Basics of Professional Mathematics

BASICS OF PROFESSIONAL MATHEMATICS

(Volume-2)

By

DR. SANAT KUMAR ADHIKARI

M.Sc. Ph.D

Professor in Mathematics Department

I.T. Pailan College of Management & Technology, Kolkata

West Bengal

UNIVERSITY SCIENCE PRESS

(An Imprint of Laxmi Publications Pvt. Ltd.)

BANGALORE • CHENNAI • COCHIN • GUWAHATI • HYDERABAD
JALANDHAR • KOLKATA • LUCKNOW • MUMBAI • RANCHI

NEW DELHI • BOSTON, USA

Published by :
UNIVERSITY SCIENCE PRESS
(An Imprint of Laxmi Publications Pvt. Ltd.)
113, Golden House, Daryaganj,
New Delhi-110002
Phone : 011-43 53 25 00
Fax : 011-43 53 25 28
info@laxmipublications.com
www.laxmipublications.com

Price : **Rs. 175.00** *Only.* *First Edition* : 2008

OFFICES

India

Bangalore	080-26 61 15 61
Chennai	044-24 34 47 26
Cochin	0484-239 70 04
Guwahati	0361-254 36 69, 251 38 81
Hyderabad	040-24 65 23 33
Jalandhar	0181-222 12 72
Kolkata	033-22 27 37 73, 22 27 52 47
Lucknow	0522-220 95 78
Mumbai	022-24 91 54 15, 24 92 78 69
Ranchi	0651-230 77 64

USA

Boston
11, Leavitt Street, Hingham,
MA 02043, USA

UPM-9213-175-BASICS OF PROFESSIONAL MATH **C—14926/07/11**

Typesetted at : Shubham Composer, N. Delhi. *Printed at :* Sudha Offset Printer, Delhi.

PREFACE

In the present scenario of globalizations in all spheres of life we are changing our educational curriculum in the light of global trend. Nowadays our education is service oriented. In the global atmosphere, the service opportunities are in the field of Management, Marketing, Information Technology, Engineering (Computer and Electronics), Communication, Multimedia etc. Keeping this in mind a large number of private colleges is coming up in all states even in West Bengal. The courses offered by these colleges are different from our traditional courses like B.A., B.Sc., B.Com. The courses that are offered to the students by these colleges are BE, IT, Computer Science, Electronics, Communications, MBA, MCA, BCA, BBA etc. These courses are basically professional in nature. Mathematics is one of the subject in all of these courses. The syllabi of mathematics for these courses are obviously applicable in character, so that learning mathematics in different semesters they can apply it in their professional lives. But for these courses there are no textbooks which contain all the chapters included in the syllabus of a semester. Due to the lack of textbooks students are facing acute problems. During my last four years of teaching in 'Pailan College of Management and Technology, Kolkata. I felt the inconvenience of the students and this motivated me to write textbooks for them. 'BASIC OF PROFESSIONAL MATHEMATICS' is the first textbook that I have written according to the syllabus of first year second semester of BCA courses of West Bengal University of Technology. Other than BCA students, the students of BE, IT and other professional courses of WBUT and similar courses of other universities including B.Sc. (Pass) course students of Calcutta University will also be benefitted through this book.

The topics covered in this book are :

Ordinary differential equations of first, second and higher orders ; Linear Algebra, in particular, Vector Spaces ; Sequences and Series.

The book is designed as a self-contained, comprehensive class room text. Each topic is treated in systematic and logical manner. The scope and depth of coverage are strictly according to the prescribed syllabus of WBUT. A large number of carefully selected illustrative examples followed by graded exercises with possible answer form an integral part of the text. Last three years questions of WBUT are included at the end of the book for the benefit of the students.

I must express my sincere gratitude to Professor D.K. Sinha, former Upacharya (Vice-Chancellor) of Visva-Bharati, and former Ghose Professor of Applied Mathematics, University of Calcutta, for his inspiration, helpful suggestions and discussion in writing the manuscript. I wish to thank Professor Saibal Kumar Mukhopadhyay, Principal, PCMT for encouraging me to write the book. I must thank all of my colleagues of PCMT for their support to prepare the manuscript. I would like to thank my dear students to motivate me for writing the book. Most of all, I would like to express my deepest gratitude to my wife, without her patience, support and encouragement, it would never have been possible to complete the manuscript.

In preparing the manuscript, I have consulted a number of literature on these topics. I express my gratitude to all the authors of the literatures.

I am sure that errors, remain to be found, however, they are mine and mine alone. I shall gratefully accept from the readers any corrections, criticism and suggestions for the improvement of the book.

—Author

SYLLABUS

Differential equations : Order, degree, solution and formation of a differential equation, standard techniques of solving a linear differential equation with constant coefficients. Cauchy's and Lagrange's linear differential equations with variable coefficients.

Linear algebra : Vector space, subspaces, bases and dimensions, co-ordinates, linear transformation, algebra of linear transformations, isomorphism, representation of transformation by matrices. Sequence and series : Bounded and unbounded sequences, convergence or divergence of a sequence, behaviour of monotone sequences, algebra of convergent sequences, Cauchy's sequence, Cauchy's general principle of convergence, infinite series—its convergence and sum, series with positive terms and standard tests of convergence (without proof), alternating series, Leibnitz test, absolute convergence, rearrangement of absolutely convergent series, test of convergence of Abel and Dirichlet (without proof).

CONTENTS

Chapter 1

Differential Equations

1.1 INTRODUCTION

We have ideas of Algebraic and Trigonometric equations. Different methods for solving these equations are also well known to us and thus by solving the equations we obtained their roots. In this chapter, we shall deal with another type of equations, known as **Differential equations**, which are involved in calculus. These equations are of fundamental importance in Engineering mathematics, I.T. mathematics and in all other branches of Science. First, we consider how these equations are formed, their order, degree and then we discuss the methods for solving the equations.

1.2 FORMATION OF DIFFERENTIAL EQUATIONS

Differential equations are formed by differentiating functional equations (without involving any arbitrary constants) of curves.

As an example, if we differentiate the equation of a curve $x^2 + xy + 3 = 0$ with respect to x, we get

$$2x + x\frac{dy}{dx} + y = 0,$$

or $$(2x + y)\, dx + x\, dy = 0$$

which is a differential equation derived from the equation of the curve $x^2 + xy + 3 = 0$.

Now, we define the formal definition of differential equation as follows :

Definition. *An equation involving differential or differential coefficient is called a differential equation. If a differential equation is formed by first derivative of a curve, it is called a first order differential equation. If it is formed by second order derivative, then it is called a second order differential equation. In general, n^{th} order derivative of a curve (if exists), is called n^{th} order differential equation.*

For example, if we differentiate the equation of the above mentioned curve namely, $x^2 + xy + 3 = 0$, twice, we get

$$x\frac{d^2y}{dx^2} + \frac{dy}{dx} + 2 + \frac{dy}{dx} = 0$$

or
$$x\frac{d^2y}{dx^2} + 2\frac{dy}{dx} + 2 = 0$$

which is a second order differential equation. Thus, we can form third order, fourth order and higher order differential equations.

The following are some differential equations :

$$\frac{dy}{dx} + y = 2x + 1 \qquad \text{... }(i)$$

$$(x^2 + y^2 - xy)\,dx + (x^2 + 3y - 1)\,dy = 0 \qquad \text{... }(ii)$$

$$3\frac{d^2y}{dx^2} - 2x\left(\frac{dy}{dx}\right)^2 + 2y = 0 \qquad \text{... }(iii)$$

$$\frac{d^3y}{dx^3} + 2\frac{d^2y}{dx^2}\cdot\frac{dy}{dx} + 3x^2\left(\frac{dy}{dx}\right)^3 = 0 \qquad \text{... }(iv)$$

$$\left[2x + \left(\frac{dy}{dx}\right)^2\right]^{\frac{3}{2}} = 5\frac{d^2y}{dx^2} \qquad \text{... }(v)$$

Remark. *Differential equations which involve only one independent variable and the differential coefficients with respect to it are called ordinary differential equations.*

In the above examples, the equations (i) to (v) are all ordinary differential equations.

1.3 ORDER AND DEGREE OF A DIFFERENTIAL EQUATION

The **order** of a differential equation is the order of the highest derivative that appears in the equation.

The order of the equations (i) and (ii) above is one. The order of the equations (iii) and (v) is two, that is, the equations (iii) and (v) are of the second order while equation (iv) is of the third order.

The **degree** of a differential equation is the degree of the highest ordered derivative that appears in the equation, provided the equation is free from the radical signs and fractional powers, as far as the derivatives are concerned. The degree of the above equations (i), (ii), (iii) and (iv) is one, while the degree of the equation (v) is two, because if we remove the fractional power in (v), the equation becomes

$$\left[2x + \left(\frac{dy}{dx}\right)^2\right]^3 = 25\left(\frac{d^2y}{dx^2}\right)^2$$, where the degree of second order derivative is two.

1.4 METHOD OF FORMATION OF ORDINARY DIFFERENTIAL EQUATION FROM THE EQUATION OF A CURVE CONSISTING OF ARBITRARY CONSTANTS

Let $$f(x, y, c_1) = 0 \quad \text{... (1)}$$

be an equation of a curve consisting an arbitrary constant c_1. Differentiating equation (1) once, we get an equation generally, containing c_1. If c_1 be eliminated between the equation (1) and the equation, from the curve, we shall get a relation involving x, y and $\frac{dy}{dx}$, which is obviously a differential equation of first order. Similarly, if we have an equation

$$f(x, y, c_1, c_2) \quad \text{... (2)}$$

containing two arbitrary constants c_1 and c_2, then differentiating the equation (2) twice, we shall get two equations. Now, between these two equations and the given equation (2), if the two arbitrary constants c_1 or c_2 be eliminated, then we shall get a differential equation of second order.

In general, if we have an equation

$$f(x, y, c_1, c_2, ..., c_n) = 0 \quad \text{... (3)}$$

containing n arbitrary constants $c_1, c_2, ..., c_n$, then differentiating the equation (3) n times. We shall get n equations consisting $c_1, c_2, ..., c_n$. Now, if the n arbitrary constants $c_1, c_2, ..., c_n$ be eliminated from equation (3) and the n equations (derived), we shall get a differential equation of nth order.

1.4.1 Illustrative Examples

1. *Find the differential equation of the curve*

$$(x^2 + 1)(y^2 + 1) = c.$$

Sol. Differentiating the given curve with respect to x, we get

$$\left(x^2+1\right) 2y \frac{dy}{dx} + \left(y^2+1\right)(2x) = 0$$

or $$\frac{dy}{dx} = -\frac{x\left(y^2+1\right)}{y\left(x^2+1\right)},$$

which is the required differential equation of the given conic.

2. *Find the differential equation of the conic $y^2 = 4ax$.*

Sol. $$y^2 = 4ax \quad \text{... (i)}$$

Differentiating w.r.t. x, we get

$$2yy_1 = 4a$$

Substituting the value of $4a$ in (1), we get

$$y^2 = 2yy_1x$$

or $y = 2xy_1$, which is the required differential equation.

3. *Find the differential equation of the curve*

$$y = a \log x + b.$$

Sol. Given $\qquad y = a \log x + b \qquad \text{... (1)}$

Differentiating (1) with respect to x, we hvae

$$y_1 = \frac{a}{x}$$

or, $\qquad xy_1 = a$

Differentiating again, w.r.t. x, we get

$$xy_2 + y_1 = 0$$

or $\qquad x\frac{d^2y}{dx^2} + \frac{dy}{dx} = 0,$

which is the required differential equation.

4. *Eliminate A and B from the equation*

$$y = Ae^x + Be^{-x} + x^2.$$

Sol. Given, $\qquad y = Ae^x + Be^{-x} + x^2 \qquad \text{... (1)}$

Differentiating (1), w.r.t. x, we get

$$\frac{dy}{dx} = Ae^x - Be^{-x} + 2x$$

Differentiating again w.r.t. x, we have

$$\frac{d^2y}{dx^2} = Ae^x + Be^{-x} + 2 = y - x^2 + 2, \text{ using (1)}$$

or $\qquad \frac{d^2y}{dx^2} - y = 2 - x^2$

which is the required eliminant.

5. *Eliminate a and b from* $y = a \tan^{-1} x + b.$

Sol. Given $\qquad y = a \tan^{-1} x + b$

Differentiating w.r.t. x, we get

$$y_1 = \frac{a}{1+x^2}$$

or $\qquad (1 + x^2)\, y_1 = a$

Differentiating again w.r.t. x, we have

$$(1 + x^2)\, y_2 + 2xy_1 = 0$$

which is the required eliminant.

6. *Form the differential equation of the curve*

$$y = ax^{n+1} + b\, x^{-n}$$

Sol. Given $\quad y = ax^{n+1} + b\, x^{-n}$

$\therefore \quad x^n\, y = ax^{2n+1} + b$

Differentiating w.r.t. x, we get

$$nx^{n-1}\, y + x^n\, y_1 = (2n + 1)\, ax^{2n}$$

or $\quad nx^{-(n+1)}\, y + x^{-n}\, y_1 = (2n + 1)\, a$

Again differentiating w.r.t. x, we have

$$-n\,(n+1)\, x^{-(n+2)}\, y + nx^{-(n+1)}\, y_1 - nx^{-(n+1)}\, y_1 + x^{-n}\, y_2 = 0$$

or $\quad -n\,(n+1)\, x^{-(n+2)}\, y + x^{-n}\, y_2 = 0$

or $\quad x^2\, y_2 = n\,(n+1)\, y.$

Hence the required differential equation is $\quad x^2\, y_2 = n\,(n+1)\, y.$

7. *Form the differential equation by eliminating A and B from $y = A \cos x + B \sin x$*

[W.B.U.T. 2006]

Sol. Given : $\quad y = \text{A} \cos x + \text{B} \sin x \quad \ldots (i)$

Differentiating (*i*) w.r.t. x, we get

$$y_1 = -\text{A} \sin x + \text{B} \cos x$$

Again differentiating w.r.t. x, we have

$$y_2 = -\text{A} \cos x - \text{B} \sin x = -(\text{A} \cos x + \text{B} \sin x)$$

or $\quad y_2 = -y \quad$ [by (*i*)]

or $\quad y_2 + y = 0.$

8. *Find the differential equation of all circles of radius a.*

Sol. Let the equation of all circles of radius a be

$$(x - h)^2 + (y - k)^2 = a^2 \quad \ldots (i)$$

where (h, k), the coordinates of the centres of the circles.

Differentiating (*i*) w.r.t. x, we get

$$2\,(x - h) + 2\,(y - k)\, \frac{dy}{dx} = 0$$

or $\quad (x - h) + (y - k)\, \dfrac{dy}{dx} = 0 \quad \ldots (ii)$

Differentiating (*ii*), again w.r.t. x, we have

$$1 + (y - k)\, .\, \frac{d^2y}{dx^2} + \left(\frac{dy}{dx}\right)^2 = 0 \quad \ldots (iii)$$

From (*iii*), $$y - k = -\frac{1+\left(\frac{dy}{dx}\right)^2}{\frac{d^2y}{dx^2}}$$

∴ From (*ii*), $$(x-h) = -(y-k)\frac{dy}{dx} = \frac{\frac{dy}{dx}\left[1+\left(\frac{dy}{dx}\right)^2\right]}{\frac{d^2y}{dx^2}}$$

Substituting the values of $(x-h)$ and $(y-k)$ in (*i*), we get

$$\frac{\left(\frac{dy}{dx}\right)\left[1+\left(\frac{dy}{dx}\right)^2\right]^2}{\left(\frac{d^2y}{dx^2}\right)^2} + \frac{\left[1+\left(\frac{dy}{dx}\right)^2\right]^2}{\left(\frac{d^2y}{dx^2}\right)^2} = a^2$$

or $$\left[1+\left(\frac{dy}{dx}\right)^2\right]^2\left[\left(\frac{dy}{dx}\right)^2+1\right] = a^2\left(\frac{d^2y}{dx^2}\right)^2$$

or $$\left[1+\left(\frac{dy}{dx}\right)^2\right]^3 = a^2\left(\frac{d^2y}{dx^2}\right)^2$$

which is the required differential equation of all circles of radius a.

9. *Find the differential equation of all circles which touch the x-axis at the origin.*

Sol. Since the circles touch the x-axis at the origin, then the diameter is the y-axis, *i.e.*, centre lies on the y-axis [Figure 1.1].

Let the equations of the circles touching the x-axis at the origin be

$$x^2 + y^2 + 2fy = 0 \qquad \text{... }(i)$$

Differentiating (*i*) w.r.t. x, we get

$$2x + 2y\frac{dy}{dx} + 2f\frac{dy}{dx} = 0$$

or $$f\frac{dy}{dx} = -\left(x + y\frac{dy}{dx}\right)$$

∴ $$f = -\frac{1}{\frac{dy}{dx}}\left(x + y\frac{dy}{dx}\right)$$

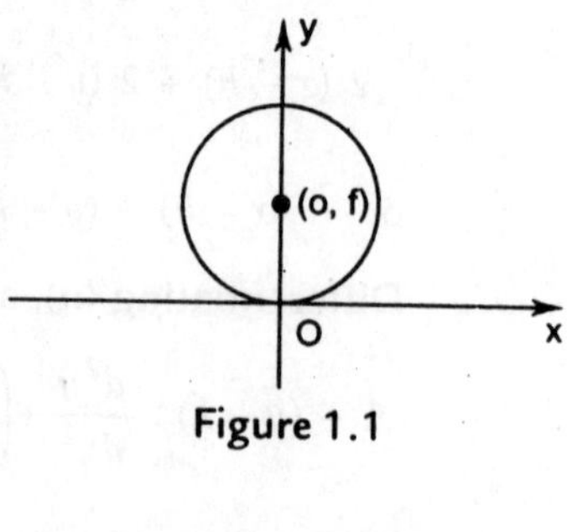

Figure 1.1

Substituting the value of f in (*i*), we obtain

$$x^2 + y^2 - 2y \frac{1}{\frac{dy}{dx}} \left(x + y \frac{dy}{dx} \right) = 0$$

or $\left(x^2 + y^2\right) \frac{dy}{dx} - 2xy - 2y^2 \frac{dy}{dx} = 0$

or $$\left(x^2 - y^2\right) \frac{dy}{dx} = 2xy,$$

which is the required differential equation.

10. *Find the differential equation of all the parabolas whose vertex is at the origin and axis is x-axis.*

Sol. Let the equation of parabolas whose axis is x-axis and vertex is at origin be

$$y^2 = 4ax \quad \text{... }(i)$$

Differentiating (i) w.r.t. x, we get

$$2y \frac{dy}{dx} = 4a$$

Substituting the value of $4a$ in (i), we have

$$y^2 = 2xy \frac{dy}{dx}$$

or, $$2x \frac{dy}{dx} = y, \text{ or } 2x \frac{dy}{dx} - y = 0.$$

which is the required differential equation.

11. *Eliminating the arbitrary constants a, b and c obtain the differential equations of the following curves :*

(a) $y = a + bx + cx^2$ *(b)* $y = ae^{2x} + be^{-3x} + ce^x$.

Sol. (*a*) Given, $y = a + bx + cx^2$

Differentiating , we get

$$\frac{dy}{dx} = b + 2cx$$

Differentiating again, we get

$$\frac{d^2y}{dx^2} = 2c$$

Further differentiating, we obtain

$$\frac{d^3y}{dx^3} = 0,$$

which is the required differential equation.

(*b*) Given, $y = ae^{2x} + be^{-3x} + ce^{x}$ ... (*i*)

Differentiating (*i*) w.r.t. *x*, we have

$$y_1 = 2ae^{2x} - 3be^{-3x} + ce^{x} \quad \text{... (ii)}$$

From (*i*), $ce^{x} = y - ae^{2x} - be^{-3x}$

Substituting the value of ce^x in (*ii*), we get

$$y_1 = 2ae^{2x} - 3be^{-3x} + y - ae^{2x} - be^{-3x}$$

or $$y_1 = ae^{2x} - 4be^{-3x} + y \quad \text{... (iii)}$$

Differentiating (*iii*), w.r.t. *x*, we get

$$y_2 = 2ae^{2x} + 12be^{-3x} + y_1 \quad \text{... (iv)}$$

Multiplying (*iii*) by 3, then adding with (*iv*), we obtain

$$2y_1 + y_2 = 5ae^{2x} + 3y \quad \text{... (v)}$$

Differentiating (*v*) w.r.t. *x*, we get

$$2y_2 + y_3 = 10ae^{2x} + 3y_1 \quad \text{... (vi)}$$

Multiplying (*v*) by 2, then subtracting from (*vi*), we have

$$y_3 - 4y_1 = 3y_1 - 6y$$

or $$y_3 - 7y_1 + 6y = 0$$

or $$\frac{d^3y}{dx^3} - 7\frac{dy}{dx} + 6y = 0$$

which is the required differential equation.

12. *Write down the order and degree of the following differential equations :*

(*i*) $x^2 - y^2 + 2xy\dfrac{dy}{dx} = 0,$ (*ii*) $(1 - x^2)\left(\dfrac{dy}{dx}\right)^2 + 1 = 0,$

(*iii*) $(x^2 - 2y^2)\left(\dfrac{dy}{dx}\right)^2 - 4xy\dfrac{dy}{dx} - x^2 = 0,$

(*iv*) $\sqrt{\left\{3y + \left(\dfrac{dy}{dx}\right)^2\right\}} = (2 + x),$ (*v*) $\dfrac{d^2y}{dt^2} + 4y = 0,$

(*vi*) $\dfrac{d^2y}{dx^2} = 4\left\{1 + \left(\dfrac{dy}{dx}\right)^2\right\},$

(*vii*) $2a\dfrac{d^2y}{dx^2} + \left(\dfrac{dy}{dx}\right)^3 = 0,$ (*viii*) $\dfrac{d^2P}{d\theta^2} = \sqrt{P + \left(\dfrac{dP}{d\theta}\right)^2},$

(*ix*) $\dfrac{d^3y}{dx^3} - 9\dfrac{dy}{dx} + 4y = 0,$ (*x*) $\dfrac{d^3y}{dx^3} = 2\left(\dfrac{dy}{dx}\right)^{\frac{2}{3}}$

Answers

Question numbers		Order	Degree	Justification
12.	*(i)*	1	1	
	(ii)	1	2	
	(iii)	1	2	
	(iv)	1	2	Simplest form of the equation is $3y+\left(\frac{dy}{dx}\right)^2=(2+x)^2$
	(v)	2	1	Removing radical sign
	(vi)	2	2	Simplifying, we have, $\left(\frac{d^2y}{dx^2}\right)^2=16\left\{1+\left(\frac{dy}{dx}\right)^2\right\}^3$
	(vii)	2	1	Removing fractional power
	(viii)	2	2	Simplifying, we get $\left(\frac{d^2P}{d\theta^2}\right)^2=P+\left(\frac{dP}{d\theta}\right)^2$, Removing radical sign
	(ix)	3	1	
	(x)	3	3	Removing fractional power, we have $\left(\frac{d^3y}{dx^3}\right)^3=8\left(\frac{dy}{dx}\right)^2$.

1.4.2 Multiple Choice Questions

A. *Choose the correct answer from the answers given below :*

(a) The order of the differential equation $\frac{d^2y}{dx^2}+1=2\frac{dy}{dx}+3\left(\frac{dy}{dx}\right)^2$ *is*

(i) 1 *(ii)* 2 *(iii)* 3 *(iv)* 0.

(W.B.U.T. 2006)

(b) The order and degree of the differential equation $\frac{d}{dx}\left(y\frac{dy}{dx}\right)=0$ *are*

(i) 1, 2 *(ii)* 2, 2 *(iii)* 1, 1 *(iv)* 2, 1.

(c) The order and degree of the differential equation $\left[1+y_1^2\right]^{\frac{3}{2}}=ay_2$ *are*

(i) 2, 1 *(ii)* 1, 2 *(iii)* 2, 2 *(iv)* 2, 3.

B. *(a) The differential equation of all the straight lines passing through the origin is*

(i) $x^2y_1=1$ *(ii)* $xy_1=y$ *(iii)* $xy_2=y$ *(iv)* $xy_1^2=y$.

(b) The differential equation of the curve y = A cos x + sin x is

(i) $\sin x\frac{dy}{dx}+y\cos x=1$ *(ii)* $\frac{dy}{dx}+x\cot x=\sec x$

(iii) $\frac{dy}{dx}+y\cot x=\sec x$ *(iv)* $\frac{d^2y}{dx^2}+y=1$.

(c) *The differential equation of the curve* $y = ax^2 + b$ *is*

(i) $y_2 - xy_1 = 0$ (ii) $y_2 = 0$ (iii) $xy_2 = y_1$ (iv) $xy_2 = y$.

Solution

A. (a) (ii), (b) (iv), (c) (iii).

B. (a) (ii). The equation of all the straight lines through the origin is

$$y = mx \quad \text{... (1)}$$

Differentiating (1) w.r.t. x, we get $y_1 = m$

$\therefore$ From (1), we have, $y = xy_1$

(b) (iii) $\because$ $y = A \cos x + \sin x$ ... (1)

$\therefore$ $y_1 = - A \sin x + \cos x$ ... (2)

Multiplying (1) by $\sin x$ and (2) by $\cos x$ and adding, we get

$$y \sin x + y_1 \cos x = \sin^2 x + \cos^2 x = 1$$

or $y_1 + y \cot x = \sec x.$

(c) (iii) $\because$ $y = ax^2 + b,$

$\therefore$ $y_1 = 2ax,$

or $\dfrac{y_1}{x} = 2a,$

Differentiating again w.r.t. x, we get

$$\frac{xy_2 - y_1}{x^2} = 0 \Rightarrow xy_2 = y_1.$$

EXERCISE 1.1

A. *Eliminating the arbitrary constants obtain the differential equations for the following curves :*

1. $y = a\sqrt{x} + \dfrac{b}{\sqrt{x}}$
2. $ax^2 + by^2 = 1$
3. $x = A \cos (nt + \alpha)$
4. $y = A \sin x + B \cos x + x \sin x$
5. $y = c_1 e^{2x} + c_2 x e^{2x} + e^x$
6. $y = a \cos (\log x) + b \sin (\log x)$
7. $y = a \sec x + b \tan x$
8. $y = ax \cos (\log x) + bx \sin (\log x)$
9. $xy = Ae^x + Be^{-x} + x^2$
10. $y^2 + x^2 = 2ay + a^2$
11. $y = ae^{2x} + be^{-3x} + ce^x$
12. $e^{2y} + 2ax\, e^y + a^2 = 0$
13. $x = (a + bx)\, e^{\frac{y}{x}}$.

B. *Find the differential equations of*

(i) all straight lines in a plane.

(ii) all circles of radius a whose centre lie on the x-axis.

(iii) all parabolas with x-axis as the axis and $(a, 0)$ as focus.

(*iv*) all the straight lines which make angle θ with the positive direction of x-axis.
(*v*) all circles passing through the origin and having centres on the x-axis.
(*vi*) all circles in the first quadrant which touch the coordinate axes.

Answers

A. **1.** $4x^2y_2 + 4xy_1 = y$ **2.** $x\{yy_2 + y_1^2\} - yy_1 = 0$

3. $\frac{d^2x}{dt^2} + n^2x = 0$ **4.** $y_2 + y = 2\cos x$

5. $y_2 - 4y_1 + 4y = e^x$ **6.** $x^2y_2 + xy_1 + y = 0$

7. $y_2 = y\sec^2 x + \tan x\, y_1$ **8.** $x^2y_2 - xy_1 + 2y = x\log x$

9. $xy_2 + 2y_1 + x^2 - xy = 0$ **10.** $(x^2 - 2y^2)\, y_1^2 - 4xyy_1 - x^2 = 0$

11. $y_3 - 7y_1 + 6y = 0$ **12.** $(1 - x^2)\, y_1^2 + 1 = 10$

13. $x^3y_2 = [xy_1 - y]^2$.

B. (*i*) $y_2 = 0$ (*ii*) $y^2[1 + y_1^2] = a^2$

(*iii*) $yy_1 = 2a$ (*iv*) $y_1 = \tan\theta$

(*v*) $x^2 - y^2 + 2xyy_1 = 0$ (*vi*) $(x - y)^2[1 + y_1^2] = (x + yy_1)^2$.

1.5 GEOMETRICAL MEANING OF A DIFFERENTIAL EQUATION OF FIRST ORDER AND FIRST DEGREE

A first order differential equation must contain the first derivative y' or (y_1) of the unknown function y, and it may contain y itself and given functions of x. Hence, we can write any first order differential equation in implicit form

$$F(x, y, y') = 0 \qquad \text{... (1)}$$

More generally, we write a first order differential equation in the explicit form.

$$y' = f(x, y) \qquad \text{... (2)}$$

Before discussing the methods of solution of differential equations, let us explain the geometrical meaning of $y' = f(x, y)$. From calculus, we know that the direction of a curve at a particular point is determined by drawing a tangent line at that point that is, its slope is given by $y' = \frac{dy}{dx}$ at that particular point.

Hence, if (2) has a solution $y(x)$ passing through a point (x_0, y_0) of the xy-plane, it must have at (x_0, y_0) the slope $f(x_0, y_0)$. This suggests the idea of plotting approximate solution curves of a given differential equation (2) without actually solving the equation, so that we obtain a picture of the general behaviour of these solution curves.

The slope $f(x, y)$ of the differential equation (2), at some points in the xy-plane, will be indicated by short segments (shown in Figures 1.2*a* and 1.2*b*) which are called *linear* **elements**. The region consists of *linear* elements is known as a **direction field** or **slope**

field. It is a field of tangent directions (slopes) of solution curves of the equation (2), which can be used for plotting (approximate) solution curves. Each such curve represents a particular solution corresponding to some initial conditions.

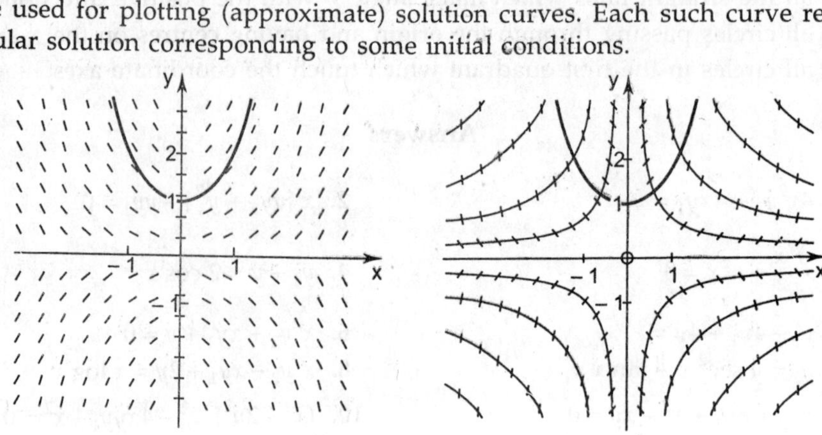

(a) Computer plotted graph (b) Hand plotted graph

Figure 1.2

1.6 CONCEPT OF SOLUTION

A solution of a given first order differential equation $y' = f(x, y)$ in some open interval $a < x < b$ is a function $y = h(x)$ that has a derivative $y' = h'(x)$ and satisfies the equation $y' = f(x, y)$ for all x in that interval; that is the equation becomes an identity if we replace the unknown function y by h and y' by h'.

As an example, $y = x^2$ is a solution of $xy' = 2y$ for all x. Because, if we substitute $y = x^2$ and $y' = 2x$ into the equation, we obtain $xy' = x(2x) = 2x^2 = 2y$, an identity in x.

Let us consider another example, $y' = \cos x$, which can be solved by integrating.

On integration, we get sine curves $y = \sin x + c$ with arbitrary c. For each c, we get a solution of the given equation and for all c's we get all possible solutions of the given differential equation.

Figure 1.3 shows some of them ; for $c = -3, -2, -1, 0, 1, 2$.

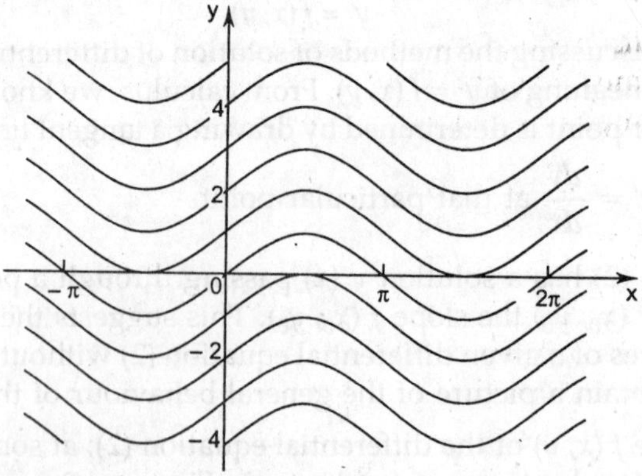

Figure 1.3

From the solutions of the above example we can conclude that all solutions of a differential equation are represented by a single formula involving an arbitrary constant c (additively or multiplicatively or in some other way). Such a function (solution) involving an arbitrary constant is called a **general solution** of a first order differential equation.

Geometrically, these are infinitely many curves, one for each c. We call this a family of curves and if we choose a particular c, the solution obtained is called a **particular solution** of that equation. A differential equation together with an initial condition, is called an **initial value problem** or **boundary value problem.**

1.7 CLASSIFICATIONS OF DIFFERENTIAL EQUATIONS OF FIRST ORDER AND FIRST DEGREE

We can classify the first order first degree differential equations in four classes :

I. Variables separable equations

II. Homogeneous equations

III. Linear equations

IV. Exact equations.

I. Variables Separable Differential Equations

If a differential equation of first order and first degree can be put in the form where dx and all terms containing x are at one place, also dy and all terms containing y are at one place, then the equation is said to be variables separable equation. In this case, the equation can be put in the form $f(x)\,dx + f(y)\,dy = 0$.

Then integrating, we get $\int f(x)\,dx + \int f(y)\,dy = c$, which is the general solution of the equation, c being an arbitrary constant. c can be taken according to the need of the simplification of the solution.

The general solution represents a family of solution curves. For each c we get a member of family curves.

As an example let us, solve the differential equation

$$9yy' + 4x = 0.$$

By separating variables, we have

$$9y\,dy = -4x\,dx.$$

Now, by integrating on both sides we obtain the general solution

$$\frac{9}{2}y^2 = -2x^2 + c_1.$$

Thus $$\frac{x^2}{9} + \frac{y^2}{4} = c \qquad \left(c = \frac{c_1}{18}\right)$$

The solution represents a family of ellipses. Some of them are shown in Figure 1.4

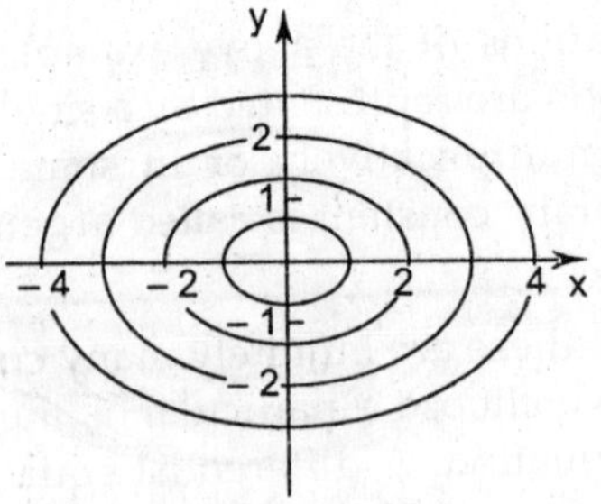

Figure 1.4

Remark 1. Any equation of the form $f_1(x)\,\phi_1(y)\,dx + f_2(x)\,\phi_2(y)\,dy = 0$ can be expressed in the variables separable form by dividing throughout the equation by $f_2(x)\,\phi_1(y)$.

Thus, $$\frac{f_1(x)}{f_2(x)}\,dx + \frac{\phi_2(y)}{\phi_1(y)}\,dy = 0$$

or $$f(x)\,dx + f(y)\,dy = 0.$$

Remark 2. Differential equation of the form

$$\frac{dy}{dx} = f(ax + by + c) \qquad \text{... (1)}$$

can be reduced to a variables separable form by substituting $ax + by + c = u$.

[As, $$a + b\frac{dy}{dx} = \frac{du}{dx} \quad \text{(Differentiating w.r.t. } x)$$

or $$\frac{dy}{dx} = \frac{1}{b}\left(\frac{du}{dx} - a\right)$$

Substituting, $\frac{dy}{dx}$ in (1), we get $\frac{1}{b}\left(\frac{du}{dx} - a\right) = f(u)$

or $$\frac{du}{dx} = a + b\,f(u) \quad \text{or} \quad \frac{du}{a + b\,f(u)} = dx]$$

1.7.1 Illustrative Examples

1. *Solve the differential equation :*

$$x\sqrt{1-y^2}\,dx + y\sqrt{1-x^2}\,dy = 0$$

Sol. Given $x\sqrt{1-y^2}\,dx + y\sqrt{1-x^2}\,dy = 0$

or, $$\frac{x\,dx}{\sqrt{1-x^2}} + \frac{y\,dy}{\sqrt{1-y^2}} = 0 \qquad \left[\text{Dividing by } \sqrt{1-x^2}\,.\,\sqrt{1-y^2}\right]$$

Integrating, we have

$$\int \frac{x\,dx}{\sqrt{1-x^2}} + \int \frac{y\,dy}{\sqrt{1-y^2}} = c \qquad \dots (i)$$

Now for $\int \frac{x\,dx}{\sqrt{1-x^2}}$ putting $1 - x^2 = z^2$, $-2x\,dx = 2z\,dz \Rightarrow x\,dx = -z\,dz$

or $$\int \frac{x\,dx}{\sqrt{1-x^2}} = \int \frac{-z\,dz}{z} = -\int dz = -z = -\sqrt{1-x^2}$$

Similarly, $$\int \frac{y}{\sqrt{1-y^2}}\,dy = -\sqrt{1-y^2}$$

$\therefore$ From (*i*), we get $\sqrt{1-x^2} + \sqrt{1-y^2} = -c$

or, $$\sqrt{1-x^2} + \sqrt{1-y^2} = c_1, \text{ where } c_1 = -c$$

2. *Solve the differential equation*

$$\sec^2 x \tan y\,dx + \sec^2 y \tan x\,dy = 0.$$

Sol. Given, $\sec^2 x \tan y\,dx + \sec^2 y \tan x\,dy = 0$

$\therefore$ $$\frac{\sec^2 x}{\tan x}dx + \frac{\sec^2 y}{\tan y}dy = 0 \qquad \text{[dividing by } \tan x \tan y]$$

Integrating, we get

$$\int \frac{\sec^2 x}{\tan x}dx + \int \frac{\sec^2 y}{\tan y}dy = c,$$

where c is an arbitrary constant

or $$\log \tan x + \log \tan y = \log c_1 \qquad \text{where } c = \log c_1$$

or $$\log(\tan x \tan y) = \log c_1$$

or $$\tan x \tan y = c_1,$$

which is the required general solution.

3. *Solve the differential equation*

$$\tan x \frac{dy}{dx} = 1 + y^2, \text{ given } y = 1 \text{ when } x = \frac{\pi}{2}.$$

Sol. Since $$\tan x \frac{dy}{dx} = 1 + y^2$$

$\therefore$ $$\cot x\,dx = \frac{dy}{1+y^2}$$

Integrating, $$\int \cot x\,dx = \int \frac{dy}{1+y^2} + c,$$ where c is an arbitrary constant.

or $$\log|\sin x| = \tan^{-1} y + c$$

Using the given conditions, *i.e.*, $y = 1$ when $x = \frac{\pi}{2}$, we get

$$\log \sin\frac{\pi}{2} = \tan^{-1} 1 + c$$

or $$\log 1 = \frac{\pi}{4} + c$$

or $$0 = \frac{\pi}{4} + c, \text{ i.e., } c = -\frac{\pi}{4}$$

Hence the particular solution is

$$\log|\sin x| = \tan^{-1} y - \frac{\pi}{4}$$

or $$y = \tan\left(\frac{\pi}{4} + \log|\sin x|\right).$$

4. *Solve the equation* $y - x\frac{dy}{dx} = a\left(y^2 + \frac{dy}{dx}\right)$, *where a is a non-zero constant.*

Sol. Given, $$y - x\frac{dy}{dx} = a\left(y^2 + \frac{dy}{dx}\right)$$

or $$(y - ay^2) = (x + a)\frac{dy}{dx}$$

$\therefore$ $$\frac{dy}{y(1 - ay)} = \frac{dx}{x + a}$$

or $$\left(\frac{1}{y} + \frac{a}{1 - ay}\right)dy = \frac{dx}{x + a}$$

Integrating, we get

$$\int\frac{1}{y}\,dy + \int\frac{a}{1 - ay}\,dy = \int\frac{dx}{x + a}$$

or $$\log|y| - \log|1 - ay| = \log|x + a| + \log|c|,$$

where c is a non-zero constant

or $$\log\left|\frac{y}{1 - ay}\right| = \log|c(x + a)|$$

or $$\frac{y}{1 - ay} = c(x + a)$$

or
$$y = c(x+a)(1-ay),$$
which is the required general solution of the given equation.

5. *Find the general solution of the differential equation*
$$\frac{dy}{dx} = \frac{x(2\log x + 1)}{\sin y + y\cos y}, \text{ where } x > 0, \text{ and } y \neq 0.$$

Sol. Since
$$\frac{dy}{dx} = \frac{x(2\log x + 1)}{\sin y + y\cos y}$$

$\therefore$
$$(\sin y + y\cos y)\, dy = (2x\log x + x)\, dx$$

Integrating, we have
$$\int \sin y\, dy + \int y\cos y\, dy = \int x\, dx + \int 2x\log x\, dx + c$$

where c is an arbitrary constant.

or
$$-\cos y + y\sin y - \int \sin y\, dy = \frac{x^2}{2} + 2\log x \cdot \frac{x^2}{2} - 2\int \frac{1}{x} \cdot \frac{x^2}{2}\, dx + c$$

or
$$-\cos y + y\sin y + \cos y = \frac{x^2}{2} + x^2\log x - \int x\, dx + c$$

or
$$y\sin y = \frac{x^2}{2} + x^2\log x - \frac{x^2}{2} + c$$

or
$$y\sin y = x^2\log x + c,$$
which is the required general solution.

6. *Find the general solution of the differential equation*
$$\frac{dy}{dx} = e^{x-y} + x^2 e^{-y}$$

Sol. Given,
$$\frac{dy}{dx} = e^{x-y} + x^2 e^{-y}$$

or
$$\frac{dy}{dx} = e^x \cdot e^{-y} + x^2 e^{-y}$$

or
$$e^y \frac{dy}{dx} = e^x + x^2$$

or
$$e^y\, dy = (e^x + x^2)\, dx$$

Integrating,
$$\int e^y\, dy = \int \left(e^x + x^2\right) dx + c,$$

where c is an arbitrary constant.

$\therefore$
$$e^y = e^x + \frac{1}{3}x^3 + c$$

or $$3e^y = 3e^x + x^3 + c_1, \text{ where } c_1 = 3c,$$

which is the required general solution of the given differential equation.

7. *Solve the equation*

$$(e^x + 1)\, y\, dy = (y^2 + 1)\, e^x\, dx,$$

given $y = 0$ when $x = 0$.

Sol. From the given equation, we have

$$\frac{y}{y^2+1}\, dy = \frac{e^x}{e^x+1}\, dx$$

Integrating, we get $\frac{1}{2} \log (y^2 + 1) = \log (e^x + 1) + c,$

where c is an arbitrary constant.

Using the given conditions, *i.e.*, $y = 0$ when $x = 0$, we get,

$$\frac{1}{2} \log 1 = \log 2 + c, \text{ or } c = -\log 2 \qquad [\because \log 1 = 0]$$

$\therefore$ The particular solution is

$$\frac{1}{2} \log (y^2 + 1) = \log (e^x + 1) - \log 2$$

or $$\log \sqrt{y^2+1} = \log\left(\frac{e^x+1}{2}\right)$$

or $$\sqrt{y^2+1} = \frac{1}{2}\left(e^x+1\right)$$

or $$4\,(y^2 + 1) = (e^x + 1)^2.$$

8. *Find the general solution of the differential equation*

$$\frac{dy}{dx} = (4x + y + 1)^2$$

Sol. The given equation is

$$\frac{dy}{dx} = (4x + y + 1)^2 \qquad \text{...}(i)$$

Let $$4x + y + 1 = u$$

Differentiating w.r.t. x, we get

$$4 + \frac{dy}{dx} = \frac{du}{dx}$$

or $$\frac{dy}{dx} = \frac{du}{dx} - 4 \qquad \text{...}(ii)$$

Replacing $\frac{dy}{dx}$, from (*ii*), in (*i*), we obtain

$$\frac{du}{dx} - 4 = u^2$$

or $$\frac{du}{dx} = u^2 + 4$$

or $$\frac{du}{u^2 + 4} = dx$$

Integrating, we get

$$\frac{1}{2}\tan^{-1}\frac{u}{2} = x + c$$

or $$u = 2\tan(2x + 2c)$$

or $$4x + y + 1 = 2\tan(2x + 2c) = 2\tan 2(x + c).$$

which is the required general solution of the given equation.

9. *Find the general solution of the differential equation*

$$sin^{-1}\left(\frac{dy}{dx}\right) = x + y$$

Sol. Given, $$\sin^{-1}\left(\frac{dy}{dx}\right) = x + y$$

or $$\frac{dy}{dx} = \sin(x + y) \qquad \text{...}(i)$$

Let $x + y = \theta$, then $1 + \frac{dy}{dx} = \frac{d\theta}{dx}$

or $$\frac{dy}{dx} = \frac{d\theta}{dx} - 1$$

Therefore, from (*i*), we get

$$\frac{d\theta}{dx} - 1 = \sin\theta$$

or $$d\theta = (1 + \sin\theta)\,dx$$

or $$\frac{d\theta}{1 + \sin\theta} = dx$$

Integrating, we get

$\int \frac{d\theta}{1 + \sin\theta} = x + c$, where c is an arbitrary constant or $$\int \frac{1-\sin\theta}{\cos^2\theta}\,d\theta = x + c$$

or $$\int \sec^2\theta\,d\theta - \int \tan\theta\sec\theta\,d\theta = x + c$$

or $$\tan\theta - \sec\theta = x + c$$

or $$\tan(x+y) - \sec(x+y) = x + c, \qquad [\because \theta = x + y)].$$

which is the required general solution of the given equation.

10. *Solve the differential equation*

$$\log\left(\frac{dy}{dx}\right) = 4x - 2y - 2, \text{ given } y = 1 \text{ when } x = 1.$$

Sol. Since $$\log\left(\frac{dy}{dx}\right) = 4x - 2y - 2$$

$\therefore$ $$\frac{dy}{dx} = e^{4x-2y-2} = e^{4x}\,.\,e^{-2(y+1)}$$

or, $$e^{2(y+1)}\,dy = e^{4x}\,dx$$

Integrating, $$\int e^{2(y+1)}\,dy = e^{4x}\,dx$$

or, $$\frac{1}{2}e^{2(y+1)} = \frac{1}{4}e^{4x} + \frac{1}{4}c, \text{ where } c \text{ is an arbitrary constant.}$$

Using the given conditions, *i.e.*, $y = 1$ when $x = 1$, we get

$$\frac{1}{2}e^4 = \frac{1}{4}e^4 + \frac{1}{4}c, \text{ or } \frac{1}{4}c = \frac{1}{4}e^4 \text{ or } c = e^4.$$

$\therefore$ $$\frac{1}{2}e^{2(y+1)} = \frac{1}{4}e^{4x} + \frac{1}{4}e^4$$

or $$2e^{2(y+1)} = e^{4x} + e^4,$$

which is the required particular solution of the given equation.

11. *Find the general solution of the differential equation*

$$\frac{dy}{dx} = \sin(x+y) + \cos(x+y).$$

Sol. Given, $$\frac{dy}{dx} = \sin(x+y) + \cos(x+y) \qquad \ldots(i)$$

Let $x + y = \theta$, then $$1 + \frac{dy}{dx} = \frac{d\theta}{dx}$$

or $$\frac{dy}{dx} = \frac{d\theta}{dx} - 1$$

$\therefore$ From (*i*), we get

$$\frac{d\theta}{dx} - 1 = \sin\theta + \cos\theta$$

or $$\frac{d\theta}{dx} = 1 + \sin\theta + \cos\theta = (1 + \cos\theta) + \sin\theta$$

$$= 2\cos^2\frac{\theta}{2} + 2\sin\frac{\theta}{2}\cos\frac{\theta}{2}$$

$$= 2\cos^2\frac{\theta}{2}\left(1 + \tan\frac{\theta}{2}\right)$$

or $$\frac{\sec^2\frac{\theta}{2}}{2\left(1+\tan\frac{\theta}{2}\right)}\,d\theta = dx$$

Integrating, $$\int\frac{\sec^2\frac{\theta}{2}}{2\left(1+\tan\frac{\theta}{2}\right)}\,d\theta = \int dx + c,$$ where c is an arbitrary constant

we get, $$\log\left(1+\tan\frac{\theta}{2}\right) = x + c$$

i.e., $$\log\left[1+\tan\frac{1}{2}(x+y)\right] = x + c \qquad (\because\ \theta = x + y)$$

or $$1 + \tan\frac{1}{2}(x+y) = c_1 e^x, \text{ where } e^c = c_1.$$

12. *Solve the equation*

$$\frac{x\,dx + y\,dy}{x\,dy - y\,dx} = \sqrt{\frac{a^2 - x^2 - y^2}{x^2 + y^2}}$$ [W.B.U.T. BCA, 2006]

Sol. Putting $x = r\cos\theta$ and $y = r\sin\theta$

we get $$\frac{dx}{d\theta} = \cos\theta\,\frac{dr}{d\theta} - r\sin\theta$$

$$\frac{dy}{d\theta} = \sin\theta\,\frac{dr}{d\theta} + r\cos\theta$$

and $$x^2 + y^2 = r^2,\ \tan\theta = \frac{y}{x},\ i.e.,\ \theta = \tan^{-1}\frac{y}{x}$$

From the given equation, we have

$$\frac{x\dfrac{dx}{d\theta}+y\dfrac{dy}{d\theta}}{x\dfrac{dy}{d\theta}-y\dfrac{dx}{d\theta}}=\sqrt{\frac{a^2-r^2}{r^2}} \qquad \text{...}(i)$$

Now substituting, $\dfrac{dx}{d\theta}$ and $\dfrac{dy}{d\theta}$ in (*i*), we get

$$\frac{r\dfrac{dr}{d\theta}\left(\cos^2\theta+\sin^2\theta\right)}{r^2\left(\cos^2\theta+\sin^2\theta\right)}=\frac{\sqrt{a^2-r^2}}{r}$$

or
$$\frac{\dfrac{dr}{d\theta}}{r}=\frac{\sqrt{a^2-r^2}}{r}$$

or
$$\frac{dr}{\sqrt{a^2-r^2}}=d\theta$$

Integrating, we get $\qquad \sin^{-1}\dfrac{r}{a}=\theta+c$, where c is an arbitrary constant.

or
$$r=a\sin(\theta+c)$$

or
$$r^2=a^2\sin^2(\theta+c)$$

or
$$x^2+y^2=a^2\sin^2\left[c+\tan^{-1}\frac{y}{x}\right],$$

which is the required general solution.

EXERCISE 1.2

A. *Solve the following differential equations :*

1. $e^y\dfrac{dy}{dx}+x^3\cos\left(x^2\right)=0$
2. $x\dfrac{dy}{dx}=y^2+y$
3. $\dfrac{ds}{dx}+x^2=x^2e^{3s}$
4. $(e^y+1)\cos x\,dx+e^y\sin x\,dy=0$
5. $e^y\left(1+x^2\right)\dfrac{dy}{dx}-2x\left(1+e^y\right)=0$
6. $(x-y^2x)\,dx-(y-x^2y)\,dy=0$
7. $3e^x\tan y\,dx+(1-e^x)\sec^2 y\,dy=0$
8. $(\sin y+y\cos y)\,dy-(2\log x+1)\,x\,dx=0$
9. $\left(x^2-yx^2\right)\dfrac{dy}{dx}+y^2+xy^2=0$
10. $xy\dfrac{dy}{dx}=1+x+y+xy$

11. $\frac{y}{x} \cdot \frac{dy}{dx} = \sqrt{1 + x^2 + y^2 + x^2y^2}$

12. $\operatorname{cosec} x \log y \, dy + x^2y^2 \, dx = 0$

13. $(3 + 2 \sin x + \cos x) \, dy = (1 + 2 \sin y + \cos y) \, dx$

14. $y^2 + \sin 2x \frac{dy}{dx} = 1$

15. $\frac{dy}{dx} + \frac{\sqrt{(x^2-1)(y^2-1)}}{xy} = 0$

16. $x(e^y + a) \, dx + e^{x+y} \, dy = 0$, where a is a constant.

B. *Solve the following differential equations :*

1. $y' = (y + 4x)^2$

2. $\frac{dy}{dx} = \sqrt{y - x}$

3. $\cot^2(x + y) \frac{dy}{dx} = 1$

4. $y' = \cos(x - y)$

5. $\frac{dy}{dx} \tan y = \sin(x + y) + \sin(x - y)$

6. $\frac{dy}{dx} - x \tan(y - x) = 1.$

C. *Find the particular solutions of the following differential equations :*

1. $(1 - x^2) \, dy = 2y \, dx$, given $y = 1$ when $x = 2$.
2. $x(y^2 + 1) \, dx + y(x^2 + 1) \, dy = 0$, given $y = -1$, when $x = 0$.
3. $8y \, dy - x \, dx + xy^2 \, dx + 2x^2y \, dy = 0$, given $y = 0$, when $x = 1$.
4. $v \frac{dv}{dx} + n^2x = 0$, ($n$ is a non-zero constant), given $v = u$, when $x = a$.
5. $x \frac{dy}{dx} + \cot y = 0$, given $y = \frac{\pi}{4}$, when $x = \sqrt{2}$.
6. $y' = (1 + y^2) \cot x$, given $y = 1$, when $x = \frac{\pi}{2}$.
7. $y - xy' = 2(1 + x^2 y')$, given $y = 1$, when $x = 1$.
8. $\tan x \, dy - \tan y \, dx = 0$, given $y = \frac{\pi}{2}$, when $x = \frac{\pi}{6}$.
9. $\log\left(\frac{dy}{dx}\right) = 4x - 2y - 2$, given $y = 1$, when $x = 1$.
10. $y' = 1 + e^{2x-y}$, given $y = 2$, when $x = 2$.

D. 1. Find the equation of the curve which satisfies the differential equation $(1 + y^2) \, dx - xy \, dy = 0$ and passes through the point (1, 0).

2. If $f'(x) = \frac{\tan x}{1 + 2\tan^2 x}$ and $f\left(\frac{\pi}{2}\right) = \sqrt{2} + \frac{1}{2} \log 2$,

show that $f(x) = \frac{1}{2} \log\left(1 + \sin^2 x\right) + \sqrt{2}$

3. *Identify the curve, which is a solution of the equation $y' = 0$, from the following :*
   (*a*) a straight line passing through the origin.
   (*b*) a straight line parallel to y-axis.
   (*c*) a straight line parallel to x-axis.
   (*d*) a circle.
4. *The curve, which is a solution of the differential equation $y' = e^{x-y}$ and passes through the origin is*
   (*a*) a straight line through the origin
   (*b*) $e^y + x = 0$
   (*c*) $e^x - e^y = 2$
   (*d*) none of these.
5. *The curve, which is a solution of the differential equation $(1 + y^2)\, dx - xy\, dy = 0$ and passes through the point (1, 0) is*
   (*a*) a circle
   (*b*) a parabola
   (*c*) an ellipse
   (*d*) a rectangular hyperbola.

Answers

A. 1. $2e^y + x^2 \sin(x^2) + \cos(x^2) = c$

2. $y = cx(1 + y)$

3. $1 - e^{-3s} = c\, e^{x^3}$

4. $(e^y + 1) \sin x = c$

5. $1 + e^y = c(1 + x^2)$

6. $(1 - y^2) = c(1 - x^2)$

7. $\tan y = c(1 - e^x)^3$

8. $y \sin y = x^2 \log x + c$

9. $\log\left(\dfrac{x}{y}\right) - \dfrac{1}{x} - \dfrac{1}{y} = c$

10. $y = x + \log\{x(1 + y)\} + c$

11. $(1+y)^{\frac{1}{2}} = \dfrac{2}{3}(1+x^2)^{\frac{3}{2}} + c$

12. $\dfrac{1}{y}(1 + \log y) + x^2 \cos x - 2x \sin x - 2 \cos x = c$

13. $2 \tan^{-1}\left(1 + \tan \dfrac{1}{2} x\right) = c + \log\left(1 + 2 \tan \dfrac{1}{2} y\right)$

14. $(1 + y) = c(1 - y) \tan x$

15. $\sqrt{x^2 - 1} + \sqrt{y^2 - 1} = \sec^{-1} x + c$

16. $(x + 1) e^{-x} = \log(e^y + a) + c$

B. 1. $\tan^{-1}\left(\dfrac{y + 4x}{2}\right) = 2(x + c)$

2. $2\sqrt{y - x} + 2 \log\left(\sqrt{y - x} - 1\right) = x + c$

3. $2(x - y) = \sin 2(x + y) + c$

4. $\cot(x - y) + \operatorname{cosec}(x - y) + x = c$

5. $2 \cos x + \sec y = c$

6. $\log \sin(y - x) = \dfrac{x^2}{2} + c$

C. 1. $3xy - x - 3y = 1$

2. $(x^2 + 1)(y^2 + 1) = 2$

3. $(1 - y^2)^2 (4 + x^2) = 5$

4. $(v^2 - u^2) = n^2 (a^2 - x^2)$

5. $x \sec y = 2$

6. $y = \tan\left(\frac{\pi}{4} + \log \sin x\right)$

7. $(2 - y)(2x + 1) = 3x$

8. $\sin y = \sqrt{2} \sin x$

9. $2e^{2(y+1)} = e^{4x} + e^4$

10. $x + \log\{e^2(e^2 - 1)^{-1}\} = 2 + \log\left\{\frac{e^{2x-y}}{e^{2x-y} - 1}\right\}$.

D. 1. $x^2 - y^2 = 1$

3. (c) $y = k$

4. (a), $x = y$

5. (d) $x^2 - y^2 = 1$.

II. Homogeneous Differential Equations

A differential equation of the form $\frac{dy}{dx} = \frac{f(x, y)}{g(x, y)}$, where $f(x, y)$ and $g(x, y)$ are both homogeneous function of x, y, that is, of the same degree, is called a homogeneous differential equation.

Method of Solution

If $f(x, y)$ and $g(x, y)$ are homogeneous functions of degree k in x, y, then

$$f(x, y) = x^k \phi_1\left(\frac{y}{x}\right) \text{ and } g(x, y) = x^k \phi_2\left(\frac{y}{x}\right).$$

Then, $\frac{dy}{dx} = \frac{f(x, y)}{g(x, y)}$ reduces to

$$\Rightarrow \quad \frac{dy}{dx} = \frac{\phi_1\left(\frac{y}{x}\right)}{\phi_2\left(\frac{y}{x}\right)} = F\left(\frac{y}{x}\right), \text{ say} \quad \text{... (1)}$$

Now, putting, $\frac{y}{x} = v$, or $y = vx$, we have

$$\frac{dy}{dx} = v + x\frac{dv}{dx}$$

Then from (1), we get

$$v + x\frac{dv}{dx} = F(v)$$

Now, separating the variables, we obtain

$$\frac{dv}{F(v) - v} = \frac{dx}{x} \quad \text{... (2)}$$

Integrating (2), we get the solution in terms of v and x.

Then replacing v by $\frac{y}{x}$, we get the required solution.

1.7.2 Illustrative Examples

1. *Solve the equation* $(x + y)\, dx + (y - x)\, dy = 0$

Sol. Since $(x + y)\, dx + (y - x)\, dy = 0$

or $$\frac{dy}{dx} = -\frac{x+y}{y-x} = \frac{x+y}{x-y} \qquad \text{...}(i)$$

Putting $y = vx$, we have

$$\frac{dy}{dx} = v + x\frac{dv}{dx}$$

$\therefore$ From (*i*), we get

$$v + x\frac{dv}{dx} = \frac{x+vx}{x-vx} = \frac{1+v}{1-v}$$

or $$x\frac{dv}{dx} = \frac{1+v}{1-v} - v = \frac{1+v^2}{1-v}$$

or $$\frac{1-v}{1+v^2}\, dv = \frac{dx}{x}$$

Integrating, $$\int \frac{dv}{1+v^2} - \frac{1}{2}\int \frac{2v}{1+v^2}\, dv = \int \frac{dx}{x}$$

or $$\tan^{-1} v - \frac{1}{2}\log(1 + v^2) = \log |x| + c$$

or $$2\tan^{-1}\frac{y}{x} - \log\left(\frac{x^2+y^2}{x^2}\right) = 2\log |x| + c_1$$

(where $c_1 = 2c$)

or $2\tan^{-1}\frac{y}{x} - \log(x^2 + y^2) + 2\log |x| = 2\log |x| + c_1$

or $$\log(x^2 + y^2) + c_1 = 2\tan^{-1}\frac{y}{x}.$$

2. *Solve the equation* $\left(\sqrt{xy} - x\right) dy + y\, dx = 0$

Sol. Given $\left(\sqrt{xy} - x\right) dy + y\, dx = 0$

or, $$\frac{dy}{dx} = -\frac{x}{\sqrt{xy} - x} \qquad \text{... (i)}$$

Putting $y = vx$, $\frac{dy}{dx} = v + x\frac{dv}{dx}$

∴ From (*i*), we get,

$$v + x\frac{dv}{dx} = -\frac{vx}{x\left(\sqrt{v} - 1\right)} = -\frac{v}{\left(\sqrt{v} - 1\right)}$$

or, $$x\frac{dv}{dx} = -\frac{v}{\sqrt{v} - 1} - v = \frac{-v\left(1 + \sqrt{v} - 1\right)}{\sqrt{v} - 1}$$

or $$x\frac{dv}{dx} = -\frac{v\sqrt{v}}{\sqrt{v} - 1}$$

or $$\frac{\sqrt{v} - 1}{v\sqrt{v}}\,dv = -\frac{dx}{x}$$

Integrating, we get

$$\int \frac{1}{v}\,dv - \int v^{-\frac{3}{2}}\,dv = -\log|x| - \log|c| \text{ (where } c \text{ is a non-zero constant)}$$

or $$\log|v| + 2v^{-\frac{1}{2}} = -\log|cx|$$

or $$\log|cvx| = -2v^{-\frac{1}{2}}$$

or $$cvx = e^{-\frac{2}{\sqrt{v}}}$$

or $$cy = e^{-2\sqrt{\frac{x}{y}}}, \text{ or } y = c_1 e^{-2\sqrt{\frac{x}{y}}}, \text{ where } c_1 = \frac{1}{c}.$$

3. *Solve the differential equation*

$$\left(x + y\cos\left(\frac{y}{x}\right)\right)dx = x\cos\left(\frac{y}{x}\right)dy$$

Sol. Given $\left(x + y\cos\left(\frac{y}{x}\right)\right)dx = x\cos\left(\frac{y}{x}\right)dy$

∴ $$\frac{dy}{dx} = \frac{x + y\cos\left(\frac{y}{x}\right)}{x\cos\left(\frac{y}{x}\right)} \qquad \text{... (i)}$$

Putting, $y = vx$, $\dfrac{dy}{dx} = v + x\dfrac{dv}{dx}$

$\therefore$ From (*i*), we get

$$v + x\frac{dv}{dx} = \frac{x[1 + v\cos v]}{x\cos v} = \frac{1 + v\cos v}{\cos v}$$

or,
$$x\frac{dv}{dx} = \frac{1}{\cos v}$$

Integrating, we get

$$\int \cos v\, dv = \frac{dx}{x}$$

or
$$\sin v = \log|x| + \log|c|,$$

where c is a non-zero constant

or
$$\sin\left(\frac{y}{x}\right) = \log|cx| \qquad (\because y = vx)$$

or
$$cx = e^{\sin\left(\frac{y}{x}\right)}$$

or
$$x = k\,e^{\sin\left(\frac{y}{x}\right)}, \text{ where } k = \frac{1}{c}.$$

4. *Solve the differential equation*

$$\left(x\tan\frac{y}{x} - y\sec^2\frac{y}{x}\right)dx + x\sec^2\frac{y}{x}\,dy = 0$$

Sol. From the given equation, we have

$$\frac{dy}{dx} = \frac{y\sec^2\frac{y}{x} - x\tan\frac{y}{x}}{x\sec^2\frac{y}{x}} = \frac{y}{x} - \frac{\tan\frac{y}{x}}{\sec^2\frac{y}{x}} \qquad \ldots(i)$$

Putting $y = vx$, we get $\dfrac{dy}{dx} = v + x\dfrac{dv}{dx}$

$\therefore$ From (*i*), we have

$$v + x\frac{dv}{dx} = v - \frac{\tan v}{\sec^2 v}$$

or
$$\frac{\sec^2 v}{\tan v}dv + \frac{dx}{x} = 0$$

Integrating, we get $\log|\tan v| + \log|x| = \log|c|$, where c is a non-zero constant.

or $$\log|x \tan v| = \log|c|$$

or $$x \tan v = c$$

or $$x \tan\left(\frac{y}{x}\right) = c.$$

5. *Solve the differential equation*

$$y\, e^{\frac{x}{y}}\, dx = \left(x\, e^{\frac{x}{y}} + y^2\right) dy$$

Sol. From the given equation, we have

$$\frac{dy}{dx} = \frac{y e^{\frac{x}{y}}}{x e^{\frac{x}{y}} + y^2} \qquad \text{... }(i)$$

Putting, $x = vy$, and differentiating w.r.t. y, we have

$$\frac{dx}{dy} = v + y\frac{dv}{dy}$$

$\therefore$ From (i), we get

$$\frac{1}{v + y\frac{dv}{dy}} = \frac{y e^v}{vy e^v + y^2} = \frac{e^v}{v e^v + y}$$

or $$e^v\left(v + y\frac{dv}{dy}\right) = ve^v + y$$

or $$ve^v + ye^v\frac{dv}{dy} = ve^v + y$$

or $$ye^v\frac{dv}{dy} = y$$

or $$e^v\, dv = dy$$

Integrating, we get

$$y = e^v + c$$

or $$y = e^{\frac{x}{y}} + c.$$

6. *Solve the differential equation*

$$y^2 + x^2\frac{dy}{dx} = xy\frac{dy}{dx}$$

Sol. From the given equation, we have

$$\left(x^2 - xy\right)\frac{dy}{dx} = -y^2$$

or $$\frac{dy}{dx} = -\frac{y^2}{x^2 - xy} \quad \text{...}(i)$$

Putting $y = vx$, we have $\frac{dy}{dx} = v + x\frac{dv}{dx}$

∴ From (*i*), we get

$$v + x\frac{dv}{dx} = -\frac{v^2x^2}{x^2 - vx^2} = -\frac{v^2}{1-v}$$

or $$x\frac{dv}{dx} = -\frac{v^2}{1-v} - v = \frac{-v^2 - v(1-v)}{1-v}$$

$$= \frac{-v^2 - v + v^2}{1-v} = -\frac{v}{1-v}$$

∴ $$\frac{1-v}{v}\,dv = \frac{dx}{x}$$

Integrating, we have

$$\int \frac{1}{v}\,dv - \int dv = \int \frac{dx}{x}$$

or $$\log|v| - v = \log|x| + \log|c|, \text{ (where } c \text{ is a non-zero constant)}$$

or $$\log\left[\frac{v}{cx}\right] = v$$

or $$v = cxe^v$$

or $$y = cx^2 e^{\frac{y}{x}} \quad \left(\because v = \frac{y}{x}\right)$$

7. *Find the solution of the differential equation*

$$x^3\,dy = (2x^2y + y^3)\,dx, \text{ when } x = 1, y = 2.$$

Sol. From the given equation, we have

$$\frac{dy}{dx} = \frac{2x^2y + y^3}{x^3} \quad \text{...}(i)$$

Putting $y = vx$, we get $\frac{dy}{dx} = v + x\frac{dv}{dx}$

$\therefore$ From (*i*), we get

$$v + x\frac{dv}{dx} = \frac{x^3\left(2v + v^3\right)}{x^3} = 2v + v^3$$

or $$x\frac{dv}{dx} = v + v^3 = v\,(1 + v^2)$$

$\therefore$ $$\frac{dv}{v\left(1+v^2\right)} = \frac{dx}{x}$$

or $$\left[\frac{1}{v} - \frac{v}{1+v^2}\right]dv = \frac{dx}{x}$$

Integrating, we get

$$\log|v| - \frac{1}{2}\log|1 + v^2| = \log|x| + \log|c|, \quad \text{(where } c \text{ is a non-zero constant)}$$

or $$\log\left|\frac{y}{x}\right| - \frac{1}{2}\log\left|1 + \frac{y^2}{x^2}\right| = \log|x| + \log|c|$$

Using the conditions, $y = 2$, when $x = 1$, we have

$$\log 2 - \frac{1}{2}\log 5 = \log 1 + \log|c|$$

or $$\log|c| = \log 2 - \frac{1}{2}\log 5$$

$\therefore$ $$\log\left|\frac{y}{x}\right| - \frac{1}{2}\log\left|\frac{x^2 + y^2}{x^2}\right| = \log|x| + \log 2 - \frac{1}{2}\log 5$$

or $$\log\frac{y^2}{x^2} - \log\frac{x^2 + y^2}{x^2} = \log x^2 + \log 4 - \log 5$$

or $$\log\left(\frac{y^2}{x^2} \times \frac{x^2}{x^2 + y^2}\right) = \log\frac{4x^2}{5}$$

or $$\frac{y^2}{x^2 + y^2} = \frac{4x^2}{5}$$

or $$5y^2 = 4x^2\,(x^2 + y^2).$$

8. *Solve the equation* $\dfrac{dy}{dx} + \dfrac{y}{x} = \dfrac{y^2}{x^2}$ [W.B.U.T. BCA, 2006]

Sol. Given $$\frac{dy}{dx} + \frac{y}{x} = \frac{y^2}{x^2} \quad \text{...}(i)$$

Putting $y = vx$, we have $\frac{dy}{dx} = v + x\frac{dv}{dx}$

$\therefore$ From (*i*), we get

$$v + x\frac{dv}{dx} + v = v^2$$

or $$x\frac{dv}{dx} = v^2 - 2v$$

or $$\frac{dv}{v^2 - 2v} = \frac{dx}{x}$$

Integrating, $$\int \frac{dv}{v^2 - 2v} = \int \frac{dx}{x}$$

or $$\int \frac{dv}{(v-1)^2 - (1)^2} = \log |x| + \log |c|,$$ (where c is a non-zero constant)

or $$\frac{1}{2}\log\left|\frac{v-1-1}{v-1+1}\right| = \log |cx|$$

or $$\log\left|\frac{v-2}{v}\right| = \log c^2x^2$$

or $$\frac{v-2}{v} = c^2x^2$$

or $$\frac{\frac{y-2x}{x}}{\frac{y}{x}} = c^2x^2 \qquad (\because y = vx)$$

or $$y - 2x = c^2yx^2$$

or $$(1 - kx^2)\, y = 2x, \text{ where } k = c^2.$$

9. *Solve the equation*

$$xy\frac{dy}{dx} - y^2 = (x+y)^2 e^{-\frac{y}{x}}$$

Sol. Given $$xy\frac{dy}{dx} - y^2 = (x+y)^2 e^{-\frac{y}{x}} \qquad \text{...}(i)$$

Putting $y = vx$, we have $\frac{dy}{dx} = v + x\frac{dv}{dx}$

$\therefore$ From (*i*), we get

$$vx^2\left[v + x\frac{dv}{dx}\right] - v^2x^2 = x^2(1+v)^2 e^{-v}$$

or $$v + x\frac{dv}{dx} = v + \frac{1}{v}(1+v)^2 e^{-v}$$

or $$x\frac{dv}{dx} = \frac{1}{v}(1+v)^2 e^{-v}$$

or $$\frac{v}{(1+v)^2 e^{-v}}\,dv = \frac{dx}{x}$$

or $$\frac{v\,e^{v}}{(1+v)^2}\,dv = \frac{dx}{x}$$

$$\frac{(1+v)\,e^{v} - e^{v}}{(1+v)^2}\,dv = \frac{dx}{x}$$

Integrating,

$$\int \frac{e^{v}}{1+v}\,dv - \int \frac{e^{v}}{(1+v)^2}\,dv = \log|x| + c$$

or $$\frac{1}{1+v}e^{v} + \int \frac{e^{v}}{(1+v)^2}\,dv - \int \frac{e^{v}}{(1+v)^2} = \log|x| + c$$

[Integrating the first integrand by parts]

or $$\frac{e^{v}}{1+v} = \log|x| + c$$

or $$\frac{e^{\frac{y}{x}}}{1+\frac{y}{x}} = \log|x| + c \qquad [\because y = vx]$$

or $$x\,e^{\frac{y}{x}} = (x+y)(c + \log|x|).$$

10. *Find the particular solution of the equation*

$$(x^2 + y^2)\,dx = 2xy\,dy, \text{ when } x = 1, y = 0.$$

Sol. From the given equation, we have

$$\frac{dy}{dx} = \frac{x^2 + y^2}{2xy} \qquad \text{...}(i)$$

Putting $y = vx$, $\frac{dy}{dx} = v + x\frac{dv}{dx}$

$\therefore$ From (*i*), we get

$$v + x\frac{dv}{dx} = \frac{x^2\left(1+v^2\right)}{2x^2 v} = \frac{1+v^2}{2v}$$

or $$x\frac{dv}{dx} = \frac{1+v^2}{2v} - v = \frac{1+v^2-2v^2}{2v} = \frac{1-v^2}{2v}$$

or $$\frac{2v}{1-v^2}dv = \frac{dx}{x}$$

Integrating, $$\int\frac{2v}{1-v^2}dv = \int\frac{dx}{x}$$

or $$-\log|1-v^2| = \log|x| + c$$

or $$-\log\left|1-\frac{y^2}{x^2}\right| = \log|x| + c$$

Using the given conditions, *i.e.*, $y = 0, x = 1$, we get

$$-\log 1 = \log 1 + c$$

or $$c = 0$$

Hence, the particular solution is

$$-\log\left|\frac{x^2-y^2}{x^2}\right| = \log|x|$$

or $$\log\left|\frac{x^2-y^2}{x}\right| = 0, \text{ or } \frac{x^2-y^2}{x} = e^0 = 1$$

or $$x^2 - y^2 = x.$$

EXERCISE 1.3

1. Solve the following differential equations :

(*i*) $x^2y\,dx - (x^3 + y^3)\,dy = 0$ (*ii*) $x\,dy - y\,dx = \sqrt{x^2+y^2}\,dx$

(*iii*) $x^2\frac{dy}{dx} + y^2 = xy$ (*iv*) $\frac{dy}{dx} = \frac{y}{x} + \sin\frac{y}{x}$

(*v*) $\frac{dy}{dx} = \frac{y}{x} + \tan\frac{y}{x}$ (*vi*) $(x^2 - 2xy)\,dy + (x^2 - 3xy + 2y^2)\,dx = 0$

(*vii*) $x\frac{dy}{dx} = y(\log y - \log x)$ (*viii*) $x\frac{dy}{dx} = y + x\cos^2\frac{y}{x}$

(ix) $x \sin\left(\frac{y}{x}\right) dy = \left[y \sin\left(\frac{y}{x}\right) - x\right] dx$ (x) $\left(1 + 3e^{\frac{x}{y}}\right) dx + 3e^{\frac{x}{y}}\left(1 - \frac{x}{y}\right) dy = 0$

(xi) $\frac{dy}{dx} = \frac{y}{x}\left[\log\left(\frac{y}{x}\right) + 1\right]$ (xii) $xy \log\left(\frac{x}{y}\right) dx + \left[y^2 - x^2 \log\left(\frac{x}{y}\right)\right] dy = 0.$

2. Find the particular solutions of the following equations :

(i) $x\, dy = (x + y)\, dx$, $y = 0$, when $x = 1$

(ii) $x\,(2y + x)\, dy = y\,(2y - x)\, dx$, $y = 1$, when $x = 1$

(iii) $(x + y)\, dy = (x - y)\, dx$, $y = 1$, when $x = 1$

(iv) $(x^2 + 2y^2)\, dx - xy\, dy = 0$, $y = 0$, when $x = 1$.

(v) $(x + 2y)\frac{dy}{dx} = x$, $y = 2$, when $x = 1$.

Answers

1. (i) $y = c\, e^{\frac{x^3}{3y^3}}$ (ii) $cx^2 = y + \sqrt{x^2 + y^2}$ (iii) $e^{\frac{x}{y}} = cx$

(iv) $y = 2x \tan^{-1}(cx)$ (v) $x = c \sin\left(\frac{y}{x}\right)$ (vi) $y = x \log\left(\frac{c}{x}\right)$

(vii) $y = x\, e^{1 + cx}$ (viii) $\tan\left(\frac{y}{x}\right) = \log(cx)$ (ix) $\log x = \cos\left(\frac{y}{x}\right) + c$

(x) $x + 3y\, e^{\frac{x}{y}} = c$ (xi) $y = x\, e^{cx}$ (xii) $\log y - \frac{x^2}{4y^2}\left[2 \log\left(\frac{y}{x}\right) + 1\right] = c.$

2. (i) $y = x \log x$ (ii) $\frac{y}{x} + \frac{1}{2}\log(xy) = 1$ (iii) $y^2 + 2xy - x^2 = 2$

(iv) $x^2 + y^2 = cx^4$ (v) $(2y - x)(2y^2 + xy - x^2)^3 = 729\,(x + y)$.

Remark Sometimes, we come across differential equations of the form

$$\frac{dy}{dx} = \frac{a_1x + b_1y + c_1}{a_2x + b_2y + c_2}$$

which are not homogeneous due to the presence of the constant terms c_1, c_2. These equations can be reduced to homogeneous form by applying some transformations. We now discuss how to transform these equations into homogeneous form.

1.7.3 Equations Reducible to Homogeneous Form

Let
$$\frac{dy}{dx} = \frac{a_1x + b_1y + c_1}{a_2x + b_2y + c_2} \qquad \text{... (1)}$$

There are two possible cases :

Case I : When $\frac{a_1}{a_2} = \frac{b_1}{b_2}$.

Let $\frac{a_1}{a_2} = \frac{b_1}{b_2} = \frac{1}{k}, (k \neq 0)$

Then, $a_2 = ka_1$ and $b_2 = kb_1$

Now, equation (1) becomes

$$\frac{dy}{dx} = \frac{(a_1x + b_1y) + c_1}{k(a_1x + b_1y) + c_2}$$

Using the transformation $a_1x + b_1y = v$, the equation is transformed into

$$\frac{dy}{dx} = \frac{v + c_1}{kv_1 + c_2}$$

which is variable separable form and using the method described in 1.7(I). it can be solved

Case II : When $\frac{a_1}{a_2} \neq \frac{b_1}{b_2}$

Let $x = X + h$, $y = Y + k$, where h and k are constants to be chosen suitably.

Then $dx = dX$, $dy = dY$ and $\frac{dy}{dx} = \frac{dY}{dX}$

Now, the equation (1) becomes,

$$\frac{dY}{dX} = \frac{(a_1X + b_1Y) + (a_1h + b_1k + c_1)}{(a_2X + b_2Y) + (a_2h + b_2k + c_2)} \quad \text{... (2)}$$

Choose h and k, such that

$$a_1h + b_1k + c_1 = 0$$

and $$a_2h + b_2k + c_2 = 0$$

Then solve the equations for h and k.

The equation (2) then reduces to

$$\frac{dY}{dX} = \frac{a_1X + b_1Y}{a_2X + b_2Y} \quad \text{... (3)}$$

Which is homogeneous in X and Y and can be solved by putting $Y = vX$, as described in 1.7 (II). Then resubstituting the values of v, h, k and X, Y in the solution, obtained by solving (3), we get the solution of the equation (1).

1.7.4 Illustrative Examples

1. *Solve the equation*

$$\frac{dy}{dx} = \frac{2x - 6y + 7}{x - 3y + 4}$$

Sol. The given equation is

$$\frac{dy}{dx} = \frac{2(x - 3y) + 7}{(x - 3y) + 4} \qquad \text{...}(i)$$

Putting, $x - 3y = v$, we have

$$1 - 3\frac{dy}{dx} = \frac{dv}{dx}$$

or $$\frac{dy}{dx} = \frac{1}{3}\left(1 - \frac{dv}{dx}\right)$$

$\therefore$ From (*i*), we get

$$\frac{1}{3}\left(1 - \frac{dv}{dx}\right) = \frac{2v + 7}{v + 4}$$

or $$1 - \frac{dv}{dx} = \frac{6v + 21}{v + 4}$$

or $$\frac{dv}{dx} = 1 - \frac{6v + 21}{v + 4} = \frac{v + 4 - 6v - 21}{v + 4} = -\frac{(5v + 17)}{v + 4}$$

or $$\frac{v + 4}{5v + 17}\,dv = -\,dx \text{ or } \frac{5v + 20}{5v + 17}\,dv = -\,dx$$

or $$\left(1 + \frac{3}{5v + 17}\right)dv = -5\,dx$$

Integrating, we have $v + \frac{3}{5}\log(5v + 17) = -5x + c$

or $$x - 3y + \frac{3}{5}\log(5x - 15y + 17) = -5x + c$$

or $$2x - y + \frac{1}{5}\log(5x - 15y + 17) = \frac{c}{3}$$

or $$10x - 5y + \log(5x - 15y + 17) = \frac{5}{3}c$$

or $$10x - 5y + \log(5x - 15y + 17) = c_1, \qquad \left(\text{where } c_1 = \frac{5}{3}c\right)$$

2. *Solve the equation* $(2x + 3y - 5)\dfrac{dy}{dx} = 5 - 3x - 2y.$

Sol. The given equation is

$$\frac{dy}{dx} = -\frac{3x + 2y - 5}{2x + 3y - 5} \qquad \text{... (i)}, \left(\textit{Here, } \frac{3}{2} \neq \frac{2}{3}\right)$$

Let $x = X + h$ and $y = Y + k$

where h and k are constants to be chosen suitably.

Then,
$$\frac{dY}{dX} = \frac{dy}{dx}$$

∴ From (*i*), we get

$$\frac{dY}{dX} = -\frac{(3X + 2Y) + (3h + 2k - 5)}{(2X + 3Y) + (2h + 3k - 5)} \qquad \text{... (ii)}$$

Let us choose h and k such that

$$3h + 2k - 5 = 0$$

and
$$2h + 3k - 5 = 0$$

∴
$$\frac{h}{-10 + 15} = \frac{k}{-10 + 15} = \frac{1}{9 - 4}$$

or
$$\frac{h}{5} = \frac{k}{5} = \frac{1}{5}.$$

i.e.,
$$h = 1, k = 1.$$

∴ From (*ii*), we have

$$\frac{dY}{dX} = -\frac{3X + 2Y}{2X + 3Y} \qquad \text{... (iii)}$$

Now putting, $Y = vX$, we get

$$\frac{dY}{dX} = v + X\frac{dv}{dX}$$

∴ From (*iii*), we get

$$v + X\frac{dv}{dX} = -\frac{X(3 + 2v)}{X(2 + 3v)} = -\frac{3 + 2v}{2 + 3v}$$

or,
$$X\frac{dv}{dX} = -\frac{3 + 2v}{2 + 3v} - v = -\left[\frac{3 + 2v + 2v + 3v^2}{2 + 3v}\right] = -\frac{3v^2 + 4v + 3}{2 + 3v}$$

or
$$\frac{2 + 3v}{3v^2 + 4v + 3}\,dv = -\frac{dX}{X}$$

Integrating, we have,

$$\log(3v^2 + 4v + 3) = -2\log X + c$$

or $$\log\left(3\frac{Y^2}{X^2} + 4\frac{Y}{X} + 3\right) + \log X^2 = c = \log k, \text{ (say)}$$

or $$X^2\left(\frac{3Y^2 + 4XY + 3X^2}{X^2}\right) = k$$

or $$3Y^2 + 4XY + 3X^2 = k \qquad \text{... } (iv)$$

$\because$ $$x = X + h, \therefore X = (x - 1)$$
$$y = Y + k, \therefore Y = (y - 1) \qquad (\because h = 1, k = 1)$$

$\therefore$ Replacing X, Y in (*iv*), we get

$$3(y-1)^2 + 4(x-1)(y-1) + 3(x-1)^2 = k$$

or $$3(x^2 + y^2) + 4xy - 10(x + y) = k_1, \text{ where } k_1 = k - 10.$$

which is the required solution of (*i*).

3. *Solve the equation* $(x + y)^2 \dfrac{dy}{dx} = 2x + 2y + 5$, *given* $y = 0$, *when* $x = 1$.

Sol. The given equation is

$$\frac{dy}{dx} = \frac{2(x+y)+5}{(x+y)^2} \qquad \text{... } (i)$$

Putting $x + y = z$, we have $1 + \dfrac{dy}{dx} = \dfrac{dz}{dx}$

or, $$\frac{dy}{dx} = \frac{dz}{dx} - 1$$

$\therefore$ From (*i*), we get

$$\frac{dz}{dx} - 1 = \frac{2z + 5}{z^2}$$

or $$\frac{dz}{dx} = 1 + \frac{2z+5}{z^2} = \frac{z^2 + 2z + 5}{z^2}$$

or $$\frac{z^2}{z^2 + 2z + 5}\, dz = dx$$

or $$\left[1 - \frac{2z+5}{z^2 + 2z + 5}\right] dz = dx$$

or $$\left[1 - \frac{2z+2}{z^2 + 2z + 5} - \frac{3}{(z+1)^2 + (2)^2}\right] dz = dx$$

Integrating, we have,

$$z - \log(z^2 + 2z + 5) - 3 \cdot \frac{1}{2}\tan^{-1}\left(\frac{z+1}{2}\right) = x + c$$

or $$(x+y) - \log[(x+y)^2 + 2(x+y) + 5] - \frac{3}{2}\tan^{-1}\left[\frac{x+y+1}{2}\right] = x + c$$

or $$y - \log[(x+y)^2 + 2(x+y) + 5] - \frac{3}{2}\tan^{-1}\left(\frac{x+y+1}{2}\right) = c \qquad \text{...}(ii)$$

Using the given conditions, *i.e.*, $y = 0, x = 1$, we get

$$-\log[1 + 2 + 5] - \frac{3}{2}\tan^{-1} 1 = c$$

or $$-\log 8 - \frac{3}{2} \cdot \frac{\pi}{4} = c$$

or $$c = -\log 8 - \frac{3}{8}\pi.$$

∴ From (*ii*), we get

$$y - \log[(x+y)^2 + 2(x+y) + 5] - \frac{3}{2}\tan^{-1}\left(\frac{x+y+1}{2}\right) = -\log 8 - \frac{3}{8}\pi$$

or $$y - \log[(x+y)^2 + 2(x+y) + 5] - \frac{3}{2}\tan^{-1}\left(\frac{x+y+1}{2}\right) + \log 8 + \frac{3}{8}\pi = 0.$$

4. *Solve the equation* $(2x + 3y - 6)\,dy + (2y - 6x + 7)\,dx = 0$, *given* $y = 0$, *when* $x = 1$.

Sol. The given equation can be written as

$$\frac{dy}{dx} = \frac{6x - 2y - 7}{2x + 3y - 6} \qquad \text{...}(i)$$

Putting $x = X + h, y = Y + k$, where h and k are constants to be chosen suitably.

∴ $$dx = dX, dy = dY \text{ and } \frac{dy}{dx} = \frac{dY}{dX}$$

∴ From (*i*), we get

$$\frac{dY}{dX} = \frac{(6X - 2Y) + (6h - 2k - 7)}{(2X + 3Y) + (2h + 3k - 6)} \qquad \text{...}(ii)$$

Let us choose h and k, such that

$$6k - 2k - 7 = 0 \text{ and}$$

$$2h + 3k - 6 = 0$$

∴ $$\frac{h}{12 + 21} = \frac{k}{-14 + 36} = \frac{1}{18 + 4}$$

or
$$\frac{h}{33} = \frac{k}{22} = \frac{1}{22}$$

$\therefore \quad h = \frac{3}{2}, k = 1$

Hence $X = x - \frac{3}{2}$ and $Y = y - 1$

Now, from (*ii*), we get

$$\frac{dY}{dX} = \frac{6X - 2Y}{2X + 3Y} \quad \text{... (iii)}$$

Let
$$Y = VX, \text{ then } \frac{dY}{dX} = V + X\frac{dV}{dX}$$

$\therefore$ from (*iii*), we get

$$V + X\frac{dV}{dX} = \frac{2X(3 - V)}{X(2 + 3V)} = \frac{2(3 - V)}{2 + 3V}$$

or,
$$X\frac{dV}{dX} = \frac{2(3 - V)}{2 + 3V} - V = \frac{6 - 2V - 2V - 3V^2}{2 + 3V}$$

or
$$X\frac{dV}{dX} = \frac{6 - 4V - 3V^2}{2 + 3V} = -\frac{3V^2 + 4V - 6}{2 + 3V}$$

or
$$\frac{2 + 3V}{3V^2 + 4V - 6}\,dV = -\frac{dX}{X}$$

$\therefore$
$$\int \frac{2 + 3V}{3V^2 + 4V - 6}\,dV = -\int \frac{dX}{X}$$

or
$$\int \frac{4 + 6V}{3V^2 + 4V - 6}\,dV = -2\int \frac{dX}{X}$$

or $\log\left|3V^2 + 4V - 6\right| + \log c = -2\log|X|$.

where c is a non-zero constant

or,
$$\log\left|c\left(3V^2 + 4V - 6\right)\right| = \log\left|\frac{1}{X^2}\right|$$

or,
$$c(3V^2 + 4V - 6) = \frac{1}{X^2}$$

Substituting the value of $V\left(= \frac{Y}{X}\right)$, we get

$$c\left[3\frac{Y^2}{X^2} + 4\frac{Y}{X} - 6\right] = \frac{1}{X^2}$$

or
$$c[3Y^2 + 4XY - 6X^2] = 1$$

Substituting $X = x - \frac{3}{2}$ and $Y = y - 1$, we obtain

$$c\left[3(y-1)^2+4\left(x-\frac{3}{2}\right)(y-1)-6\left(x-\frac{3}{2}\right)^2\right]=1$$

or $$c[6(y-1)^2+4(2x-3)(y-1)-3(2x-3)^2]=2.$$

5. *Solve the equation*

$$\frac{x+y-a}{x+y-b}\cdot\frac{dy}{dx}=\frac{x+y+a}{x+y+b}$$

Sol. Putting $x+y=z$, we get $1+\dfrac{dy}{dx}=\dfrac{dz}{dx}$

or $$\frac{dy}{dx}=\frac{dz}{dx}-1$$

$\therefore$ From the given equation, we get

$$\frac{z-a}{z-b}\left(\frac{dz}{dx}-1\right)=\frac{z+a}{z+b}$$

or $$\frac{dz}{dx}-1=\frac{(z+a)(z-b)}{(z+b)(z-a)}$$

or $$\frac{dz}{dx}=1+\frac{z^2+az-bz-ab}{z^2+bz-az-ab}$$

$$=\frac{z^2+bz-az-ab+z^2+az-bz-ab}{z^2+bz-az-ab}=\frac{2z^2-2ab}{z^2-az+bz-ab}$$

or $$\frac{z^2-az+bz-ab}{z^2-ab}\,dz=2\,dx$$

or $$\left[1-\frac{(a-b)\,z}{z^2-ab}\right]dz=2\,dx$$

On integration, we get

$$z-\frac{a-b}{2}\log\left(z^2-ab\right)=2x+\frac{c}{2}$$

or $$2z-(a-b)\log(z^2-ab)=4x+c$$

or $$2(x+y)-(a-b)\log[(x+y)^2-ab]=4x+c$$

or $$2(x-y)+(a-b)\log(x^2+y^2+2xy-ab)=c.$$

6. *Show that in an equation of the form*

$$y\,f_1(xy)\,dx+x\,f_2(xy)\,dy=0$$

the variables can be separated by the substitution $xy=v$.

Sol. Since $$xy=v,\ \therefore\ y=\frac{v}{x}$$

and $$d(xy)=dv$$

i.e., $$y\,dx + x\,dy = dv$$

$\because$ $$y = \frac{v}{x},\ \therefore\ dy = \frac{x\,dv - v\,dx}{x^2}$$

or $$x\,dy = dv - \frac{v}{x}\,dx \qquad \text{...}(i)$$

$\therefore$ from $\quad y f_1(xy)\,dx + x f_2(xy)\,dy = 0$

we get,

$$\frac{v}{x} f_1(v)\,dx + f_2(v)\left\{dv - \frac{v}{x}\,dx\right\} = 0 \qquad [\because xy = v \text{ and using } (i)]$$

or $$\frac{f_2(v)\,dv}{v\{f_1(v) - f_2(v)\}} + \frac{dx}{x} = 0$$

which is the variables separated equation.

EXERCISE 1.4

1. Solve the following differential equations :
   (*i*) $(2x - 2y + 5)\,dy = (x - y + 3)\,dx$ (*ii*) $(2x + 4y + 3)\,dy = (x + 2y + 1)\,dx$
   (*iii*) $\dfrac{dy}{dx} = \dfrac{x+y+1}{2x+2y+1}$ (*iv*) $\dfrac{dy}{dx} = \dfrac{6x-4y+3}{3x-2y+1}$
   (*v*) $(x + y - 1)\,dy = (x + y + 1)\,dx$, given $y = \dfrac{1}{3}$, when $x = \dfrac{2}{3}$
   (*vi*) $(3y + 2x + 4)\,dx - (4x + 6y + 5)\,dy = 0$, given $y = 0$, when $x = 1$.
2. Solve the following differential equations :
   (*i*) $\dfrac{dy}{dx} = \dfrac{x+2y-3}{2x+y-3}$
   (*ii*) $\dfrac{dy}{dx} = \dfrac{2x+9y-20}{6x+2y-10}$
   (*iii*) $(6x - 5y + 4)\,dy + (y - 2x - 1)\,dx = 0$
   (*iv*) $(2x + 3y - 6)\,dy + (2y - 6x + 7)\,dx = 0$
   (*v*) $(4x + 3y + 1)\,dx + (3x + 2y + 1)\,dy = 0$.

Answers

1. (*i*) $x - y + 2 = c\,e^{2y-x}$ (*ii*) $4x - 8y = \log(4x + 8y + 5) + c$
   (*iii*) $6y - 3x = \log(3x + 3y + 2) + c$ (*iv*) $2x - y = \log(3x - 2y + 3) + c$
   (*v*) $3(x - y) + 3\log(x + y) = 1$ (*vi*) $7(x - 2y) + 3\log(14x + 21y + 22) = 7 + 6\log 6$.
2. (*i*) $(x + y - 2) = c(x - y)^3$ (*ii*) $(2x - y)^2 = c(x + 2y - 5)$
   (*iii*) $(5y - 2x - 3)^4 = c(4y - 4x - 3)$ (*iv*) $3y^2 + 4xy - (x^2 - 12y + 14x) = c$
   (*v*) $2(x + 1)^2 + (y - 1)^2 + 3(x + 1)(y - 1) = c$.

III. Linear Differential Equations

An equation of the form

$$\frac{dy}{dx} + Py = Q \qquad \text{... }(i)$$

where P and Q are functions of x alone (or constant) is called a **linear differential equation of first order.** Another form of linear differential equation is

$$\frac{dx}{dy} + Px = Q \qquad \text{... }(ii)$$

where P and Q are functions of y alone (or constant).

Remark. The equation (*i*) is also called **Leibnitz's linear equation.**

Method of Solution

To solve equation (*i*), we multiply the equation throughout by a function of x, $e^{\int P\,dx}$, called **integrating factor (I.F.),**

Then from (*i*), we get

$$e^{\int P\,dx}\frac{dy}{dx} + e^{\int P\,dx}\,Py = e^{\int P\,dx}.Q$$

or

$$\frac{d}{dx}\left[y\,e^{\int P\,dx}\right] = e^{\int P\,dx}.Q$$

On integration, we have

$$y\,e^{\int P\,dx} = \int Q\,e^{\int P\,dx}\,dx + c, \text{ where } c \text{ is an arbitrary constant.}$$

or

$$y = e^{-\int P\,dx}\int Q\,e^{\int P\,dx}\,dx + c\,e^{-\int P\,dx}$$

which is the required solution.

Remark. To solve equation (*ii*), we multiply the equation throughout by the I.F. $e^{\int P\,dy}$; then the equation becomes

$$\frac{d}{dy}\left(x\,e^{\int P\,dy}\right) = Q\,e^{\int P\,dy}$$

Integrating, we get

$$x\,e^{\int P\,dy} = \int Q\,e^{\int P\,dy}\,dy + c \text{ which is the required solution.}$$

Thus, the solution of equation (*i*) is y (I.F.) $= \int Q.(\text{I.F.})\,dx + c$

and the solution of equation (*ii*) is x (I.F.) $= \int Q.(\text{I.F.})\,dy + c$.

1.7.5 Illustrative Examples

1. *Solve the equation*

$$\frac{dy}{dx} + y \cot x = \cos x$$

Sol. Given $$\frac{dy}{dx} + y \cot x = \cos x \qquad \text{... }(i)$$

Here $$\text{I.F.} = e^{\int \cot x \, dx} = e^{\log \sin x} = \sin x$$

Multiplying equation (*i*) throughout by $\sin x$, we get

$$\sin x \frac{dy}{dx} + y \cos x = \sin x \cos x$$

or $$\frac{d}{dx}(y \sin x) = \frac{1}{2} \sin 2x$$

Integrating, we get

$$y \sin x = \frac{1}{2} \int \sin 2x \, dx + c$$

or $$y \sin x = -\frac{1}{4} \cos 2x + c = -\frac{1}{4}(1 - 2\sin^2 x) + c$$

$$= \frac{1}{2} \sin^2 x + c_1, \text{ where } c_1 = c - \frac{1}{4}$$

or $$y = \frac{1}{2} \sin x + c_1 \operatorname{cosec} x.$$

2. *Solve the equation*

$$(x+1)\frac{dy}{dx} - y = e^x (x+1)^2.$$

Sol. Dividing the given equation throughout by $(x + 1)$, we get

$$\frac{dy}{dx} - \frac{y}{x+1} = e^x (x+1) \qquad \text{... }(i)$$

Here $$\text{I.F.} = e^{-\int \frac{1}{x+1} dx} = e^{-\log(x+1)} = \frac{1}{x+1}$$

Multiplying the equation (*i*), throughout by $\frac{1}{x+1}$ and integrating, we get

$$y \cdot \frac{1}{x+1} = \int e^x \, dx + c, \text{ where } c \text{ is an arbitrary constant.}$$

or $$\frac{y}{x+1} = e^x + c \text{ or, } y = (x+1)(e^x + c).$$

3. *Solve the equation*

$$\frac{dy}{dx} + 2y \tan x = \sin x, \text{ given } y = 0, \text{ when } x = \frac{\pi}{3}.$$

Sol. Given $\quad \frac{dy}{dx} + 2y \tan x = \sin x \qquad \ldots (i)$

Here $\quad \text{I.F.} = e^{\int 2 \tan x \, dx} = e^{2 \log \sec x} = \sec^2 x$

Multiplying equation (*i*) throughout by $\sec^2 x$ and integrating, we get

$$y \sec^2 x = \int \sin x \sec^2 x \, dx + c$$

$$= \int \sec x \tan x \, dx + c = \sec x + c$$

Using the given conditions, *i.e.*, $y = 0$, when $x = \frac{\pi}{3}$, we get

$$0 = \sec \frac{\pi}{3} + c$$

or $\quad 0 = 2 + c$, or $c = -2$

Hence, $\quad y \sec^2 x = \sec x - 2$

or $\quad y = \cos x - 2 \cos^2 x,$

is the required solution.

4. *Integrating factor of* $x\frac{dy}{dx} - y = 1$ *is*

(*i*) x $\qquad$ (*ii*) $\frac{1}{x}$ $\qquad$ (*iii*) $-x$ $\qquad$ (*iv*) $-\frac{1}{x}$.

(W.B.U.T. BCA, 2006)

Sol. Dividing the given equation throughout by x, we get

$$\frac{dy}{dx} - \frac{1}{x} y = \frac{1}{x}.$$

$\therefore \quad \text{I.F.} = e^{-\int \frac{1}{x} dx} = e^{-\log x} = e^{\log x^{-1}} = \frac{1}{x}.$

Hence, $\quad \text{I.F.} = \frac{1}{x}$

[Solution of the equation is

$$y \cdot \frac{1}{x} = \int \frac{1}{x^2} dx = -\frac{1}{x} + c$$

or $\quad y = -1 + cx$].

5. *Solve the equation*

$$y e^y dx = (y^3 + 2x e^y) dy$$

Sol. The given equation can be written as

$$\frac{dx}{dy} = \frac{y^3 + 2x e^y}{y e^y} = y^2 e^{-y} + 2\frac{x}{y}$$

or $$\frac{dx}{dy} - 2\frac{x}{y} = y^2 e^{-y} \qquad \text{... } (i)$$

Here, $$\text{I.F.} = e^{-2\int \frac{1}{y} dy} = e^{-2\log y} = e^{\log y^{-2}} = \frac{1}{y^2}$$

Multiplying equation (*i*) throughout by $\frac{1}{y^2}$ and integrating, we get

$$x \cdot \frac{1}{y^2} = \int e^{-y}\, dy + c$$

or $$\frac{x}{y^2} = -e^{-y} + c$$

or $$x = (c - e^{-y})\, y^2, \text{ where } c \text{ is an arbitrary constant.}$$

6. *Solve the equation*

$$(1 + y^2)\, dx = (\tan^{-1} y - x)\, dy, \text{ given } y = \frac{\pi}{4}, \text{ when } x = 0.$$

Sol. The given equation can be written as

$$\left(1 + y^2\right)\frac{dx}{dy} = \tan^{-1} y - x$$

or $$(1 + y^2)\frac{dx}{dy} + x = \tan^{-1} y$$

or $$\frac{dx}{dy} + \frac{x}{1 + y^2} = \frac{1}{\left(1 + y^2\right)} \tan^{-1} y \qquad \text{... } (i)$$

Here, $$\text{I.F.} = e^{\int \frac{1}{1+y^2} dy} = e^{\tan^{-1} y}$$

Multiplying equation (*i*) throughout by $e^{\tan^{-1} y}$ and integrating, we get

$$x\, e^{\tan^{-1} y} = \int \frac{\tan^{-1} y}{1 + y^2} e^{\tan^{-1} y}\, dy + c$$

$$= \int z\, e^z\, dz + c \qquad \text{Putting } \tan^{-1} y = z,$$

$$= e^z (z - 1) + c \qquad \text{we have } \frac{1}{1 + y^2} dy = dz$$

or $$x\,e^{\tan^{-1}y} = (\tan^{-1}y - 1)\,e^{\tan^{-1}y} + c$$

Using the given conditions, *i.e.*, when $x = 0, y = \dfrac{\pi}{4}$, we get

$$0 = \left(\tan^{-1}\frac{\pi}{4} - 1\right)e^{\tan^{-1}\frac{\pi}{4}} + c$$

$$= (1-1)\,e + c = c$$

i.e., $c = 0$

Hence the required solution is

$$x\,e^{\tan^{-1}y} = \left(\tan^{-1}y - 1\right)e^{\tan^{-1}y}$$

or $$x + 1 = \tan^{-1}y, \text{ or } y = \tan(x+1).$$

7. *Solve the differential equation*

$$x\cos x\,\frac{dy}{dx} + y\,(x\sin x + \cos x) = 1.$$

Sol. The given equation can be written as

$$\frac{dy}{dx} + \left(\tan x + \frac{1}{x}\right)y = \frac{1}{x\cos x} \qquad \text{[Dividing by } x\cos x\text{]}$$

Here, $$\text{I.F.} = e^{\int\left(\tan x + \frac{1}{x}\right)dx} = e^{\log\sec x}\,.\,e^{\log x} = x\sec x$$

Multiplying throughout the equation by $x\sec x$ and integrating, we get

$$yx\sec x = \int \frac{1}{x\cos x}\,.\,x\sec x\,dx + c$$

$$= \int \sec^2 x\,dx + c = \tan x + c$$

or $$xy = \sin x + c\cos x.$$

8. *Solve the equation*

$$\sqrt{1-y^2}\,dx = (\sin^{-1}y - x)\,dy$$

Sol. The given equation can be written as

$$\sqrt{1-y^2}\,\frac{dx}{dy} = \sin^{-1}y - x$$

or $$\sqrt{1-y^2}\,\frac{dx}{dy} + x = \sin^{-1}y$$

or $$\frac{dx}{dy} + \frac{x}{\sqrt{1-y^2}} = \frac{1}{\sqrt{1-y^2}}\sin^{-1}y \qquad \ldots(i)$$

Here I.F. $= e^{\int \frac{1}{\sqrt{1-y^2}} dy} = e^{\sin^{-1} y}$

Multiplying equation (*i*) throughout by $e^{\sin^{-1} y}$ and integrating, we get

$$x \cdot e^{\sin^{-1} y} = \int \frac{1}{\sqrt{1-y^2}} \cdot \sin^{-1} y \cdot e^{\sin^{-1} y}\, dy$$

Putting $\sin^{-1} y = z$, we have $y = \sin z$

and $$\frac{1}{\sqrt{1-y^2}}\, dy = dz$$

$\therefore$ $$x\, e^{\sin^{-1} y} = \int z\, e^z\, dz = e^z (z-1) + c$$

or $$x \sin^{-1} y = e^{\sin^{-1} y} (\sin^{-1} y - 1) + c.$$

or $$x = \left(\sin^{-1} y - 1\right) + c e^{-\sin^{-1} y}$$

EXERCISE 1.5

Solve the following differential equations :

1. $(1+x)\dfrac{dy}{dx} - xy = 1 - x$
2. $\sqrt{a^2+x^2}\,\dfrac{dy}{dx} + y = \sqrt{a^2+x^2} - x$
3. $\dfrac{dy}{dx} + \dfrac{y}{\left(1-x^2\right)^{\frac{3}{2}}} = \dfrac{x+\sqrt{1-x^2}}{\left(1-x^2\right)^2}$
4. $x(x-1)\dfrac{dy}{dx} - y = x^2 (x-1)^2$
5. $x \log x \dfrac{dy}{dx} + y = 2 \log x$
6. $\left(x + 2y^3\right) dy = y\, dx$
7. $x\left(\dfrac{dy}{dx} + y\right) = 1 - y$
8. $x(1-x^2)\, dy + (2x^2 y - y - ax^3)\, dx = 0$
9. $(1+x^3)\dfrac{dy}{dx} + 6x^2 y = 1 + x^2$
10. $\dfrac{dy}{dx} = \dfrac{y}{2y \log y + y - x}$
11. $e^{-y} \sec^2 y\, dy = dx + x\, dy$
12. $\left(1+y^2\right) + \left(x - e^{-\tan^{-1} y}\right)\dfrac{dy}{dx} = 0.$
13. $\dfrac{dy}{dx} - y \tan x = 3e^{-\sin x}$, given $y = 4$, when $x = 0$.
14. $\dfrac{dy}{dx} + y \cot x = 5e^{\cos x}$, given $y = -4$, when $x = \dfrac{\pi}{2}$
15. $x\dfrac{dy}{dx} - y = \cos\dfrac{1}{x}$
16. $\left(1-x^2\right)\dfrac{dy}{dx} - 2xy = x - x^3$
17. $y \log y \dfrac{dx}{dy} + x - \log y = 0.$

Answers

1. $y(1+x) = x + ce^{x}$

2. $y\left(\dfrac{\sqrt{a^2+x^2}+x}{a}\right) = a\log\left(x+\sqrt{a^2+x^2}\right)+c$

3. $y = \dfrac{x}{\sqrt{1-x^2}} + ce^{-\frac{x}{\sqrt{1-x^2}}}$

4. $\dfrac{xy}{x-1} = \dfrac{1}{3}x^3 + c$

5. $y \log x = (\log x)^2 + c$

6. $x = y^3 + cy$

7. $y(x\,e^{x}) = e^{x} + c$

8. $y + ax = \dfrac{bx}{\sqrt{1-x^2}}$

9. $12y(1+x^3)^2 = 12x + 4x^3 + 3x^4 + 2x^6 + c_1$

10. $xy = c + y^2 \log y$

11. $x\,e^{y} = \tan y + c$

12. $x = e^{-\tan^{-1} y}(\tan^{-1} y + c)$

13. $y\cos x = 7 - 3e^{-\sin x}$

14. $y \sin x = 1 - 5e^{\cos x}$

15. $\dfrac{y}{x} + \sin\dfrac{1}{x} = c$

16. $y(1-x^2) = \dfrac{1}{2}x^2 - \dfrac{1}{4}x^4 + c$

17. $x \log y = \dfrac{1}{2}(\log y)^2 + c.$

Remark. The differential equations of the form

$$\frac{dy}{dx} + Py = Qy^n.$$

where P and Q are functions of x only or constants, are not linear differential equations. These equations can be reduced to linear form. The above mentioned equation is known as **Bernoulli's equation.**

1.7.6 Bernoulli's Equation (equation reducible to linear form)

Let $$\frac{dy}{dx} + Py = Qy^n \qquad \text{... (1)}$$

Dividing both sides of equation (1) by y^n, we have

$$y^{-n}\frac{dy}{dx} + P\,y^{1-n} = Q \qquad \text{... (2)}$$

Substituting $y^{1-n} = z$, and differentiating w.r.t. x, we get

$$(1-n)\,y^{-n}\frac{dy}{dx} = \frac{dz}{dx}$$

or $$y^{-n}\frac{dy}{dx} = \frac{1}{(1-n)}\frac{dz}{dx}$$

Replacing the value of $y^{-n}\dfrac{dy}{dx}$ in (2), we have

$$\frac{1}{(1-n)}\frac{dz}{dx}+\text{P}z=\text{Q}$$

or $$\frac{dz}{dx}+(1-n)\,\text{P}z=(1-n)\,\text{Q} \qquad \text{... (3)}$$

Equation (3) is the linear differential equation with z as the dependent variable. Equation (3) can be solved by the method described in 1.7 (III).

1.7.7 Illustrative Examples

1. *Solve the equation* $\dfrac{dy}{dx}+\dfrac{y}{x}=\dfrac{y^2}{x^2}$ (W.B.U.T., BCA, 2006)

Sol. Dividing the given equation by y^2, we get

$$y^{-2}\frac{dy}{dx}+\frac{y^{-1}}{x}=\frac{1}{x^2} \qquad \text{... }(i)$$

Putting, $y^{-1}=z$, and differentiating w.r.t. x, we have

$$-y^{-2}\frac{dy}{dx}=\frac{dz}{dx} \text{ or } y^{-2}\frac{dy}{dx}=-\frac{dz}{dx}$$

Replacing $y^{-2}\dfrac{dy}{dx}$ in (i), we have

$$-\frac{dz}{dx}+\frac{z}{x}=\frac{1}{x^2}$$

or $$\frac{dz}{dx}-\frac{z}{x}=-\frac{1}{x^2} \qquad \text{... }(ii)$$

Here $$\text{I.F.}=e^{-\int\frac{1}{x}dx}=e^{-\log x}=\frac{1}{x}$$

Multiplying the equation (ii) throughout by $\dfrac{1}{x}$ and integrating, we obtain

$$z\,.\,\frac{1}{x}=-\int\frac{1}{x^2}\cdot\frac{1}{x}\,dx=-\int\frac{1}{x^3}\,dx=\frac{1}{2x^2}+c$$

or $$z=\frac{1}{2x}+cx \text{ or } \frac{1}{y}=\frac{1}{2x}+cx \qquad \left(\because z=\frac{1}{y}\right)$$

which is the required solution of the given equation.

[This problem was solved earlier by the method of homogeneous differential equation. See Example 8 of 1.7.2 Illustrative Examples]

2. *Solve the equation* $x\dfrac{dy}{dx} + y\log y = xy\,e^x$.

Sol. Dividing the given equation by xy, we obtain

$$\frac{1}{y}\frac{dy}{dx} + \frac{1}{x}\log y = e^x \qquad \text{...}(i)$$

Putting $\log y = z$ and differentiating w.r.t. x, we have

$$\frac{1}{y}\frac{dy}{dx} = \frac{dz}{dx}$$

Therefore, from (*i*), we get

$$\frac{dz}{dx} + \frac{1}{x}z = e^x \qquad \text{...}(ii)$$

Here $\qquad \text{I.F.} = e^{\int \frac{1}{x}dx} = e^{\log x} = x$

Multiplying equation (*ii*) by x and integrating, we get

$$z\,.\,x = \int x\,e^x\,dx = e^x(x-1) + c$$

or $\qquad x\log y = e^x(x-1) + c$

which is the required general solution of the given equation.

3. *Solve the equation* $x^2y - x^3\dfrac{dy}{dx} = y^4\cos x$.

Sol. The given equation is

$$-x^3\frac{dy}{dx} + x^2y = y^4\cos x$$

Dividing the equation throughout by x^3y^4, we obtain

$$-\frac{1}{y^4}\frac{dy}{dx} + \frac{1}{xy^3} = \frac{1}{x^3}\cos x \qquad \text{...}(i)$$

Putting, $y^{-3} = z$ and differentiating w.r.t. x,

we get $\qquad -3\dfrac{1}{y^4}\dfrac{dy}{dx} = \dfrac{dz}{dx}$ or $-\dfrac{1}{y^4}\dfrac{dy}{dx} = \dfrac{1}{3}\dfrac{dz}{dx}$

$\therefore$ from (*i*), we get $\dfrac{1}{3}\dfrac{dz}{dx} + \dfrac{z}{x} = \dfrac{1}{x^3}\cos x$

or

$$\frac{dz}{dx} + \frac{3}{x}z = \frac{3}{x^3}\cos x \qquad \text{...}(ii)$$

Here $\qquad \text{I.F.} = e^{\int \frac{3}{x}dx} = e^{3\log x} = x^3$

Multiplying equation (*ii*) throughout by x^3 and integrating, we get

$$z \cdot x^3 = \int 3 \cos x \, dx = 3 \sin x + c$$

or $$\frac{x^3}{y^3} = 3 \sin x + c$$

or $$x^3 = y^3 (3 \sin x + c).$$

which is the required general solution of the given equation.

4. *Solve the equation* $xy\,(1 + xy^2)\,\dfrac{dy}{dx} = 1.$

Sol. The given equation can be written as

$$\frac{dx}{dy} - xy = x^2y^3$$

Dividing by x^2, we get

$$\frac{1}{x^2}\frac{dx}{dy} - \frac{y}{x} = y^3 \qquad \text{... }(i)$$

Putting $\dfrac{1}{x} = z$ and differentiating w.r.t. y, we obtain

$$-\frac{1}{x^2}\frac{dx}{dy} = \frac{dz}{dy}$$

or $$\frac{1}{x^2}\frac{dx}{dy} = -\frac{dz}{dy}$$

Then from (*i*), we get

$$-\frac{dz}{dy} - yz = y^3$$

or $$\frac{dz}{dy} + yz = -y^3 \qquad \text{... }(ii)$$

Here, $$\text{I.F.} = e^{\int y\,dy} = e^{\frac{y^2}{2}}$$

Multiplying equation (*ii*) throughout by $e^{\frac{y^2}{2}}$, and integrating, we get

$$z.e^{\frac{y^2}{2}} = -\int y^3 e^{\frac{y^2}{2}} dy + c \qquad \text{... }(iii)$$

Let $$\text{I} = \int y^3 e^{\frac{y^2}{2}} dy$$

Putting $\frac{y^2}{2} = t$, we have $2 \cdot \frac{1}{2} y\, dy = dt$ or $y\, dy = dt$

$$\therefore \qquad I = \int 2t \cdot e^t\, dt = 2e^t (t-1) = 2e^{\frac{1}{2}y^2}\left(\frac{1}{2}y^2 - 1\right)$$

$$= e^{\frac{1}{2}y^2}\left(y^2 - 2\right)$$

$\therefore$ From (*iii*), we have

$$z \cdot e^{\frac{y^2}{2}} = -e^{\frac{y^2}{2}}(y^2 - 2) + c$$

or
$$\frac{1}{x} = -(y^2 - 2) + c\, e^{-\frac{y^2}{2}}$$

or
$$\frac{1}{x} = 2 - y^2 + c\, e^{-\frac{y^2}{2}}.$$

which is the required general solution of the given equation.

5. *Solve the equation*

$$\sin y \frac{dy}{dx} = \cos x\,(2 \cos y - \sin^2 x)$$

Sol. The given equation can be written as

$$\sin y \frac{dy}{dx} - 2 \cos x \cos y = -\cos x \sin^2 x \qquad \text{... (i)}$$

Putting $\cos y = z$ and differentiating w.r.t. x, we get

$$-\sin y \frac{dy}{dx} = \frac{dz}{dx} \text{ or } \sin y \frac{dy}{dx} = -\frac{dz}{dx}$$

Therefore, from (*i*), we obtain

$$-\frac{dz}{dx} - 2 \cos x \cdot z = -\cos x \sin^2 x$$

or
$$\frac{dz}{dx} + 2z \cos x = \cos x \sin^2 x \qquad \text{... (ii)}$$

Here
$$\text{I.F.} = e^{\int 2 \cos x\, dx} = e^{2 \sin x}$$

Multiplying equation (*ii*) throughout by $e^{2 \sin x}$ and integrating, we get

$$z \cdot e^{2 \sin x} = \int \cos x \sin^2 x \cdot e^{2 \sin x}\, dx + c$$

$$\therefore \qquad z e^{2 \sin x} = \int t^2 e^{2t}\, dt + c$$

| Putting $\sin x = t$, we have $\cos x\, dx = dt$

$$= \left[t^2 \frac{1}{2} e^{2t} - \int 2t \, . \frac{1}{2} e^{2t} \, dt\right] + c \quad \text{[Integrating by parts]}$$

$$= \frac{1}{2} t^2 e^{2t} - \int t \, e^{2t} \, dt + c$$

$$= \frac{1}{2} t^2 e^{2t} - \frac{1}{2} t \, e^{2t} + \int \frac{1}{2} e^{2t} \, dt + c$$

$$= \frac{1}{2} t^2 e^{2t} - \frac{1}{2} t \, e^{2t} + \frac{1}{4} e^{2t} + c$$

Thus the required general solution is

$$\cos y \, e^{2 \sin x} = \frac{1}{2} \sin^2 x \, e^{2 \sin x} - \frac{1}{2} \sin x \, e^{2 \sin x} + \frac{1}{4} e^{2 \sin x} + c$$

[Resubstituting the values of z and t]

or $$\cos y = \frac{1}{2} \sin^2 x - \frac{1}{2} \sin x + \frac{1}{4} + c \, e^{-2 \sin x}$$

or $$4 \cos y = 2 \sin^2 x - 2 \sin x + 1 + 4c \, e^{-2 \sin x}$$

6. *Solve the equation* $y - \cos x \dfrac{dy}{dx} = y^2 (1 - \sin x) \cos x$, *given* $y = 2$ *when* $x = 0$.

Sol. Dividing the given equation by $y^2 \cos x$, we get

$$-\frac{1}{y^2} \frac{dy}{dx} + \frac{1}{y} \sec x = 1 - \sin x \qquad \text{... (i)}$$

Putting $\dfrac{1}{y} = z$ and differentiating w.r.t. x, we obtain

$$-\frac{1}{y^2} \frac{dy}{dx} = \frac{dz}{dx}$$

Therefore, we get from (*i*)

$$\frac{dz}{dx} + z \sec x = 1 - \sin x \qquad \text{... (ii)}$$

Here $$\text{I.F.} = e^{\int \sec x} = e^{\log (\sec x + \tan x)}$$
$$= \sec x + \tan x$$

Multiplying equation (*ii*) throughout by $\sec x + \tan x$ and integrating, we get

$$z \, . (\sec x + \tan x) = \int (\sec x + \tan x)(1 - \sin x) \, dx + c$$

or $$\frac{1}{y} (\sec x + \tan x) = \int \frac{(1 + \sin x)(1 - \sin x)}{\cos x} dx + c$$

or $$\frac{1}{y}(\sec x + \tan x) = \int \frac{1 - \sin^2 x}{\cos x}\, dx + c = \int \cos x\, dx + c$$

$\therefore$ $$\frac{1}{y}(\sec x + \tan x) = \sin x + c \qquad \text{... (iii)}$$

Using the given conditions, $y = 2$, when $x = 0$, we get from (*iii*)

$$\frac{1}{2}(1+0) = 0 + c, \text{ or } c = \frac{1}{2}$$

Therefore, from (*iii*), we get

$$\frac{1}{y}(\sec x + \tan x) = \sin x + \frac{1}{2}$$

or $$2(\sec x + \tan x) = y(2 \sin x + 1)$$

which is the required solution of the given equation.

7. *Solve the equation* $\dfrac{dy}{dx} + \dfrac{x}{1-x^2} y = x\sqrt{y}$.

Sol. Dividing the given equation throughout by $\sqrt{y}$, we get

$$y^{-\frac{1}{2}} \frac{dy}{dx} + \frac{x}{1-x^2} \cdot y^{\frac{1}{2}} = x \qquad \text{... (i)}$$

Putting $y^{\frac{1}{2}} = z$, and differentiating w.r.t. x, we have

$$\frac{1}{2} y^{-\frac{1}{2}} \frac{dy}{dx} = \frac{dz}{dx}$$

or $$y^{-\frac{1}{2}} \frac{dy}{dx} = 2\frac{dz}{dx}$$

Therefore, from (*i*) we get

$$\frac{dz}{dx} + \frac{x}{2(1-x^2)} z = \frac{1}{2} x \qquad \text{... (ii)}$$

Here, $$\text{I.F.} = e^{\frac{1}{2}\int \frac{x}{1-x^2} dx} = e^{-\frac{1}{4}\log(1-x^2)}$$

$$= e^{\log(1-x^2)^{-\frac{1}{4}}} = (1-x^2)^{-\frac{1}{4}}$$

Multiplying (*ii*) by $(1-x^2)^{-\frac{1}{4}}$ and integrating, we have

$$z.(1-x^2)^{-\frac{1}{4}} = \frac{1}{2}\int x.(1-x^2)^{-\frac{1}{4}}\, dx + c$$

$$= \frac{1}{2}\int \frac{x}{(1-x^2)^{\frac{1}{4}}}\, dx + c = -\frac{1}{3}(1-x^2)^{\frac{3}{4}} + c$$

Substituting $z = y^{\frac{1}{2}}$, we have

$$\frac{\sqrt{y}}{\left(1-x^2\right)^{\frac{1}{4}}} = -\frac{1}{3}\left(1-x^2\right)^{\frac{3}{4}} + c$$

or
$$\sqrt{y} = -\frac{1}{3}\left(1-x^2\right) + c\left(1-x^2\right)^{\frac{1}{4}}$$

which is the required general solution of the given equation.

8. *Solve the equation* $x\dfrac{dy}{dx} + y = y^2 \log x.$

Sol. Dividing the given equation throughout by xy^2, we have

$$\frac{1}{y^2}\frac{dy}{dx} + \frac{1}{xy} = \frac{1}{x}\log x \qquad \text{... (i)}$$

Putting $\dfrac{1}{y} = z$ and differentiating w.r.t. x, we get

$$-\frac{1}{y^2}\frac{dy}{dx} = \frac{dz}{dx} \text{ or } \frac{1}{y^2}\frac{dy}{dx} = -\frac{dz}{dx}$$

Therefore from (*i*), we get

$$-\frac{dz}{dx} + \frac{z}{x} = \frac{1}{x}\log x$$

or
$$\frac{dz}{dx} - \frac{z}{x} = -\frac{1}{x}\log x \qquad \text{... (ii)}$$

Here,
$$\text{I.F.} = e^{-\int \frac{1}{x} dx} = e^{-\log x} = \frac{1}{x}$$

Multiplying equation (*ii*) throughout by $\dfrac{1}{x}$ and integrating, we get

$$z \cdot \frac{1}{x} = \int -\frac{1}{x^2}\log x \, dx + c$$

$$= -\int e^{-t}\, t\, dt + c, \qquad \text{Putting } \log x = t, \ \frac{1}{x}dx = dt \text{ and } x = e^t$$

$$= -\left[-t e^{-t} + \int e^{-t}\, dt\right] + c$$

$$= -[-t e^{-t} - e^{-t}] + c$$

$$= te^{-t} + e^{-t} + c$$

$$= e^{-t}(1 + t) + c$$

Substituting the values of z and t, we obtain

$$\frac{1}{xy} = \frac{1}{x}(1 + \log x) + c$$

or
$$\frac{1}{y} = (1 + \log x) + cx$$

which is the required general solution.

EXERCISE 1.6

Solve the following differential equations :

1. $2\frac{dy}{dx} = \frac{y}{x} + \frac{y^2}{x^2}$

2. $\frac{dy}{dx} + \frac{y\log y}{x} = \frac{y(\log y)^2}{x^2}$

3. $3\frac{dy}{dx} + 3\frac{y}{x} = 2x^4y^4$

4. $\frac{dy}{dx} = y\tan x - y^2\sec x$

5. $\tan y\frac{dy}{dx} + \tan x = \cos y\cos^2 x$

6. $\frac{dy}{dx} = 2y\tan x + y^2\tan^2 x$, given $y = 1$, when $x = 0$.

7. $\frac{dy}{dx} + \tan x\tan y = \cos x\sec y$

8. $\left(xy^2 - e^{x^{\frac{1}{3}}}\right)dx - x^2y\,dy = 0$

9. $\cos x\frac{dy}{dx} + 4y\sin x = 4\sqrt{y}\sec x$

10. $x(3x + 2y^2)\,dx + 2y(1 + x^2)\,dy = 0$

11. $(x^2y^2 + xy)\,y\,dx + (x^2y^2 - 1)\,x\,dy = 0$

12. $dy + y\tan x\,dx = y^2\sec x\,dx$

13. $(x - y^2)\,dx + 2xy\,dy = 0$

14. $(2x - 10y^3)\frac{dy}{dx} + y = 0.$

15. $(x^2 + y^2 + x)\,dx + xy\,dy = 0$

16. $\frac{1}{y^2}\frac{dy}{dx} - \frac{1}{y} = 2x\,e^{-x}$

17. $\frac{dy}{dx} + x\sin 2y = x^3\cos^2 y$

18. $\frac{dy}{dx} - \frac{\tan y}{1+x} = (1 + x)\,e^x\sec y.$

19. $e^y\left(\frac{dy}{dx} + 1\right) = e^x$

20. $\sec^2 y\frac{dy}{dx} + x\tan y = x^3$

21. $\frac{dy}{dx} = e^{x-y}(e^x - e^y).$

22. $\frac{dy}{dx} + \left(2x\tan^{-1} y - x^3\right)\left(1 + y^2\right) = 0$

23. $x\frac{dy}{dx} = y(\log y - \log x + 1)$

24. $\cos t\frac{dx}{dt} + x\sin t = 1$, given $x = -1$ when $t = \pi$.

Answers

1. $x = y\left(1+c\sqrt{x}\right)$

2. $\dfrac{1}{x\log y} = \dfrac{1}{2x^2}+c$

3. $\dfrac{1}{y^3} = -x^5 + cx^3$

4. $\sec x = (\tan x + c)\, y$

5. $\sec y = (\sin x + c)\cos x$

6. $\dfrac{1}{y}\sec^2 x = -\dfrac{1}{3}\tan^3 x + 1$

7. $\sin y \sec x = x + c$

8. $3y^2 = x^2\left(2e^{x^{\frac{1}{3}}} + c\right)$

9. $\sqrt{y}\sec^2 x = 2\left[\tan x + \dfrac{1}{3}\tan^3 x\right] + c$

10. $y^2(1 + x^2) = -x^3 + c$

11. $xy = \log(cy)$

12. $y(x + c) + \cos x = 0$

13. $y^2 + x\log x = cx$

14. $x = 2y^3 + c\,y^{-2}$

15. $6x^2y^2 = 3x^4 - 4x^2 + c$

16. $e^x + x^2y + cy = 0$

17. $\tan y = \dfrac{1}{2}(x^2 - 1) + c\,e^{-x^2}$

18. $\sin y = (1 + x)(e^x + c)$

19. $e^{x+y} = \dfrac{1}{2}e^{2x} + c$

20. $\tan y = x^2 - 2 + c\,e^{-\frac{x^2}{2}}$

21. $e^y = (e^x - 1) + c\,e^{-e^x}$

22. $2\tan^{-1} y = x^2 - 1 + 2c\,e^{-x^2}$

23. $\log y = \log x + cx$ or $y = xe^{cx}$.

24. $x = \cos t + \sin t$.

IV. Exact Differential Equations

A differential equation obtained from its primitive directly by differentiation, without any other operations such as multiplication, elimination or reduction etc., is said to be an *exact differential equation.*

Thus a differential equation of the form $M(x, y)\,dx + N(x, y)\,dy = 0$ is an exact differential equation if it is obtained directly by differentiating the equation $u(x, y) = c$, which is its primitive.

That is, if $$du = M\,dx + N\,dy.$$

An exact differential equation can often be solved by *inspection*, by picking out the terms of $M\,dx + N\,dy$ that may form a perfect differential and then integrating it. While grouping the terms suitably to form a *perfect differential,* it will be convenient to remember the following perfect differentials :

(*i*) $dx \pm dy = d(x \pm y)$

(*ii*) $x\,dy + y\,dx = d(xy)$

(*iii*) $x\,dx \pm y\,dy = \dfrac{1}{2}d(x^2 \pm y^2)$

(*iv*) $\dfrac{x\,dy - y\,dx}{x^2} = d\left(\dfrac{y}{x}\right)$

(v) $\dfrac{x\,dy - y\,dx}{y^2} = -d\left(\dfrac{x}{y}\right)$ (vi) $\dfrac{x\,dy - y\,dx}{xy} = d\left[\log\left(\dfrac{y}{x}\right)\right]$

(vii) $\dfrac{y\,dx - x\,dy}{x^2+y^2} = d\left[\tan^{-1}\left(\dfrac{x}{y}\right)\right]$ (viii) $\dfrac{x\,dy - y\,dx}{x^2+y^2} = d\left[\tan^{-1}\left(\dfrac{y}{x}\right)\right]$

(ix) $\dfrac{x\,dy + y\,dx}{\sqrt{1-x^2y^2}} = d\,[\sin^{-1}(xy)]$ (x) $\dfrac{2xy\,dy - y^2\,dx}{x^2} = d\left(\dfrac{y^2}{x}\right)$

(xi) $\dfrac{y^2\,dx + 2xy\,dy}{x^2y^4} = -d\left(\dfrac{1}{xy^2}\right)$.

1.7.8 Illustrative Examples

1. *Solve* $x\,dx + y\,dy + (x^2 + y^2)\,dy = 0$.

Sol. $x\,dx + y\,dy + (x^2+y^2)\,dy = 0$ can be written as $\frac{1}{2}\,d\,(x^2+y^2) + (x^2+y^2)\,dy = 0$

or
$$\frac{d\left(x^2+y^2\right)}{x^2+y^2} + 2\,dy = 0$$

Integrating, we obtain
$$\log(x^2+y^2) + 2y = c$$
which is the required solution of the given equation.

2. *Solve the differential equation*
$$y\,dx + x\,dy = xy\,(dy - dx).$$

Sol. The given equation can be written as
$$d\,(xy) = xy\,d\,(y - x)$$
or
$$\frac{d\,(xy)}{xy} = d\,(y - x).$$
On integration, we have
$$\log(xy) = y - x + c$$
which is the required solution of the given equation.

3. *Solve the equation*
$$x^2\frac{dy}{dx} + xy + 2\sqrt{1-x^2y^2} = 0.$$

Sol. From the given equation, we have
$$x^2\,dy + xy\,dx + 2\sqrt{1-x^2y^2}\,dx = 0$$

or $$\frac{x\,(x\,dy + y\,dx)}{\sqrt{1-x^2y^2}} + 2\,dx = 0$$

or $$\frac{x\,dy + y\,dx}{\sqrt{1-x^2y^2}} + \frac{2}{x}\,dx = 0$$

or $$d\,[\sin^{-1}(xy)] + \frac{2}{x}\,dx = 0$$

On integration, we have

$$\sin^{-1}(xy) + 2\log x = c$$

which is the required solution of the given equation.

4. *Solve the equation* $y\dfrac{dy}{dx}\sin x = \cos x\left(\sin x - \dfrac{y^2}{2}\right)$, *given* $y = 1$, *when* $x = \dfrac{\pi}{2}$.

Sol. $\because$ $$y\frac{dy}{dx}\sin x = \cos x\left(\sin x - \frac{y^2}{2}\right)$$

$\therefore$ $$y\frac{dy}{dx}\sin x + \frac{y^2}{2}\cos x = \cos x \sin x$$

or $$2y\frac{dy}{dx}\sin x + y^2\cos x = \sin 2x$$

or $2y \sin x\, dy + y^2 \cos x\, dx = \sin 2x\, dx$

or $$d\,(y^2 \sin x) = \sin 2x\, dx$$

Integrating, we get

$$y^2 \sin x = -\frac{1}{2}\cos 2x + c$$

Using the given conditions, $y = 1$, when $x = \dfrac{\pi}{2}$, we get

$$1 \,.\, \sin\frac{\pi}{2} = -\frac{1}{2}\cos \pi + c$$

or $$1 = \frac{1}{2} + c \;\therefore\; c = \frac{1}{2}$$

Hence, the required solution is

$$y^2 \sin x = \frac{1}{2}(1 - \cos 2x) = \frac{1}{2}\,.\,2\sin^2 x$$

or $$y^2 = \sin x.$$

5. *Solve the equation $x^2 (x\,dx + y\,dy) + 2y (x\,dy - y\,dx) = 0$.*

Sol. The given equation can be written as

$$x^2 (x\,dx + y\,dy) + 2x (x\,dx + y\,dy) - 2 (x^2 + y^2)\,dx = 0$$

or $$(x\,dx + y\,dy)(x^2 + 2x) - 2 (x^2 + y^2)\,dx = 0$$

or $$\frac{1}{2} d(x^2 + y^2)(x^2 + 2x) - 2 (x^2 + y^2)\,dx = 0$$

or $$\frac{d\left(x^2+y^2\right)}{x^2+y^2} - 4\frac{dx}{x(x+2)} = 0$$

or $$\frac{d\left(x^2+y^2\right)}{x^2+y^2} + 2\left[\frac{1}{x+2} - \frac{1}{x}\right]dx = 0$$

Integrating, we get

$$\log (x^2 + y^2) + 2 [\log (x + 2) - \log x] = \log c$$

or $$\log (x^2 + y^2) + \log \left(\frac{x+2}{x}\right)^2 = \log c$$

or $$\log\left[\left(x^2+y^2\right).\left(\frac{x+2}{x}\right)^2\right] = \log c$$

or $$(x^2 + y^2)(x + 2)^2 = cx^2.$$

which is the required solution of the given equation.

EXERCISE 1.7

Find the solutions of the following differential equations :

1. $x\,dy - y\,dx - 2\sqrt{x^2 - y^2}\,dx = 0$
2. $x\,dy - y\,dx = \cos\left(\frac{1}{x}\right)dx$
3. $(x + x^5 + 2x^3y^2 + xy^4)\,dx + y\,dy = 0$
4. $x\,dx + y\,dy + \dfrac{x\,dy - y\,dx}{x^2 + y^2} = 0$
5. $y\,dx - x\,dy = xy\,dx$
6. $x\,dx + y\,dy + (x^2 + y^2)\,dy = 0$, given $y = 0$, when $x = 1$
7. $(1 + xy)\,y\,dx + (1 - xy)\,x\,dy = 0$
8. $y\,(2x^2y + e^x)\,dx - (e^x + y^3)\,dy = 0$
9. $x^2y_1 + xy + 2\sqrt{1 - x^2y^2} = 0$
10. $y\,(1 + xy)\,dx - x\,dy = 0.$

Answers

1. $y = x \sin [\log (cx^2)]$
2. $\frac{y}{x} + \sin\left(\frac{1}{x}\right) = c$
3. $(x^2 + y^2)(x^2 + c) = 1$
4. $\frac{1}{2}(x^2 + y^2) + \tan^{-1}\left(\frac{y}{x}\right) = c$
5. $y = kx\, e^{-x}$
6. $y + \frac{1}{2}\log(x^2 + y^2) = 0$
7. $x = ky\, e^{\frac{1}{xy}}$
8. $\frac{e^x}{y} + \frac{2x^3}{3} - \frac{y^2}{2} = c$
9. $\sin^{-1}(xy) + 2 \log x = c$
10. $2x + x^2 y = 2cy.$

Remark. If an equation of the form M dx + N dy = 0 cannot be solved by inspection then we shall test whether the equation is exact or not. If the equation is exact, we can solve the equation by applying a method that will be discussed here. Before describing the method of solution we state and prove a theorem on the necessary and sufficient condition for the equation M dx + N dy = 0 to be exact.

Theorem 1.1. *The necessary and sufficient condition for the differential equation M dx + N dy = 0 to be exact is* $\frac{\partial M}{\partial y} = \frac{\partial N}{\partial x}$.

Proof. *The condition is necessary.*

The equation $\quad M\,dx + N\,dy = 0$ will be exact, if

$$du = M\,dx + N\,dy \qquad \dots (i)$$

But,

$$du = \frac{\partial u}{\partial x}dx + \frac{\partial u}{\partial y}dy \qquad \dots (ii)$$

$\therefore$ From (*i*) and (*ii*), we have

$$M\,dx + N\,dy = \frac{\partial u}{\partial x}dx + \frac{\partial u}{\partial y}dy$$

Equating the coefficients of dx and dy from both sides, we get

$$M = \frac{\partial u}{\partial x} \text{ and } N = \frac{\partial u}{\partial y}$$

Again,

$$\frac{\partial M}{\partial y} = \frac{\partial^2 u}{\partial y \partial x} \text{ and } \frac{\partial N}{\partial x} = \frac{\partial^2 u}{\partial x \partial y}$$

But

$$\frac{\partial^2 u}{\partial y \partial x} = \frac{\partial^2 u}{\partial x \partial y}$$

$\therefore$

$$\frac{\partial M}{\partial y} = \frac{\partial N}{\partial x}$$

which is the necessary condition of exactness

The condition is sufficient :

Let $u = \int M\,dx$, (in M y is constant)

$$\therefore \qquad \frac{\partial u}{\partial x} = M \text{ and } \frac{\partial^2 u}{\partial y \partial x} = \frac{\partial M}{\partial y}$$

$$\text{But} \qquad \frac{\partial^2 u}{\partial y \partial x} = \frac{\partial^2 u}{\partial x \partial y} \text{ and } \frac{\partial M}{\partial y} = \frac{\partial N}{\partial x}$$

$$\therefore \qquad \frac{\partial N}{\partial x} = \frac{\partial^2 u}{\partial x \partial y} = \frac{\partial}{\partial x}\left(\frac{\partial u}{\partial y}\right)$$

Integrating both sides w.r.t. x, treating y as constant, we have

$$N = \frac{\partial u}{\partial y} + f(y)$$

$$\therefore \qquad M\,dx + N\,dy = \frac{\partial u}{\partial x}dx + \left[\frac{\partial u}{\partial y} + f(y)\right]dy, \qquad \left[\because M = \frac{\partial u}{\partial x} \text{ and } N = \frac{\partial u}{\partial y} + f(y)\right]$$

$$= \left(\frac{\partial u}{\partial x}dx + \frac{\partial u}{\partial y}dy\right) + f(y)\,dy$$

$$= du + f(y)\,dy = d\left[u + \int f(y)\,dy\right]$$

which shows that M dx + N dy is an exact differential and hence M dx + N dy = 0 is an exact differential equation.

Hence the theorem is proved.

Thus, $\frac{\partial M}{\partial y} = \frac{\partial N}{\partial x}$ is the required condition of exactness of the equation M dx + N dy = 0.

Method for solving M dx + N dy = 0 when it is exact.

Step I. First integrate the terms in M w.r.t. x, treating y as constant.

Step II. Then integrate, those terms of N which do not contain x, w.r.t. y.

Step III. Equate to some constant the sum of the results of Step I and Step II,

i.e., $$\int M\,dx + \int N\,dy = c$$

(Treating y as constant)(Only those terms which do not contain x).

Step IV. If N has no term which is free from x, then the solution will be

$$\int_{(y \text{ as constant})} M\,dx = c$$

Thus, the solution of $M\,dx + N\,dy = 0$ is

$$\int_{(y \text{ as constant})} M\,dx + \int (\text{terms of N not containing } x)\,dy = c$$

1.7.9 Illustrative Examples

1. *Solve the equation* $(x^2 - 4xy - 2y^2)\,dx + (y^2 - 4xy - 2x^2)\,dy = 0$

Sol. The given equation is of the form $M\,dx + N\,dy = 0$

Here, $M = x^2 - 4xy - 2y^2$ and $N = y^2 - 4xy - 2x^2$

$\therefore$ $\dfrac{\partial M}{\partial y} = -4x - 4y$ and $\dfrac{\partial N}{\partial x} = -4y - 4x$

$\therefore$ $\dfrac{\partial M}{\partial y} = \dfrac{\partial N}{\partial x}$

Hence the given equation is exact.

Therefore, its solution is

$$\int \left(x^2 - 4xy - 2y^2\right) dx + \int y^2\,dy = c$$

or $\dfrac{x^3}{3} - 2x^2y - 2xy^2 + \dfrac{y^3}{3} = c$

or $x^3 - 6x^2y - 6xy^2 + y^3 = 3c = k$, when $k = 3c$

or $x^3 + y^3 - 6xy\,(x + y) - k = 0.$

2. *Solve the equation*

$\cos x\,(\cos x - \sin a \sin y)\,dx + \cos y\,(\cos y - \sin a \sin x)\,dy = 0$

Sol. The given equation is of the form

$M\,dx + N\,dy = 0$

where $M = \cos x\,(\cos x - \sin a \sin y)$ and $N = \cos y\,(\cos y - \sin a \sin x)$

$\therefore$ $\dfrac{\partial M}{\partial y} = -\sin a \cos x \cos y$ and $\dfrac{\partial N}{\partial x} = -\sin a \cos y \cos x$

$\therefore$ $\dfrac{\partial M}{\partial y} = \dfrac{\partial N}{\partial x}$

Hence the given equation is exact.

Therefore, the general solution of the given equation is

$$\int \left(\cos^2 x - \sin a \cos x \sin y\right) dx + \int \cos^2 y\,dy = \frac{c}{2}$$

or $\int (1 + \cos 2x - 2 \sin a \cos x \sin y)\, dx + \int (1 + \cos 2y)\, dy = c$

or $$x + \frac{\sin 2x}{2} - 2 \sin a \sin x \sin y + y + \frac{\sin 2y}{2} = c$$

or $$2(x + y) + (\sin 2x + \sin 2y) - 4 \sin a \sin x \sin y = 2c$$

or $2(x + y) + (\sin 2x + \sin 2y) = 4 \sin a \sin x \sin y + k$, (where $k = 2c$)

3. *Solve the equation* $(e^y + 1) \cos x\, dx + e^y \sin x\, dy = 0$.

Sol. The given equation is of the form $M\,dx + N\,dy = 0$

Here, $$M = (e^y + 1) \cos x, N = e^y \sin x$$

$\therefore$ $$\frac{\partial M}{\partial y} = e^y \cos x \text{ and } \frac{\partial N}{\partial x} = e^y \cos x$$

$\because$ $$\frac{\partial M}{\partial y} = \frac{\partial N}{\partial x}.$$

therefore, the given equation is exact.

Hence its solution is $\int (e^y + 1) \cos x\, dx = c$

[Since N has no term of y only]

or $$(e^y + 1) \sin x = c.$$

4. *Find the value of* λ, *for which the differential equation*

$(xy^2 + \lambda x^2 y)\, dx + (x + y)\, x^2\, dy = 0$ *is exact, and solve the equation for this value of* λ.

Sol. $(xy^2 + \lambda x^2 y)\, dx + (x + y)\, x^2\, dy = 0$ ... (*i*)

Comparing the equation (*i*) with $M\,dx + N\,dy = 0$, we have

$$M = xy^2 + \lambda x^2 y \text{ and } N = (x + y)\, x^2$$

$\therefore$ $$\frac{\partial M}{\partial y} = 2xy + \lambda x^2 \text{ and } \frac{\partial N}{\partial x} = (3x^2 + 2xy)$$

The equation (*i*) will be exact,

if $$\frac{\partial M}{\partial y} = \frac{\partial N}{\partial x}$$

i.e., if $$2xy + \lambda x^2 = 3x^2 + 2xy$$

i.e., if $$\lambda x^2 = 3x^2$$

i.e., if $$\lambda = 3.$$

Hence, the equation (*i*) will be exact if $\lambda = 3$.

$\therefore$ The solution of the equation (*i*) for $\lambda = 3$ is

$$\int (xy^2 + 3x^2 y)\, dx = c$$

or $$\frac{1}{2}x^2y^2 + x^3y = c$$

or $$x^2y^2 + 2x^3y = k, \text{ where } k = 2c.$$

5. *Find the solution of the equation*
(2x − y + 1) dx + (2y − x − 1) dy = 0.

Sol. Comparing the given equation with M dx + N dy = 0, we have M = $2x - y + 1$ and N = $2y - x - 1$

$$\therefore \quad \frac{\partial M}{\partial y} = -1, \frac{\partial N}{\partial x} = -1.$$

$$\because \quad \frac{\partial M}{\partial y} = \frac{\partial N}{\partial x}$$

therefore, the equation is exact, and its solution is

$$\int (2x - y + 1)\, dx + \int (2y - 1)\, dy = c$$

or $$x^2 - xy + x + y^2 - y = c$$

or $$x^2 + y^2 - xy + x - y = c.$$

6. *The differential equation*
$(ay^2 + x + x^8)\, dx + (y^8 - y + bxy)\, dy = 0$ is exact if

(i) $b = a$ (ii) $b = 2a$ (iii) $a = 1, b = 3$ (iv) $b \neq 2a$

Sol. Comparing the equation with M dx + N dy = 0, we have M = $ay^2 + x + x^8$ and N = $y^8 - y + bxy$

$$\therefore \quad \frac{\partial M}{\partial y} = 2ay \text{ and } \frac{\partial N}{\partial x} = by$$

The equation will be exact, if $\frac{\partial M}{\partial y} = \frac{\partial N}{\partial x}$

i.e., if $2ay = by$

i.e., if $2a = b$

Hence, the answer is (*ii*) $b = 2a$.

7. *Solve the equation*
$(\sec x \tan x \tan y - e^x)\, dx + \sec x \sec^2 y\, dy = 0$

Sol. The given equation is

$$(\sec x \tan x \tan y - e^x)\, dx + \sec x \sec^2 y\, dy = 0 \quad \text{...}(i)$$

Here, M = $\sec x \tan x \tan y - e^x$, N = $\sec x \sec^2 y$

$$\therefore \quad \frac{\partial M}{\partial y} = \sec x \tan x \sec^2 y, \frac{\partial N}{\partial x} = \sec x \tan x \sec^2 y$$

$$\because \quad \frac{\partial M}{\partial y} = \frac{\partial N}{\partial x},$$

therefore equation (*i*) is exact and its solution is $\int \left(\sec x \tan x \tan y - e^x\right) dx = c$

or $$\sec x \tan y - e^x = c$$

8. *Solve* : $y\, e^{xy}\, dx + (x\, e^{xy} + 2y)\, dy = 0$

Sol. The given equation is of the form $M\,dx + N\,dy = 0$

where $$M = y\, e^{xy} \text{ and } N = x\, e^{xy} + 2y$$

Now, $$\frac{\partial M}{\partial y} = e^{xy} + xy\, e^{xy} \text{ and } \frac{\partial N}{\partial x} = e^{xy} + xy\, e^{xy}$$

Here, $\frac{\partial M}{\partial y} = \frac{\partial N}{\partial x}$, so the equation is exact and its solution is

$$\int y\, e^{xy}\, dx + \int 2y\, dy = c$$

or $$e^{xy} + y^2 = c.$$

EXERCISE 1.8

Solve the following differential equations :

1. $(3x^2 + 6xy^2)\, dx + (6x^2y + 4y^3)\, dy = 0$
2. $(x^3 - 3xy^2)\, dx + (y^3 - 3x^2y)\, dy = 0$ when $y(0) = 1$
3. $\left(y^2 e^{xy^2} + 4x^3\right) dx + \left(2xy\, e^{xy^2} - 3y^2\right) dy = 0$
4. $(y \sec^2 x + \sec x \tan x)\, dx + (\tan x + 2y)\, dy = 0$
5. $(2xy \cos x^2 - 2xy + 1)\, dx + (\sin x^2 - x^2)\, dy = 0$
6. $\left[\cos x \log(2y - 8) + \frac{1}{x}\right] dx + \frac{\sin x}{y - 4}\, dy = 0$
7. $\left[y\left(1 + \frac{1}{x}\right) + \cos y\right] dx + (x + \log x - x \sin y)\, dy = 0$
8. $(y \cos x + 1)\, dx + \sin x\, dy = 0.$
9. $(x^4 e^x - 2m\, xy^2)\, dx + 2m\, x^2y\, dy = 0$
10. $\left(1 + 3e^{\frac{x}{y}}\right) dx + 3e^{\frac{x}{y}}\left(1 - \frac{x}{y}\right) dy = 0$
11. $(\sin x \cos y + e^{2x})\, dx + (\cos x \sin y + \tan y)\, dy = 0$
12. $\left(1 + e^{\frac{x}{y}}\right) dx + \left(1 - \frac{x}{y}\right) e^{\frac{x}{y}}\, dy = 0.$

Answers

1. $x^3 + 3x^2y^2 + y^4 = c$
2. $x^4 - 6x^2y^2 + y^4 = 1$
3. $e^{xy^2} + x^4 - y^3 = c$
4. $y \tan x + \sec x + y^2 = c$
5. $y \sin x^2 - x^2y + x = c$
6. $\sin x \log (2y - 8) + \log x = c$
7. $y (x + \log x) + x \cos y = c$
8. $y \sin x + x = c$
9. $x^2 e^x + my^2 = cx^2$
10. $x + 3e^{\frac{x}{y}} = c$
11. $\frac{1}{2} e^{2x} - \cos x \cos y + \log \sec y = c$
12. $x + y e^{\frac{x}{y}} = c.$

Remark. Linear differential equations of the form M dx + N dy = 0 which are not exact can sometimes be reduced to exact form after multiplying by a suitable factor of x or y or x and y, called integrating factor (I.F.). These equations are reducible to exact equations. We mention below some rules for finding integrating factors.

1.7.10 Some rules for finding integrating factors for equations of the form M dx + N dy = 0 to make it exact (when it is not exact)

Rule 1. If $Mx + Ny \neq 0$ and the equation M dx + Ndy = 0 is homogeneous in x and y, then $\frac{1}{Mx + Ny}$ is an I.F. of the equation.

As an example, the equation $(x^2y - 2xy^2)\, dx - (x^3 - 3x^2y)\, dy = 0$ or $(x^2y - 2xy^2)\, dx + (3x^2y - x^3)\, dy = 0$ is not exact,

because $\frac{\partial M}{\partial y} = x^2 - 4xy$ and $\frac{\partial N}{\partial x} = 6xy - 3x^2$

i.e., $\frac{\partial M}{\partial y} \neq \frac{\partial N}{\partial x}$, but the equation is homogeneous in x, y.

Now,
$$Mx + Ny = (x^2y - 2xy^2)\,x + (3x^2y - x^3)\,y$$
$$= x^3y - 2x^2y^2 + 3x^2y^2 - x^3y = x^2y^2 \neq 0$$

$\therefore$
$$\text{I.F.} = \frac{1}{x^2y^2}.$$

Multiplying the equation throughout by $\frac{1}{x^2y^2}$, we get

$$\left(\frac{1}{y} - \frac{2}{x}\right) dx + \left(3\frac{1}{y} - \frac{x}{y^2}\right) dy = 0 \qquad \ldots (i)$$

Now
$$M = \frac{1}{y} - \frac{2}{x} \text{ and } N = \frac{3}{y} - \frac{x}{y^2}$$

and
$$\frac{\partial M}{\partial y} = -\frac{1}{y^2}, \frac{\partial N}{\partial x} = -\frac{1}{y^2}$$

$$\therefore \quad \frac{\partial M}{\partial y} = \frac{\partial N}{\partial x}$$

Thus the equation (*i*) is now exact and its solution is

$$\int\left(\frac{1}{y}-\frac{2}{x}\right)dx+\int\frac{3}{y}\,dy = c$$

or
$$\frac{x}{y} - 2\log x + 3\log y = c$$

or
$$\frac{x}{y} + \log\left(\frac{y^3}{x^2}\right) = c.$$

Rule 2. If the equation M dx + N dy = 0 is of the form $y f_1(x, y)\, dx + x f_2(x, y)\, dy = 0$, and $Mx - Ny \neq 0$, then $\frac{1}{Mx - Ny}$ is the integrating factor.

As an example, the equation $y(xy + 2x^2y^2)\,dx + x(xy - x^2y^2)\,dy = 0$ is not exact because here $\frac{\partial M}{\partial y} = 2xy + 6y^2x^2$ is not equal to $\frac{\partial N}{\partial x} = 2xy - 3x^2y^2$.

But the equation is of the form $y f_1(x, y)\, dx + x f_2(x, y)\, dy = 0$

and
$$Mx - Ny = xy(xy + 2x^2y^2) - xy(xy - x^2y^2) = 3x^3y^3 \neq 0$$

Hence, $\frac{1}{x^3y^3}$ is an integrating factor of the equation.

Multiplying the equation throughout by $\frac{1}{x^3y^3}$, we get

$$\left(\frac{1}{3x^2y}+\frac{2}{3x}\right)dx+\left(\frac{1}{3xy^2}-\frac{1}{3y}\right)dy = 0 \quad \text{... }(i)$$

Now,
$$\frac{\partial M}{\partial y} = -\frac{1}{3x^2y^2} \text{ and } \frac{\partial N}{\partial x} = -\frac{1}{3x^2y^2}$$

$$\therefore \quad \frac{\partial M}{\partial y} = \frac{\partial N}{\partial x}$$

The equation (*i*) now becomes exact and hence its solution is

$$\int\left(\frac{1}{3x^2y}+\frac{2}{3x}\right)dx+\int-\frac{1}{3y}\,dy=c$$

or $$-\frac{1}{3xy}+\frac{2}{3}\log x-\frac{1}{3}\log y=c$$

or, $$\frac{1}{3}\log\left(\frac{x^2}{y}\right)-\frac{1}{3xy}=c$$

or $$\log\left(\frac{x^2}{y}\right)-\frac{1}{xy}=3c=K$$

Rule 3. If $\dfrac{\dfrac{\partial M}{\partial y}-\dfrac{\partial N}{\partial x}}{N}$ is a function of x (or constant) alone say $f(x)$, then I.F. $= e^{\int f(x)\,dx}$.

Rule 4. If $\dfrac{\dfrac{\partial N}{\partial x}-\dfrac{\partial M}{\partial y}}{M}$ is a function of y (or constant) alone say $f(y)$, then I.F. $= e^{\int f(y)\,dy}$

As for examples : 1. The equation $(x^2+y^2+2x)\,dx+2y\,dy=0$ is not exact, because, here $\dfrac{\partial M}{\partial y}(=2y)\neq\dfrac{\partial N}{\partial x}(=0)$ and the equation is not of the form as defined in Rule 1 and Rule 2, so we have to find I.F. by Rule 3 or Rule 4. First, we apply Rule 3 (if it fails we apply Rule 4).

$$\frac{\dfrac{\partial M}{\partial y}-\dfrac{\partial N}{\partial x}}{N}=\frac{2y-0}{2y}=1$$

(which may be regarded as a function of x). It satisfies Rule 3,

$\therefore$ $$\text{I.F.}=e^{\int dx}=e^x$$

Multiplying the equation $(x^2+y^2+2x)\,dx+2y\,dy=0$ throughout by e^x, we get

$$e^x(x^2+y^2+2x)\,dx+2e^x y\,dy=0 \quad \ldots (i)$$

Now, $$M=e^x(x^2+y^2+2x) \text{ and } N=2e^x y$$

$\therefore$ $$\frac{\partial M}{\partial y}=2y\,e^x \text{ and } \frac{\partial N}{\partial x}=2y\,e^x$$

$\therefore$ $$\frac{\partial M}{\partial y}=\frac{\partial N}{\partial x}$$

Hence the equation (*i*) is exact and its solution is

$$\int e^x\left(x^2+y^2+2x\right)dx=c$$

or $$\int x^2 e^x\,dx + \int y^2 e^x\,dx + \int 2x\,e^x\,dx = c$$

or $$x^2 e^x - \int 2x\,e^x\,dx + y^2 e^x + \int 2x\,e^x\,dx = c$$

or $$e^x (x^2 + y^2) = c$$

2. The equation $(y^4 + 2y)\,dx + (xy^3 + 2y^4 - 4x)\,dy = 0$ is not exact, since $\frac{\partial M}{\partial y}\ (= 4y^3 + 2) \neq \frac{\partial N}{\partial x}$ $(= y^3 - 4)$.

The equation is also not of the form of the equations under Rule 1 and Rule 2.

Now, $$\frac{\frac{\partial M}{\partial y} - \frac{\partial N}{\partial x}}{N} = \frac{4y^3 + 2 - y^3 + 4}{xy^3 + 2y^4 - 4x} = \frac{3y^3 + 6}{xy^3 + 2y^4 - 4x}$$

which is not a function of x or y alone.

Again, $$\frac{\frac{\partial N}{\partial x} - \frac{\partial M}{\partial y}}{M} = \frac{y^3 - 4 - 4y^3 - 2}{y^4 + 2y} = \frac{-3y^3 - 6}{y\left(y^3 + 2\right)}$$

$$= \frac{-3\left(y^3 + 2\right)}{y\left(y^3 + 2\right)} = -\frac{3}{y}$$ which is a function of y alone,

Therefore, $$\text{I.F.} = e^{-\int \frac{3}{y}dy} = e^{-3\log y} = e^{\log y^{-3}} = \frac{1}{y^3}$$

Multiplying the equation throughout by $\frac{1}{y^3}$, we get

$$\left(y + \frac{2}{y^2}\right)dx + \left(x + 2y - \frac{4x}{y^3}\right)dy = 0$$

Now, $$\frac{\partial M}{\partial y} = 1 - \frac{4}{y^3} \text{ and } \frac{\partial N}{\partial x} = 1 - \frac{4}{y^3}.$$

$\therefore$ $$\frac{\partial M}{\partial y} = \frac{\partial N}{\partial x}$$

Hence, the equation is exact and its solution is

$$\int \left(y + \frac{2}{y^2}\right)dx + \int 2y\,dy = c$$

or $$xy + \frac{2x}{y^2} + y^2 = c$$

or $$\left(y + \frac{2}{y^2}\right)x + y^2 = c.$$

EXERCISE 1.9

Solve the following differential Equations :

1. $x\,dy - y\,dx = xy^2\,dx$
2. $x\,dy - y\,dx = (x^2 + y^2)\,dx$
3. $(y - 2x^3)\,dx - x(1 - xy)\,dy = 0$
4. $y(x^2y + e^x)\,dx - e^x\,dy = 0$
5. $(x \sec^2 y - x^2 \cos y)\,dy = (\tan y - 3x^4)\,dx$
6. $(x^2y^2 + xy + 1)\,y\,dx + (x^2y^2 - xy + 1)\,x\,dy = 0$
7. $(3x^2y^4 + 2xy)\,dx + (2x^3y^3 - x^3)\,dy = 0$
8. $(y^2 + 2yx^2)\,dx + (2x^2 - xy)\,dy = 0$
9. $y \log y\,dx + (x - \log y)\,dy = 0$
10. $(x - y^2)\,dx + 2xy\,dy = 0$
11. $\left(xy\,e^{\frac{x}{y}} + y^2\right)dx - x^2\,e^{\frac{x}{y}}\,dy = 0$
12. $(2x^2y - 3y^4)\,dx + (3x^3 + 2xy^3)\,dy = 0$
13. $(xy^2 + 2x^2y^3)\,dx + (x^2y - x^3y^2)\,dy = 0$
14. $y(1 + xy)\,dx + x(1 - xy)\,dy = 0$
15. $y(1 + xy)\,dx + x(1 + xy + x^2y^2)\,dy = 0$
16. $(xy \sin xy + \cos xy)\,y\,dx + (xy \sin xy - \cos xy)\,x\,dy = 0$
17. $y(2xy + 1)\,dx + x(1 + 2xy - x^3y^3)\,dy = 0$
18. $(3xy^2 - y^3)\,dx - (2x^2y - xy^2)\,dy = 0$
19. $(3xy - 2ay^2)\,dx + (x^2 - 2axy)\,dy = 0$
20. $(2x \log x - xy)\,dy + 2y\,dx = 0.$

Answers

1. $\dfrac{x^2}{2} + \dfrac{x}{y} = c$
2. $y = x \tan(x + c)$
3. $\dfrac{y^2}{2} - x^2 - \dfrac{y}{x} = c$
4. $\dfrac{x^3}{3} + \dfrac{e^x}{y} = c$
5. $-\dfrac{1}{x}\tan y - x^3 + \sin y = c$
6. $xy + \log \dfrac{x}{y} - \dfrac{1}{xy} = c$
7. $x^3y^2 + \dfrac{x^2}{y} = c$
8. $\dfrac{6}{x^2} - x^{-\frac{3}{2}}\,y^{\frac{3}{2}} = c$
9. $2x \log y = c + (\log y)^2$
10. $y^2 = (x - x \log x)$
11. $e^{\frac{x}{y}} + \log x = c$
12. $5x^{-\frac{36}{13}} . y^{\frac{24}{13}} - 12x^{-\frac{10}{3}} . y^{-\frac{15}{13}} = c$
13. $-\dfrac{1}{xy} + 2\log x - \log y = c$
14. $xy \log xy = cxy - 1$
15. $\dfrac{1}{2x^2y^2} + \dfrac{1}{xy} - \log y = c$
16. $y \cos xy = cx$

17. $\frac{1}{x^2y^2} + \frac{1}{3x^3y^3} + \log y = c$ 18. $3 \log x - 2 \log y + \frac{y}{x} = c$

19. $x^2 (ay^2 - xy) = c$ 20. $2y \log x - \frac{1}{2} y^2 = c.$

Remark. In the foregoing articles we discussed about the different forms of first order and first degree differential equations along with their method of solutions. We shall now deal with differential equations of first order and higher degree along with their method of solutions.

1.8 DIFFERENTIAL EQUATIONS OF FIRST ORDER AND HIGHER DEGREE

In first order and higher degree differential equation, for convenience, we denote $\frac{dy}{dx}$ by $p\left(\text{or } \frac{dx}{dy} = \frac{1}{p}\right)$.

A general differential equation of first order and n^{th} degree is of the form

$$p^n + P_1 p^{n-1} + P_2 p^{n-2} + \ldots + P_n = 0 \qquad \ldots (1)$$

where $P_1, P_2, \ldots, P_n$ are functions of x and y.

Since (1) is a differential equation of first order, its general solution will contain only one arbitrary constant.

The first order and higher degree differential equations are generally of the following forms :

I. Equations Solvable for p

In this case the left side of the equation (1) can be resolved into n linear factors and the equation becomes

$$[p - f_1(x, y)]\,[p - f_2(x, y)] \ldots [p - f_n(x, y)] = 0$$

which is equivalent to

$$p - f_1(x, y) = 0, p - f_2(x, y) = 0, \ldots, p - f_n(x, y) = 0$$

Each of these equations is of first order and first degree and can be solved by the methods that have already been discussed.

If the solutions of the above n component equations are $F_1(x, y, c) = 0$,

$$F_2(x, y, c) = 0, \ldots, F_n(x, y, c) = 0$$

then the general solution of the equation (1) is given by

$$F_1(x, y, c).\, F_2(x, y, c) \ldots F_n(x, y, c) = 0.$$

1.8.1 Illustrative Examples

1. *Solve the equation* $\left(\frac{dy}{dx}\right)^2 + 2x\left(\frac{dy}{dx}\right) + y\frac{dy}{dx} + 2xy = 0$

Sol. Substituting $\frac{dy}{dx} = p$, the equation becomes

$$p^2 + 2px + py + 2xy = 0$$

or $$(p + 2x)(p + y) = 0$$

$\therefore$ either $$p + 2x = 0 \quad \text{...}(i)$$

or $$p + y = 0 \quad \text{...}(ii)$$

From *(i)*, $$\frac{dy}{dx} + 2x = 0$$

or $$dy + 2x\,dx = 0$$

Integrating, we have

$$(y + x^2 - c) = 0$$

From *(ii)*, we get $$\frac{dy}{dx} + y = 0$$

or $$\frac{dy}{y} + dx = 0$$

Integrating, we get $\log|y| + x - c = 0$

$\therefore$ The required solution is

$$(x^2 + y - c)(x + \log|y| - c) = 0.$$

2. *Solve the equation* $x^2 = 1 + p^2$.

Sol. $\because$ $$x^2 = 1 + p^2 \therefore p^2 = x^2 - 1$$

or $$p = \pm\sqrt{x^2 - 1}$$

or $$dy = \pm\sqrt{x^2 - 1}\,dx \qquad \left(\because p = \frac{dy}{dx}\right)$$

Integrating, we get

$$y = \pm\frac{x}{2}\sqrt{x^2 - 1} \mp \frac{1}{2}\log\left(x + \sqrt{x^2 - 1}\right) + c$$

or $$2y = \pm x\sqrt{x^2 - 1} \mp \log\left(x + \sqrt{x^2 - 1}\right) + k, \text{ when } 2c = k$$

which is the required general solution.

3. *Solve the equation* $p^2 - p(e^x + e^{-x}) + 1 = 0$

Sol. Given: $p^2 - p(e^x + e^{-x}) + 1 = 0$

or $p(p - e^x) - e^{-x}(p - e^x) = 0$

or $(p - e^x)(p - e^{-x}) = 0$

$\therefore$ $$p - e^x = 0 \quad \text{...}(i)$$

and $$p - e^{-x} = 0 \quad \text{...}(ii)$$

From *(i)*,

$$dy - e^x\,dx = 0$$

Integrating, we get $y - e^x - c = 0$

From (*ii*),

$$dy - e^{-x}\,dx = 0$$

Integrating, we obtain $y + e^{-x} - c = 0$

Hence, the general solution of the given equation is

$$(y - e^{x} - c)\,(y + e^{-x} - c) = 0.$$

4. *Solve the equation $p\,(p^2 + xy) = p^2\,(x + y)$.*

Sol. The given equation is

$$p\,(p^2 + xy) = p^2\,(x + y)$$

or $$p^3 + pxy - p^2x - p^2y = 0$$

or $$p^2\,(p - x) - py\,(p - x) = 0$$

or $$p\,(p - x)\,(p - y) = 0$$

$\therefore$ $$p = 0 \quad \text{...(i)}$$

$$p - x = 0 \quad \text{...(ii)}$$

and $$p - y = 0 \quad \text{...(iii)}$$

From (*i*), $\frac{dy}{dx} = 0$, or $$dy = 0$$

Integrating, $$y - c = 0$$

From (*ii*), $$dy - x\,dx = 0$$

On integration, we have $y - \frac{x^2}{2} - c = 0$

or $$2y - x^2 - 2c = 0$$

From (*iii*), $$dy - y\,dx = 0$$

or $$\frac{dy}{y} - dx = 0$$

Integrating, we get $\log y - x - c = 0$

or $$\log y = x + c, \text{ or } y = e^c\,.\,e^x$$

or $$(y - e^c e^x) = 0$$

Hence, the general solution of the given equation is

$$(y - c)\,(2y - x^2 - 2c)\,(y - e^c e^x) = 0.$$

5. *Solve the equation $xy\,p^2 + p\,(3x^2 - 2y^2) - 6xy = 0$.*

Sol. Given $xy\,p^2 + p\,(3x^2 - 2y^2) - 6xy = 0$

Solving the equation for p, we have

$$p = \frac{-\left(3x^2 - 2y^2\right) \pm \sqrt{\left(3x^2 - 2y^2\right)^2 + 24x^2y^2}}{2xy}$$

$$= \frac{\left(2y^2 - 3x^2\right) \pm \sqrt{\left(3x^2 + 2y^2\right)^2}}{2xy}$$

$$= \frac{\left(2y^2 - 3x^2\right) \pm \left(3x^2 + 2y^2\right)}{2xy}$$

$$\therefore \qquad p = \frac{2y}{x} \qquad \text{(considering positive sign)}$$

and $$p = -\frac{3x}{y} \qquad \text{(considering negative sign)}$$

From $p = \frac{2y}{x}$, we have $\frac{dy}{y} - 2\frac{dx}{x} = 0$

Integrating, we get $\log y - 2\log x = \log c$

or $$\log\left(\frac{y}{x^2}\right) = \log c$$

or $$y - cx^2 = 0$$

From $$p = -\frac{3x}{y}, \text{ we get } y\,dy = -3x\,dx$$

Integrating, we have $$\frac{y^2}{2} = -\frac{3x^2}{2} + \frac{c}{2}$$

or $$y^2 + 3x^2 - c = 0$$

Therefore, the general solution of the given equation is $(y - cx^2)(y^2 + 3x^2 - c) = 0$.

EXERCISE 1.10

Solve the following differential equations :

1. $p^2 - p - 6$
2. $p(p + y) = x(x + y)$
3. $p^2 - 7p + 12 = 0$
4. $yp^2 + (x - y)p - x = 0$
5. $p^2 + 2xp - 3x^2 = 0$
6. $p^2 + 2py \cot x = y^2$
7. $p(p + x) = y(x + y)$
8. $xyp^2 - (x^2 - y^2)p - xy = 0$
9. $p - \frac{1}{p} = \frac{x^2 - y^2}{xy}$
10. $xyp^2 - (x^2 + y^2)p + xy = 0$
11. $p^3 - p(x^2 + xy + y^2) + x^2y + xy^2 = 0$
12. $p^3 - (x^2 + xy + y^2)p^2 + (x^3y + x^2y^2 + xy^3)p - x^3y^3 = 0.$

Answers

1. $(y - 3x - c)(y + 2x - c) = 0$
2. $(2y - x^2 + c)(y + x + ce^{-x} - 1) = 0$
3. $(y - 4x - c)(y - 3x - c) = 0$
4. $(y - x - c)(x^2 + y^2 - c) = 0$
5. $(2y + 3x^2 - c)(2y - x^2 - c) = 0$
6. $y(1 \pm \cos x) = c$
7. $(y - ce^x)(y + x - ce^{-x} - 1) = 0$
8. $(xy - c)(x^2 - y^2 - c) = 0$
9. $(xy - c)(x^2 - y^2 - c) = 0$
10. $(y^2 - x^2 - c)(y - cx) = 0$
11. $(y - ce^x)(2y - x^2 - c)(y + x - 1 - ce^{-x}) = 0$
12. $(x^2 - 3y + c)\left(e^{\frac{1}{2}x^2} + cy\right)(xy + cy + 1) = 0.$

II. Equations Solvable for y

If an equation is solvable for y, then it can be expressed in the form.

$$y = f(x, p) \qquad \text{... }(i)$$

Differentiating equation (*i*) w.r.t. x and denoting $\frac{dy}{dx}$ by p we shall have another relation

$$p = F\left(x, p, \frac{dp}{dx}\right) \qquad \text{... }(ii)$$

The equation (*ii*) when solved will give us a relation of the form

$$\phi(x, p, c) = 0 \qquad \text{... }(iii)$$

Eliminating p between (*i*) and (*iii*), we obtain a relation between x, y and c which will be the required solution.

However, it is not possible always to eliminate p between (*i*) and (*iii*), in this case we express the values of x and y in terms of p and c and they together will constitute the solutions of the given differential equation in terms of the parameter p.

1.8.2 Illustrative Examples

1. *Solve the equation $y + px = x^4p^2$.*

Sol. The given equation can be written as

$$y = -px + x^4p^2 \qquad \text{... }(i)$$

which is solvable for y.

Differentiating equation (*i*) w.r.t. x and writing $\frac{dy}{dx} = p$, we get

$$p = -p - x\frac{dp}{dx} + x^4 \,.\, 2p\frac{dp}{dx} + p^2 \,.\, 4x^3$$

or
$$2p + x\frac{dp}{dx} = 2px^3\left(2p + x\frac{dp}{dx}\right)$$

or
$$\left(2p + x\frac{dp}{dx}\right)(1 - 2px^3) = 0$$

$\therefore$
$$2p + x\frac{dp}{dx} = 0 \qquad \text{[discarding the factor } 1 - 2px^3]$$

or
$$\frac{2}{x}dx + \frac{dp}{p} = 0$$

Integrating, we get

$$2\log x + \log p = \log c$$

or
$$\log(px^2) = \log c$$

or
$$px^2 = c$$

$$\therefore \qquad p = \frac{c}{x^2} \qquad \text{...}(ii)$$

Eliminating p between (i) and (ii), we obtain

$$y = -x \cdot \frac{c}{x^2} + x^4 \cdot \frac{c^2}{x^4}$$

or $$xy = -c + c^2 x.$$

which is the required solution.

2. *Solve the equation* $y = (x - a)\, p - p^2$.

Sol. The given equation is

$$y = (x - a)\, p - p^2 \qquad \text{...}(i)$$

Differentiating equation (i) w.r.t. x, we get

$$\frac{dy}{dx} = p - (x - a)\frac{dp}{dx} - 2p\frac{dp}{dx}$$

or $$0 = -\left[x - a + 2p\right]\frac{dp}{dx} \qquad \left(\because \frac{dy}{dx} = p\right)$$

$$\therefore \qquad \frac{dp}{dx} = 0 \qquad \text{[discarding the factor } (x - a + 2p)]$$

Integrating, we have $\qquad p = c$

Putting $p = c$ in (i), we get

$$y = (x - a)\, c - c^2.$$

which is the required solution.

3. *Solve the equation*

$$4xp^2 - 8yp - x = 0.$$

Sol. From the given equation, we have

$$y = \frac{1}{8}x\left(4p - \frac{1}{p}\right) \qquad \text{...}(i)$$

Differentiating (i) w.r.t. x, we get

$$p = \frac{1}{8}\left\{\left(4p - \frac{1}{p}\right) + x\left(4 + \frac{1}{p^2}\right)\frac{dp}{dx}\right\} \qquad \left(\because \frac{dy}{dx} = p\right)$$

or $$8p = 4p - \frac{1}{p} + x\left(4 + \frac{1}{p^2}\right)\frac{dp}{dx}$$

or $$4p + \frac{1}{p} = x\left(4 + \frac{1}{p^2}\right)\frac{dp}{dx}$$

or $$p\left(4+\frac{1}{p^2}\right)=x\left(4+\frac{1}{p^2}\right)\frac{dp}{dx}$$

or $$p=x\frac{dp}{dx}$$ $\left[\text{discarding the factor}\left(4+\frac{1}{p^2}\right)\right].$

or $$\frac{dx}{x}=\frac{dp}{p}$$

Integrating, we get

$$\log p=\log x+\log c=\log cx$$

$\therefore$ $$p=cx.$$

Putting $p=cx$ in (*i*), we get

$$y=\frac{1}{8}x\left(4cx-\frac{1}{cx}\right)=\frac{1}{8}x\left(\frac{4c^2x^2-1}{cx}\right)$$

$$=\frac{1}{8c}\left(4c^2x^2-1\right)$$

or $$8cy-4c^2x^2+1=0$$

which is the required solution.

4. *Solve the equation*

$$y=2px-p^2.$$

Sol. Given $$y=2px-p^2 \qquad \ldots(i)$$

Differentiating w.r.t. x, we get

$$\frac{dy}{dx}=2p+2x\frac{dp}{dx}-2p\frac{dp}{dx}$$

or $$0=p+(2x-2p)\frac{dp}{dx}. \qquad \left(\because \frac{dy}{dx}=p\right)$$

or $$p.\frac{dx}{dp}+2x-2p=0$$

or $$\frac{dx}{dp}+\frac{2}{p}x=2 \qquad \ldots(ii)$$

which is a linear equation.

Here $$\text{I.F.}=e^{2\int\frac{1}{p}dp}=e^{2\log p}=p^2.$$

Multiplying equation (*ii*) by p^2 and integrating, we get

$$x.p^2=\int 2p^2\,dp=\frac{2}{3}p^3+c$$

or $$x = \frac{2}{3}p + cp^{-2} \qquad \text{... (iii)}$$

Putting the value of x in (*i*), we get

$$y = 2p\left(\frac{2}{3}p + cp^{-2}\right) - p^2$$

or $$y = \frac{1}{3}p^2 + 2cp^{-1} \qquad \text{... (iv)}$$

Equation (*iii*) and (*iv*) together constitute the general solution of (*i*).

5. *Solve the equation* $y = p^2x + p$.

Sol. $$y = p^2x + p \qquad \text{... (i)}$$

Differentiating (*i*) w.r.t. x, we get

$$p = 2px\frac{dp}{dx} + p^2 + \frac{dp}{dx} \qquad \left(\because p = \frac{dy}{dx}\right)$$

or $$p = p^2 + (2px + 1)\frac{dp}{dx}$$

or $$(2px + 1)\frac{dp}{dx} = p - p^2 = p\,(1 - p)$$

or $$p\,(1 - p)\frac{dx}{dp} = 2px + 1$$

or $$p\,(1 - p)\frac{dx}{dp} - 2px = 1$$

or $$\frac{dx}{dp} - \frac{2x}{1 - p} = \frac{1}{p\,(1 - p)} \qquad \text{... (ii)}$$

which is a linear form.

$\therefore$ $$\text{I.F.} = e^{-\int \frac{2}{1-p}dp} = e^{2\log(1-p)} = (1 - p)^2$$

Multiplying equation (*ii*) throughout by $(1 - p)^2$ and integrating, we get

$$x\,.\,(1 - p)^2 = \int \frac{1 - p}{p}\,dp = \log p - p + c$$

or $$x = \frac{1}{(1 - p)^2}[\log p - p + c] \qquad \text{... (iii)}$$

Equations (*i*) and (*iii*) together will constitute the solution.

III. Equations Solvable for x

In this case the equation will be of the form

$$x = f(y, p)$$

Differentiating this equation w.r.t. y, and replacing $\frac{dx}{dy}$ by $\frac{1}{p}$, we shall get an equation in two variables y and p. If p be eliminated between the solution of the latter equation (which contains an arbitrary constant c) and the given equation, we shall get the required solution.

IV. Equations not Containing *x*

In this case the equation will be of the form

$$p = f(y)$$

or of the form $$y = f(p).$$

If the equation is of the form $p = f(y)$, it can be solved by separating the variables and if it is of the form $y = f(p)$, it can be solved by the method explained in 2 (Equations solvable for y).

V. Equations not Containing *y*

In this case the equation will be of the forms

$$p = f(x) \text{ or } x = f(p).$$

If the equation is of the $p = f(x)$, it can be solved by separating the variables and if it is of the form $x = f(p)$, it can be solved by the method explained in 3 (Equations solvable for x).

We shall illustrate the cases 3, 4 and 5 by taking suitable examples.

1.8.3 Illustrative Examples

1. *Solve the equation $x + yp = ap^2$.*

Sol. From the given equation, we have

$$x = -yp + ap^2 \qquad \text{...}(i)$$

Differentiating w.r.t. y, we get

$$\frac{dx}{dy} = -p - y\frac{dp}{dy} + 2ap\frac{dp}{dy}$$

or $$\frac{1}{p} = -p - y\frac{dp}{dy} + 2ap\frac{dp}{dy} \qquad \left(\because \frac{dx}{dy} = \frac{1}{p}\right)$$

or $$(2ap - y)\frac{dp}{dy} = \frac{1+p^2}{p}$$

or $$\frac{1+p^2}{p}\frac{dy}{dp} = 2ap - y$$

or $$\frac{1+p^2}{p}\frac{dy}{dp} + y = 2ap$$

or $$\frac{dy}{dp} + \frac{p}{1+p^2}y = \frac{2ap^2}{1+p^2} \qquad \text{...}(ii)$$

which is linear differential equation.

$$\therefore \qquad \text{I.F.} = e^{\int \frac{p}{1+p^2} dp} = e^{\frac{1}{2}\log(1+p^2)} = \sqrt{1+p^2}$$

Multiplying (*ii*) throughout by $\sqrt{1+p^2}$ and integrating, we get

$$y \cdot \sqrt{1+p^2} = \int \frac{2ap^2}{\sqrt{1+p^2}} dp + c$$

$$= 2a \int \left[\sqrt{1+p^2} - \frac{1}{\sqrt{1+p^2}} \right] dp + c$$

$$= 2a \left[\frac{p\sqrt{1+p^2}}{2} + \frac{1}{2}\log\left(p+\sqrt{1+p^2}\right) - \log\left(p+\sqrt{1+p^2}\right) \right] + c$$

$$= 2a \left[\frac{p\sqrt{1+p^2}}{2} - \frac{1}{2}\log\left(p+\sqrt{1+p^2}\right) \right] + c$$

$$= a\left[p\sqrt{1+p^2} - \log\left(p+\sqrt{1+p^2}\right) \right] + c$$

or $$y = \frac{a}{\sqrt{1+p^2}} \left[p\sqrt{1+p^2} - \log\left(p+\sqrt{1+p^2}\right) \right] + c \qquad \text{... (iii)}$$

The equations (*i*) and (*iii*) will constitute the general solution.

2. *Solve the equation*

$$y = 2px + y^2p^3.$$

Sol. Solving for x, the equation can be written as

$$x = \frac{1}{2}\left(\frac{y}{p} - y^2p^2 \right)$$

Differentiating w.r.t. y, we get

$$\frac{dx}{dy} = \frac{1}{2}\left[\frac{1}{p} - \frac{y}{p^2}\frac{dp}{dy} - 2yp^2 - 2y^2p\frac{dp}{dy} \right]$$

or $$\frac{1}{p} = \frac{1}{2p^2}\left[p - y\frac{dp}{dy} - 2yp^4 - 2y^2p^3\frac{dp}{dy} \right]$$

or $$2p = p - y\frac{dp}{dy} - 2yp^4 - 2y^2p^3\frac{dp}{dy}$$

or $p + 2yp^4 + y\dfrac{dp}{dy} + 2y^2p^3\dfrac{dp}{dy} = 0$

or $p\left(1+2yp^3\right) + y\dfrac{dp}{dy}\left(1+2yp^3\right) = 0$

or $$\left(p+y\frac{dp}{dy}\right)\left(1+2yp^3\right)=0$$

Discarding the factor $(1+2yp^3)$, we have

$$p+y\frac{dp}{dy}=0$$

or $$\frac{dy}{y}+\frac{dp}{p}=0$$

Integrating, we obtain $\log y+\log p=\log c$

or $$py=c \text{ or } p=\frac{c}{y}$$

Putting $p=\frac{c}{y}$ in the given equation, we get

$$y=\frac{2cx}{y}+\frac{c^3}{y} \text{ or } y^2=2cx+c^3$$

which is the required solution.

3. *Solve the equation $y = Ap^3 + Bp^2$*

Sol. Given $$y=Ap^3+Bp^2 \qquad \text{...}(i)$$

Differentiating w.r.t. x, we get

$$p=\left(3Ap^2+2Bp\right)\frac{dp}{dx} \qquad \left(\because \frac{dy}{dx}=p\right)$$

or $$(3Ap+2B)\,dp=dx$$

Integrating, we have

$$\frac{3}{2}Ap^2+2Bp+c=x$$

or $$x=\frac{3}{2}Ap^2+2Bp+c \qquad \text{...}(ii)$$

Equations (*i*) and (*ii*) will provide the solution.

4. *Solve the equation $x = 4p + 4p^2$*

Sol. Given $$x=4p+4p^2 \qquad \text{...}(i)$$

Differentiating (*i*) w.r.t. y, we get

$$\frac{dx}{dy}=4\frac{dp}{dy}+8p\frac{dp}{dy}$$

or $$\frac{1}{p}=4(1+2p)\frac{dp}{dy}$$

or $$dy=4(p+2p^2)\,dp$$

Integrating, we have

$$y=4\left(\frac{p^2}{2}+\frac{2}{3}p^3\right)+c \qquad \text{...}(ii)$$

Equations (*i*) and (*ii*) will constitute the solution.

EXERCISE 1.11

Solve the following differential equations :

1. $p = \tan\left(x - \dfrac{p}{1+p^2}\right)$

2. $y = 3px + 6p^2y^2$

3. $x^3p^2 + x^2py + a^3 = 0$

4. $y^2 \log y = xyp + p^2$

5. $x^2 + p^2x = yp$

6. $y = 2px + y^2p^3$

7. $y = 4p^3 + 3px$

8. $y = yp^2 + 2px$

9. $p^2 + py = x^2 + xy$

10. $p^3 - 4xyp + 8y^2 = 0$

11. $p^3x - p^2y - 1 = 0$

12. $\sin y \cos px - \cos y \sin px - p = 0$

13. $e^y - p^3 - p = 0$

14. $y = p \cos p - \sin p$

15. $p^2 - 2xp + 1 = 0.$

Answers

1. $\left.\begin{aligned} x &= \tan^{-1} p + \frac{p}{1+p^2} \\ y &= c - \frac{1}{1+p^2} \end{aligned}\right\}$

2. $y^2 = 3cx + 6c^2$

3. $c^2 + cxy + a^3x = 0$

4. $\log y = cx + c^2$

5. $y = \dfrac{\left(c\sqrt{p} - \frac{1}{3}p^2\right)^2}{p} + p\left(c\sqrt{p} - \frac{1}{3}p^2\right)$

6. $y^2 = 2cx + c^3$

7. $\left.\begin{aligned} y &= -\frac{8}{7}p^3 + cp^{-\frac{1}{2}} \\ x &= -\frac{12}{7}p^2 + \frac{c}{3}p^{-\frac{3}{2}} \end{aligned}\right\}$

8. $y^2 = 2cx + c^2$

9. $(c + x^2 - 2y)\{c - (y + x - 1)e^x\} = 0$

10. $64y = c(c - 4x)^2$

11. $c^3x - c^2y - 1 = 0$

12. $y = cx + \sin^{-2} c$

13. $\left.\begin{aligned} x &= 2\tan^{-1} p - p^{-1} + c \\ y &= \log\left(p^3 + p\right) \end{aligned}\right\}$

14. $\left.\begin{aligned} x &= c \cos p \\ y &= p \cos p - \sin p \end{aligned}\right\}$

15. $\left.\begin{aligned} x &= \frac{1}{2}\left(p + p^{-1}\right) \\ y &= \frac{1}{4}p^2 - \frac{1}{2}\log p + c \end{aligned}\right\}.$

Remark. While solving problems in 1.8.2. Illustrative examples, we discarded some factors containing derivatives (p) in few problems, which give rise to other solution of the differential equations, which are not included in the general solution. Such solutions are called *singular solutions* and these solutions will be considered in *Clairaut's Equations* in the next-article.

1.9 CLAIRAUT'S EQUATIONS

Any differential equation of the form

$$y = px + f(p), \left(p = \frac{dy}{dx}\right) \text{ is known as Clairaut's equation.} \quad \dots (1)$$

To solve the equation (1), we have to differentiate (1) w.r.t. x.

Differentiating (1) w.r.t. x, we get

$$\frac{dy}{dx} = p + x\frac{dp}{dx} + f'(p)\frac{dp}{dx}$$

or $$\left[x + f'(p)\right]\frac{dp}{dx} = 0$$

Therefore, either $$\frac{dp}{dx} = 0 \quad \dots (2)$$

or $$x + f'(p) = 0 \quad \dots (3)$$

Integrating (2), we get $p = c$.

Replacing $p = c$ in (1), we obtain the general solution (or complete solution),

$$y = cx + f(c).$$

If p be eliminated between (3) and the original equation (1), we shall obtain a relation between x and y which satisfies the equation (1) and so it is also a solution of the equation (1). Since this solution does not contain any arbitrary constant nor can it be derived from the complete solution by giving any particular value to the arbitrary constant, it is called the **Singular Solution** of the differential equation.

Thus, a Clairaut's equation has two kinds of solutions : (i) **a complete solution** (linear in x, y) containing one arbitrary constant ;

and (ii) **a singular solution** containing no arbitrary constant.

1.9.1 Illustrative Examples

1. *Solve the equation* $y = px + ap(1 - p)$.

Sol. Given $$y = px + ap(1 - p) \quad \dots (i)$$

Differentiating (i) w.r.t. x, we get

$$p = p + x\frac{dp}{dx} + a(1 - 2p)\frac{dp}{dx}, \quad \left(\because \frac{dy}{dx} = p\right)$$

or $$\left[x + a(1 - 2p)\right]\frac{dp}{dx} = 0$$

$\therefore$ either $\qquad \dfrac{dp}{dx} = 0$ ... (ii)

or $\qquad x + a(1 - 2p) = 0$ ... (iii)

Integrating, (ii) we get $p = c$

Replacing $p = c$ in (i), we have

$$y = cx + ac(1 - c),$$

which is the complete solution.

From (iii), we get $\qquad \dfrac{x}{a} + 1 = 2p$

or $\qquad p = \dfrac{x+a}{2a}$

Substituting, $p = \dfrac{x+a}{2a}$ in (i), we obtain

$$y = \frac{x+a}{2a}.x + a.\frac{x+a}{2a}\left(1 - \frac{x+a}{2a}\right)$$

or $\qquad y = \dfrac{x(x+a)}{2a} + \dfrac{(x+a)(a-x)}{4a}$

or $\qquad 4ay = 2x^2 + 2ax + a^2 - x^2 = (x + a)^2$

or $\qquad (x + a)^2 = 4ay$

which is the singular solution.

2. *Solve the equation sin y cos px – cos y sin px – p = 0.*

Sol. From the given equation, we get

$$\sin(y - px) = p$$

or $\qquad y - px = \sin^{-1} p$

or $\qquad y = px + \sin^{-1} p$ ... (i)

which is the Clairaut's equation.

Differentiating equation (i), we get

$$p = p + x\frac{dp}{dx} + \frac{1}{\sqrt{1-p^2}}\frac{dp}{dx} \qquad \left(\because p = \frac{dy}{dx}\right)$$

or $\qquad \left[x + \dfrac{1}{\sqrt{1-p^2}}\right]\dfrac{dp}{dx} = 0$

$\therefore$ either $\qquad \dfrac{dp}{dx} = 0$ ... (ii)

or $\qquad x + \dfrac{1}{\sqrt{1-p^2}} = 0$ ... (iii)

Integrating (*ii*), we get $\quad p = c$

Replacing $p = c$ in (*i*), we get

$$y = cx + \sin^{-1} c$$

which is the general solution.

From (*iii*),

$$\sqrt{1-p^2} = -\frac{1}{x}$$

or
$$1 - p^2 = \frac{1}{x^2}$$

or
$$p^2 = \frac{x^2-1}{x^2} \text{ or } p = \pm\frac{\sqrt{x^2-1}}{x}$$

Replacing
$$p = \pm\frac{\sqrt{x^2-1}}{x} \text{ in } (i), \text{ we have}$$

$$y = \pm\frac{\sqrt{x^2-1}}{x}x + \sin^{-1}\left(\pm\frac{\sqrt{x^2-1}}{x}\right)$$

or
$$y = \pm\left[\sqrt{x^2-1} + \sin^{-1}\frac{\sqrt{x^2-1}}{x}\right]$$

which is the singular solution.

3. *Solve the equation*

$$p = \log (px - y).$$

Sol. Given
$$p = \log (px - y)$$

$\therefore$
$$px - y = e^p$$

or
$$y = px - e^p \qquad \text{... } (i)$$

which is a Clairaut's equation.

Differentiating (*i*) w.r.t. x, we have

$$p = p + x\frac{dp}{dx} - e^p \cdot \frac{dp}{dx} \qquad \left(\because p = \frac{dy}{dx}\right)$$

or
$$\left[x - e^p\right]\frac{dp}{dx} = 0$$

$\therefore$ either
$$\frac{dp}{dx} = 0 \qquad \text{... } (ii)$$

and
$$x - e^p = 0 \qquad \text{... } (iii)$$

Integrating (*ii*), we get $\quad p = c$

Replacing $p = c$ in (*i*), we obtain

$$y = cx - e^c$$

which is the complete solution.

From (*iii*), we get

$$e^p = x$$

$$\therefore \quad p = \log|x|$$

Substituting p in (*i*), we get

$$y = x\log|x| - x$$

$$= x[\log|x| - 1]$$

which is the particular solution.

4. *Solve the equation* $y = px + \sqrt{a^2p^2 + b^2}$ (W.B.U.T. BCA, 2006)

Sol. Given

$$y = px + \sqrt{a^2p^2 + b^2} \quad \text{... }(i)$$

Differentiating w.r.t. x, we get

$$p = p + x\frac{dp}{dx} + \frac{1}{2\sqrt{a^2p^2 + b^2}} \times 2a^2p\frac{dp}{dx} \quad \left(\because p = \frac{dy}{dx}\right)$$

or

$$\left[x + \frac{a^2p}{\sqrt{a^2p^2 + b^2}}\right]\frac{dp}{dx} = 0$$

$\therefore$ Either

$$\frac{dp}{dx} = 0 \quad \text{... }(ii)$$

or

$$x + \frac{a^2p}{\sqrt{a^2p^2 + b^2}} = 0 \quad \text{... }(iii)$$

From (*i*), integrating we get $p = c$

Replacing $p = c$ in (*i*), we get

$$y = cx + \sqrt{a^2c^2 + b^2}$$

which is the complete solution.

From (*iii*), we get $\sqrt{a^2p^2 + b^2} = -\dfrac{a^2p}{x}$

or

$$a^2p^2 + b^2 = \frac{a^4p^2}{x^2}$$

or

$$a^2p^2x^2 + b^2x^2 = a^4p^2$$

or

$$a^2p^2(x^2 - a^2) = -b^2x^2$$

or $$p^2 = -\frac{b^2x^2}{a^2\left(x^2-a^2\right)} \qquad \text{... (iv)}$$

Again, since $\sqrt{a^2p^2+b^2} = -\frac{a^2p}{x}.$

Therefore from (*i*), we have

$$y = px - \frac{a^2p}{x} = p\left(\frac{x^2-a^2}{x}\right)$$

or $$y^2 = p^2\frac{\left(x^2-a^2\right)^2}{x^2} = -\frac{b^2x^2}{a^2\left(x^2-a^2\right)}\cdot\frac{\left(x^2-a^2\right)^2}{x^2}, \text{ [Using (iv)]}$$

or $$y^2 = -\frac{b^2}{a^2}\left(x^2-a^2\right)$$

or $$a^2y^2 = -b^2x^2 + a^2b^2$$

or $$b^2x^2 + a^2y^2 = a^2b^2$$

or $$\frac{x^2}{a^2}+\frac{y^2}{b^2} = 1 \qquad \text{[Dividing by } a^2b^2]$$

which is the required singular solution.

5. *Solve the equation* $(y + 1)\,p - xp^2 + 2 = 0$.

Sol. The given equation can be written as

$$y + 1 = xp - \frac{2}{p}$$

or $$y = px - \frac{2}{p} - 1 \qquad \text{... (i)}$$

which is a Clairaut's equation.

Differentiating equation (*i*) w.r.t. *x*, we get

$$p = p + x\frac{dp}{dx} + \frac{2}{p^2}\frac{dp}{dx} \qquad \left(\because p = \frac{dy}{dx}\right)$$

or $$\left[x + \frac{2}{p^2}\right]\frac{dp}{dx} = 0$$

∴ either $$\frac{dp}{dx} = 0 \qquad \text{... (ii)}$$

or $$x + \frac{2}{p^2} = 0 \qquad \text{... (iii)}$$

From (*ii*), integrating, we have

$$p = c$$

Replacing $p = c$ in (*i*), we get

$$y = cx - \frac{2}{c} - 1$$

which is the complete solution.

From (*iii*), we have $\quad p^2 = -\frac{2}{x}$

$$\because \qquad (y+1) = \left(xp - \frac{2}{p}\right) \qquad \text{[given equation]}$$

$$\therefore \qquad (y+1)^2 = x^2p^2 + \frac{4}{p^2} - 4x$$

$$= x^2\left(-\frac{2}{x}\right) + \frac{4}{-\frac{2}{x}} - 4x \qquad \left(\because p^2 = -\frac{2}{x}\right)$$

$$= -2x - 2x - 4x = -8x$$

or $\qquad (y+1)^2 + 8x = 0$

which is the required singular solution.

1.10 LAGRANGE'S EQUATION

An equation of the form

$$y = x\,\phi(p) + f(p) \qquad \text{... (1),}$$

where $\qquad p = \frac{dy}{dx}$

is called *Lagrange's equation*.

To solve the equation, we have to differentiate the equation (1) w.r.t. x.

Differentiating (1) w.r.t. x, we get

$$\frac{dy}{dx} = \phi(p) + x\phi'(p)\frac{dp}{dx} + f'(p)\frac{dp}{dx}$$

or

$$p - \phi(p) = \left[x\phi'(p) + f'(p)\right]\frac{dp}{dx}$$

or

$$\frac{dx}{dp} = \frac{x\phi'(p) + f'(p)}{p - \phi(p)}$$

or

$$\frac{dx}{dp} - \frac{\phi'(p)}{p - \phi(p)}x = \frac{f'(p)}{p - \phi(p)}$$

The equation can be written as

$$\frac{dx}{dp} - \alpha(p)\,x = \beta(p) \qquad \text{... (2)}$$

where $$\alpha(p) = \frac{\phi'(p)}{p - \phi(p)} \text{ and } \beta(p) = \frac{f'(p)}{p - \phi(p)}$$

Equation (2) is a linear equation in x. It can be solved by finding the I.F. $= e^{-\int \alpha(p)\,dp}$ and multiplying the equation by the I.F. and integrating.

The solution of the equation (2) is

$$x\,e^{-\int \alpha(p)\,dp} = \int \beta(p)\,e^{-\int \alpha(p)\,dp}\,dp + c \qquad \text{... (3)}$$

Then eliminating p between (1) and (3), we get the required solution of the equation (1).

1.10.1 Illustrative Examples

1. *Solve the equation* $y = (p + p^2)\,x + p^{-1}$. (W.B.U.T., 2003)

Sol. Given : $$y = (p + p^2)\,x + p^{-1} \qquad \text{... (i)}$$

Differentiating (*i*) w.r.t. x, we get

$$\frac{dy}{dx} = (p + p^2) + x(1 + 2p)\frac{dp}{dx} - \frac{1}{p^2}\frac{dp}{dx} \qquad \left(\because p = \frac{dy}{dx}\right)$$

or $$-p^2 = \left[x(1 + 2p) - \frac{1}{p^2}\right]\frac{dp}{dx}$$

or $$\frac{dx}{dp} + \frac{1 + 2p}{p^2}.x = \frac{1}{p^4} \qquad \text{... (ii)}$$

Equation (*ii*) is a linear equation in x.

Here $$\text{I.F.} = e^{\int \frac{1+2p}{p^2}\,dp} = e^{\int \left(\frac{1}{p^2} + \frac{2}{p}\right)dp} = e^{-\frac{1}{p} + 2\log p}$$

$$= e^{-\frac{1}{p}}.e^{\log p^2} = p^2\,e^{-\frac{1}{p}}$$

Multiplying the equation (*ii*) throughout by $p^2\,e^{-\frac{1}{p}}$ and integrating, we get

$$x.p^2.e^{-\frac{1}{p}} = \int \frac{1}{p^2}\,e^{-\frac{1}{p}}\,dp + c$$

Substituting $-\frac{1}{p} = t,\ \frac{1}{p^2}\,dp = dt$

$$\therefore \qquad x.p^2.e^{-\frac{1}{p}} = \int e^t\,dt + c = e^t + c$$

or
$$x p^2.e^{-\frac{1}{p}} = e^{-\frac{1}{p}} + c$$

or
$$x = \frac{1}{p^2} + c\frac{1}{p^2}.e^{\frac{1}{p}} = \frac{1}{p^2}\left(1 + c.e^{\frac{1}{p}}\right) \qquad \text{... (iii)}$$

$\therefore$ From (*i*),
$$y = \frac{1}{p}(1+p)\left(1 + c\,e^{\frac{1}{p}}\right) + \frac{1}{p} = \frac{1}{p}\left[(1+p)\left(1 + c\,e^{\frac{1}{p}}\right) + 1\right] \text{... (iv)}$$

Equations (*iii*) and (*iv*) will give the solution of (*i*).

2. *Solve the equation $xp^2 + x = 2yp$.*

Sol. Given
$$xp^2 + x = 2yp$$

or
$$(p^2 + 1)\,x = 2yp$$

or
$$y = \frac{1}{2}\left(p + \frac{1}{p}\right)x \qquad \text{... (i)}$$

Equation (*i*) is a Lagrange's equation.

Differentiating (*i*), we get

$$\frac{dy}{dx} = \frac{1}{2}\,x\left[1 - \frac{1}{p^2}\right]\frac{dp}{dx} + \frac{1}{2}\left(p + \frac{1}{p}\right)$$

or
$$2p = x\left(1 - \frac{1}{p^2}\right)\frac{dp}{dx} + p + \frac{1}{p} \qquad \left[\because \frac{dy}{dx} = p\right]$$

or
$$p - \frac{1}{p} = x\left(1 - \frac{1}{p^2}\right)\frac{dp}{dx}$$

or
$$\frac{p^2 - 1}{p} = x\left(\frac{p^2 - 1}{p^2}\right)\frac{dp}{dx}$$

or
$$\frac{dx}{x} = \frac{1}{p}\,dp$$

Integrating, we get $\log|x| = \log p + \log c$

or
$$x = cp$$

or
$$p = \frac{x}{c}$$

Putting the value of p in (*i*), we get

$$y = \frac{1}{2}\left(\frac{x}{c} + \frac{c}{x}\right)x = \frac{1}{2}\left(\frac{x^2 + c^2}{c}\right)$$

or $$y = \frac{x^2 + c^2}{2c}$$

which is the required solution of the equation (*i*).

EXERCISE 1.12

A. *Solve the following differential equations. Also find the singular solutions.*

1. $y = px + \frac{a}{p}$
2. $y = px + p - p^2$
3. $y = x\frac{dy}{dx} + \left(\frac{dy}{dx}\right)^2$
4. $(y - px)(p - 1) = p$
5. $y = px + p^n$
6. $y = px + ap(1 - p)$
7. $py = p^2(x - b) + a$
8. $y = px + p \log p$
9. $y = px + p^3$
10. $y = px + p^2 - p^3$.

B. *Solve the following equations :*

1. $y = 3px + 4p^3$
2. $y = yp^2 + 2px$
3. $y = (1 + p)x + ap^2$
4. $p^3x - p^2y - 1 = 0$
5. $y = p^2x + p$
6. $x + yp = ap^2$
7. $y = 2px + p^2$
8. $x = py - p^2$.

Answers

A. 1. $y = cx + \frac{a}{c}$; $y^2 = 4ax$

2. $y = cx + c - c^2$; $4y = (x + 1)^2$

3. $y = cx + c^2$; $4y = x^4 - 2x^2$

4. $y = cx - \frac{c}{c-1}$; $y = \sqrt{x}\left(\sqrt{x} \pm 1\right) \pm \left(\sqrt{x} \pm 1\right)$

5. $y = cx + c^n$; $n^n y^{n-1} + x^n (n - 1)^{n-1} = 0$

6. $y = cx + ac(1 - c)$; $(x + a)^2 = 4ay$

7. $cy = c^2(x - b) + a$; $y^2 = 4a(x - b)$

8. $y = cx + c \log c$; $y = -e^{-(x+1)}$

9. $y = cx + c^3$; $27y^2 + 4x^3 = 0$

10. $y = cx + c^2 - c^3$; $y = \frac{1}{3}\left(1 \pm \sqrt{1+3x}\right)\left[x + \frac{1}{3}\left(1 \pm \sqrt{1+3x}\right) - \frac{1}{9}\left(1 \pm \sqrt{1+3x}\right)^2\right]$.

B. 1. $$\left.\begin{aligned} y &= -\frac{8}{7}p^3 + cp^{-\frac{1}{2}}; \\ x &= -\frac{12}{7}p^2 + \frac{1}{3}cp^{-\frac{3}{2}} \end{aligned}\right\}$$

2. $y^2 = 2cx + c^2$

3. $x = 2a(1-p) + ce^{-p};\ y = 2a - ap^2 + (1+p)ce^{-p}$

4. $c^3x - c^2y - 1 = 0$

5. $\left.\begin{aligned} y &= p^2x + p \\ x &= (\log p - p + c)/(p-1)^2 \end{aligned}\right\}$

6. $x + yp = ap^2$;

$$x\left(1+p^2\right)^{\frac{1}{2}} = p\left[c + a\log\left(p + \sqrt{1+p^2}\right)\right]$$

7. $(3xy + 2x^2 + c)^2 - 4(x^2 + y^2) = 0$

8. $x = py - p^2;\ x = \dfrac{p}{\sqrt{p^2-1}}\log\left[p + \sqrt{p^2-1}\right]$

1.11 EQUATIONS REDUCIBLE TO CLAIRAUT'S FORM

When the differential equations in x, y, p are not directly in Clairaut's form, it is often possible to reduce the equation in Clairaut's form with the help of suitable substitution. It will be shown in the following illustrative examples.

1.11.1 Illustrative Examples

1. *Using the transformations $x^2 = u$ and $y^2 = v$, solve the equation $x^2(y - px) = p^2y$.*

Sol. Since $x^2 = u$ and $y^2 = v$

$\therefore$ $2x\,dx = du$ and $2y\,dy = dv$

Also,
$$\frac{y}{x}\frac{dy}{dx} = \frac{dv}{du} \text{ or } \frac{y}{x}p = \frac{dv}{du}$$

or
$$p = \frac{x}{y}\frac{dv}{du}$$

Putting the value of p in the given equation, we get

$$x^2\left(y - x\cdot\frac{x}{y}\frac{dv}{du}\right) = \frac{x^2}{y^2}\left(\frac{dv}{du}\right)^2\cdot y$$

or
$$\left(y^2 - x^2\frac{dv}{du}\right) = \left(\frac{dv}{du}\right)^2$$

or
$$\left(v - u\frac{dv}{du}\right) = \left(\frac{dv}{du}\right)^2$$

or $$v = u\frac{dv}{du} + \left(\frac{dv}{du}\right)^2$$

If we assume $\frac{dv}{du} = P$, then the equation becomes

$$v = Pu + P^2$$

which is Clairaut's form and its solution is

$$v = cu + c^2$$

or $$y^2 = cx^2 + c^2$$

which is the required solution of the given equation.

2. *Reduce the equation $y = 2px - p^2y$ to Clairaut's form by the substitutions $x = u$ and $y^2 = v$ and then obtain the complete primitive and singular solution, if any*

Sol. Given equation is

$$y = 2px - p^2y \quad \text{...}(i)$$

$\because$ $$x = u \text{ and } y^2 = v$$

$\therefore$ $$dx = du \text{ and } 2y\frac{dy}{dx} = \frac{dv}{du}$$

$\therefore$ $2yp = P$, say, where $\frac{dy}{dx} = p$ and $\frac{dv}{du} = P$.

or $$p = \frac{P}{2y}$$

$\therefore$ The equation (i) reduces to

$$y = 2x\frac{P}{2y} - \frac{P^2}{4y^2}y$$

or $$y^2 = xP - \frac{P^2}{4}$$

or $$v = uP - \frac{P^2}{4} \quad \text{...}(ii)$$

Equation (ii) is in Clairaut's form.

Differentiating (ii) w.r.t. u, we get

$$\frac{dv}{du} = P + u\frac{dP}{du} - \frac{P}{2}\frac{dP}{du}$$

or $$P = P + \left[u - \frac{P}{2}\right]\frac{dP}{du} \quad \left[\because P = \frac{dv}{du}\right]$$

or $$\left[u-\frac{P}{2}\right]\frac{dP}{du}=0$$

$\therefore$ $$\frac{dP}{du}=0 \qquad \text{... (iii)}$$

or $$u-\frac{P}{2}=0 \qquad \text{... (iv)}$$

Integrating (*iii*), we have P = c

Replacing P = c in (*ii*), we get $v = uc - \frac{1}{4}c^2$

or $$y = cx - \frac{1}{4}c^2$$

which is the complete primitive of (*i*).

From (*iv*), we get P = $2u$

Replacing P = 2u in (*ii*), we obtain

$$v = 2u^2 - u^2 = u^2$$

or $$y^2 = x^2 \qquad (\because v = y^2 \text{ and } u = x)$$

or $$y = \pm x$$

which is the singular solution of (*i*).

3. *Reduce the equation* $xy(y - px) = x + py$ *to Clairaut's form by the substitutions* $x^2 = u$ *and* $y^2 = v$ *and find its complete primitive.*

Sol. Given $$xy(y - px) = x + py \qquad \text{... (i)}$$

$\because$ $$x^2 = u \text{ and } y^2 = v$$

$\therefore$ $$2x\,dx = du \text{ and } 2y\,dy = dv$$

Hence $$\frac{y\,dy}{x\,dx} = \frac{dv}{du}$$

If we assume $\frac{dv}{du}$ = P, then $\frac{y}{x}.p$ = P, where $p = \frac{dy}{dx}$

or $$p = \frac{Px}{y}$$

Substituting p in (*i*), we get

$$xy\left(y-\frac{Px^2}{y}\right) = x + Px$$

or $$(y^2 - Px^2) = 1 + P$$

or $$v - Pu = 1 + P$$

or $$v = Pu + (1 + P)$$

which is in Clairaut's form.

The complete primitive is

$$v = cu + (1 + c), \text{ where } c \text{ is a constant.}$$

or $$y^2 = cx^2 + 1 + c = 1 + c\,(x^2 + 1)$$

is the required complete solution of equation (*i*).

4. *Reduce the equation $x^2p^2 + y\,(2x + y)\,p + y^2 = 0$ to Clairaut's form by the substitution $y = u$, $xy = v$.*
Hence, solve the equation and prove that $y + 4x = 0$ is a singular solution of the equation.

Sol. The given equation is

$$x^2p^2 + y\,(2x + y)\,p + y^2 = 0 \qquad \text{... }(i)$$

$\because$ $$y = u \text{ and } xy = v$$

$\therefore$ $$\frac{du}{dx} = \frac{dy}{dx} = p$$

and $$\frac{dv}{dx} = y + x\frac{dy}{dx} = y + xp$$

$\therefore$ $$\frac{dv}{du} = \frac{y + xp}{p} = x + \frac{y}{p} \qquad \text{... }(ii)$$

Considering, $\dfrac{dv}{du} = P$, we have from (*ii*)

$$P = x + \frac{y}{p} \text{ or } p = \frac{y}{P - x}$$

Putting $$p = \frac{y}{P - x} \text{ in } (i), \text{ we get}$$

$$x^2 \cdot \frac{y^2}{(P - x)^2} + y\,(2x + y)\frac{y}{P - x} + y^2 = 0$$

or $$x^2y^2 + y^2\,(2x + y)\,(P - x) + y^2\,(P - x)^2 = 0$$

or $$v^2 + u^2\left(2\frac{v}{u} + u\right)\left(P - \frac{v}{u}\right) + u^2\left(P - \frac{v}{u}\right)^2 = 0$$

or $$v^2 + (u^2 + 2v)\,(Pu - v) + (Pu - v)^2 = 0$$

or $$v^2 + Pu^3 - u^2v + 2uvP - 2v^2 + P^2u^2 - 2uvP + v^2 = 0$$

or $$Pu^3 - u^2v + P^2u^2 = 0$$

or $$v = Pu + P^2 \qquad \text{... }(iii)$$

which is in Clairaut's form.

Hence, its complete primitive is

$$v = cu + c^2, \; c \text{ is a constant.}$$

or $$xy = cy + c^2.$$

which is the complete solution of equation (*i*).

Differentiating (*iii*) w.r.t. u, we get

$$P = P + u\frac{dP}{du} + 2P\frac{dP}{du} \qquad \left[\because \frac{dv}{du} = P\right]$$

or $$(u + 2P)\frac{dP}{du} = 0$$

From it, we have $$u + 2P = 0$$

or $$P = -\frac{u}{2}$$

Substituting $P = -\frac{u}{2}$ in (*iii*), we obtain

$$v = -\frac{u^2}{2} + \frac{u^2}{4} = -\frac{u^2}{4}$$

or $$4v = -u^2 \text{ or } 4v + u^2 = 0$$

or $$4xy + y^2 = 0 \qquad [\because y = u,\ xy = v]$$

or $$y + 4x = 0$$

which is the required singular solution of equation (*i*).

5. *Reduce the equation $xp^2 - 2yp + x + 2y = 0$ to Clairaut's form by using the substitution $y - x = v$ and $x^2 = u$. Hence, find the complete primitive of the equation.*

Sol. Given $xp^2 - 2yp + x + 2y = 0$ ... (*i*)

$\because$ $$y - x = v \text{ and } x^2 = u$$

$\therefore$ $$\frac{dy}{dx} - 1 = \frac{dv}{dx} \text{ and } 2x = \frac{du}{dx}$$

$\therefore$ $$\frac{dv}{du} = \frac{p-1}{2x} \qquad \left(\because \frac{dy}{dx} = p\right)$$

or $$p = 2x\frac{dv}{du} + 1$$

Substituting the value of p in (*i*), we get

$$x\left(2x\frac{dv}{du} + 1\right)^2 - 2y\left(2x\frac{dv}{du} + 1\right) + x + 2y = 0$$

or $$4x^3\left(\frac{dv}{du}\right)^2 + 4x^2\frac{dv}{du} + 2x - 4xy\frac{dv}{du} = 0$$

or $$4x^2\left(\frac{dv}{du}\right)^2 + 4x\frac{dv}{du} + 2 - 4y\frac{dv}{du} = 0$$

or $$4u\left(\frac{dv}{du}\right)^2 - 4v\frac{dv}{dx} + 2 = 0 \qquad [\because x^2 = u \text{ and } y - x = v]$$

or
$$v = u\frac{dv}{du} + \frac{1}{2\frac{dv}{du}}$$

Taking $\frac{dv}{du} = P$, the above equation becomes

$$v = uP + \frac{1}{2P}$$

which is in Clairaut's form.

Hence, its complete primitive is

$$v = cu + \frac{1}{2c},\ c \text{ is a constant}$$

or
$$y - x = cx^2 + \frac{1}{2c}$$

or $\quad 2c^2x^2 - 2c\,(y - x) + 1 = 0$

which is the complete primitive of the equation (*i*).

EXERCISE 1.13

1. Reduce the equation $(px^2 + y^2)\,(px + y) = (p + 1)^2$ to Clairaut's form by using the substitutions $x + y = u$ and $xy = v$, and hence find its complete primitive.
2. Using the substitution $u = \frac{1}{x}$ and $v = \frac{1}{y}$, solve the equation $y^2\,(y - px) = x^4p^2$.
3. Solve the equation $4yp^2 + 2xp - y = 0$ by putting $y^2 = v$.
4. Transform the equation $e^{3x}\,(p - 1) + p^3\,e^{2y} = 0$ to Clairaut's form by substituting $e^x = u$ and $e^y = v$ and hence solve the equation.
5. Transform the equation $(x^2 + y^2)\,(1 + p)^2 - 2\,(x + y)\,(1 + p)\,(x + yp) + (x + yp)^2 = 0$ to Clairaut's form by substituting $x^2 + y^2 = v$ and $x + y = u$ and hence find the complete primitive of the equation.
6. Solve the equation $y = 2px + y^2p^3$.
7. Solve the equation $x^2\,(y - px) = p^2y$.
8. Solve the equation $(px - y)\,(x - 2y) = 2p$ by reducing it to Clairaut's form by using suitable transformations.

Answers

1. $c^2\,(x + y) - cxy + 1 = 0$
2. $c^2xy + cy - x = 0$
3. $y^2 = cx + c^2$
4. $e^y = ce^x + c^3$
5. $x^2 + y^2 = c\,(x + y) - \frac{1}{4}\,c^2$
6. $y^2 = cx + \frac{1}{8}\,c^3$
7. $y^2 = cx^2 + c^2$
8. $c^2x^2 - c\,(x^2 + y^2 - 2) + y^2 = 0.$

1.12 TRAJECTORIES OF A FAMILY OF CURVES

Definition. A curve which cuts every member of a given family of curves according to a given law is called a *trajectory* of the given family.

Here we shall consider only the case when each trajectory cuts every member of a given family at a constant angle. If the constant angle is a right angle then the trajectory is called *orthogonal.*

We shall find the orthogonal trajectory to a given family of curves when its equation is given in Cartesian or in polar co-ordinates.

1. Cartesian co-ordinates. We know that when two curves intersect, the angle between the curves is equal to the angle between the tangents to the curves at a common point of their intersection. It is also known that the slope of a tangent at any point (x, y) is $\frac{dy}{dx}$.

Let $\left(\frac{dy}{dx}\right)_{\text{I}}$ and $\left(\frac{dy}{dx}\right)_{\text{II}}$ be the slopes of the tangents to the two curves at their common point of intersection, and if the two curves are orthogonal, then

$$\left(\frac{dy}{dx}\right)_{\text{I}} \times \left(\frac{dy}{dx}\right)_{\text{II}} = -1$$

or

$$\left(\frac{dy}{dx}\right)_{\text{I}} = -\left(\frac{dx}{dy}\right)_{\text{II}}$$

Thus, replacing $\left(\frac{dy}{dx}\right)$ in the differential equation of the given curve by $-\left(\frac{dx}{dy}\right)$, we get the differential equation of the orthogonal trajectory and on solving the differential equation thus obtained, we obtain the equation to the orthogonal trajectory.

2. Polar co-ordinates. In polar co-ordinates angle between two tangents is $\phi_1 \sim \phi_2$, and

$\tan\phi = r\frac{d\theta}{dr}$.

If the two curves are orthogonal, then

$$\phi_1 \sim \phi_2 = \frac{\pi}{2}$$

or

$$\tan\phi_1 \times \tan\phi_2 = -1.$$

or

$$\left(r\frac{d\theta}{dr}\right)_{\text{I}} \times \left(r\frac{d\theta}{dr}\right)_{\text{II}} = -1$$

or

$$\left(r\frac{d\theta}{dr}\right)_{\text{I}} = -\left(\frac{1}{r}\cdot\frac{dr}{d\theta}\right)_{\text{II}}$$

or $$\left(\frac{dr}{d\theta}\right)_I = -\left(r^2\frac{d\theta}{dr}\right)_{II}$$

Thus replacing $r\frac{d\theta}{dr}$ by $-\frac{1}{r}\frac{dr}{d\theta}$ or replacing $\frac{dr}{d\theta}$ by $-r^2\frac{d\theta}{dr}$ in the differential equation of the given curve, we get a differential equation of the orthogonal trajectory and on solving this differential equation, we get the equation to the orthogonal trajectory.

Remark. For the given equation of the curve, we first find the differential equation of the curve by eliminating the parameter and then we proceed as mentioned above to obtain the orthogonal trajectory.

1.12.1 Illustrative Examples

1. *Find the orthogonal trajectories of the family of coaxial circles $x^2 + y^2 + 2gx + c = 0$, where g is the parameter.*

Sol. Given $x^2 + y^2 + 2gx + c = 0$ ...(*i*)

Differentiating (*i*) w.r.t. x, we get

$$2x + 2y\frac{dy}{dx} + 2g = 0$$

or $$x + y\frac{dy}{dx} + g = 0 \quad \text{...(ii)}$$

Eliminating, g, from (*i*) and (*ii*), we have

$$x^2 + y^2 - 2x\left(x + y\frac{dy}{dx}\right) + c = 0$$

or $$y^2 - x^2 - 2xy\frac{dy}{dx} + c = 0$$

Replacing $\frac{dy}{dx}$ by $-\frac{dx}{dy}$, we get

$$y^2 - x^2 + 2xy\frac{dx}{dy} + c = 0$$

or $$2xy\frac{dx}{dy} - x^2 = -c - y^2 \quad \text{...(iii)}$$

which is the differential equation of the orthogonal trajectory.

From (*iii*) $$2x\frac{dx}{dy} - \frac{x^2}{y} = -\frac{c}{y} - y.$$

Putting $x^2 = t$, we get $2x\frac{dx}{dy} = \frac{dt}{dy}$

$$\therefore \qquad \frac{dt}{dy} - \frac{t}{y} = -\frac{c}{y} - y \qquad ...(iv)$$

Here $$\text{I.F.} = e^{-\int \frac{1}{y} dy} = e^{-\log y} = \frac{1}{y}.$$

Multiplying (*iv*) by $\frac{1}{y}$ and integrating, we get

$$t.\frac{1}{y} = -\int \frac{1}{y}\left(\frac{c}{y} + y\right) dy + f \text{ where } f \text{ is a constant}$$

$$= -\int \left(\frac{c}{y^2} + 1\right) dy + f = -\left(-\frac{c}{y} + y\right) + f$$

or $$\frac{x^2}{y} = \frac{c}{y} - y + f$$

or $$x^2 = c - y^2 + fy$$

or $$x^2 + y^2 - fy - c = 0$$

which is a coaxial system of circles orthogonal to the given system with f, as parameter.

2. *Find the orthogonal trajectories to the family of parabolas* $y = ax^2$

Sol. Given $$y = ax^2 \qquad ...(i)$$

Differentiating (*i*), w.r. t. x, we get

$$\frac{dy}{dx} = 2ax$$

or $$\frac{dy}{dx} = 2x . \frac{y}{x^2} \qquad \left[\because a = \frac{y}{x^2}\right]$$

or $$\frac{dy}{dx} = \frac{2y}{x}$$

which is the equation of the trajectory. Replacing $\frac{dy}{dx}$ by $-\frac{dx}{dy}$, we get the differential equation of the orthogonal trajectory, which is $-\frac{dx}{dy} = \frac{2y}{x}$

or $$2y\ dy + x\ dx = 0$$

Integrating, we have

$$y^2 + \frac{x^2}{2} = \frac{c^2}{2}$$

or $$2y^2 + x^2 = c^2$$

which represents the orthogonal trajectories of the family of the given parabolas.

3. *Find the orthogonal trajectories of the curves*

$$\frac{x^2}{a^2+\lambda} + \frac{y^2}{b^2+\lambda} = 1,$$

λ being the parameter of the family.

Sol. Given $$\frac{x^2}{a^2+\lambda} + \frac{y^2}{b^2+\lambda} = 1 \quad ...(i)$$

Differentiating (*i*) w.r.t. x, we get

$$\frac{x}{a^2+\lambda} + \frac{y}{b^2+\lambda} \cdot \frac{dy}{dx} = 0 \quad ...(ii)$$

we have to eliminate λ from (*i*) and (*ii*).

From (*ii*), we have

$$\lambda = \frac{b^2x + a^2y\frac{dy}{dx}}{x + y\frac{dy}{dx}}$$

$$\therefore \quad a^2 + \lambda = \frac{(a^2-b^2)x}{x + y\frac{dy}{dx}}, \; b^2 + \lambda = -\frac{(a^2-b^2)y\frac{dy}{dx}}{x + y\frac{dy}{dx}}$$

Substituting these values in (*i*), we get

$$\left(x - y\frac{dx}{dy}\right)\left(x + y\frac{dy}{dx}\right) = a^2 - b^2 \quad ...(iii)$$

as the differential equation of the given family.

Changing $\frac{dy}{dx}$ to $-\frac{dx}{dy}$ in (*iii*), we obtain

$$\left(x + y\frac{dy}{dx}\right)\left(x - y\frac{dx}{dy}\right) = a^2 - b^2 \quad ...(iv)$$

which is the same as (*iii*).

Thus, we see that the family (*i*) is self-orthogonal, that is, every member of the family. (*i*) cuts every member of the same family orthogonally.

4. *Find the orthogonal trajectories of the rectangular hyperbola* $xy = a^2$.

Sol. Given
$$xy = a^2 \quad ...(i)$$
Differentiating (*i*) w.r. t. *x*, we get
$$x\frac{dy}{dx} + y = 0 \quad ...(ii)$$
Replacing $\frac{dy}{dx}$ by $-\frac{dx}{dy}$, we get
$$-x\frac{dx}{dy} + y = 0$$
which is the differential equation of the orthogonal trajectory.

or
$$-x\,dx + y\,dy = 0$$
Integrating, we get
$$\frac{y^2}{2} - \frac{x^2}{2} = \frac{c^2}{2}$$
or
$$y^2 - x^2 = c^2$$
which is the required equation of the orthogonal trajectories.

5. *Find the orthogonal trajectories of the cardioides*
$$r = a(1 - \cos\theta).$$

Sol. Given
$$r = a(1 - \cos\theta).$$
Taking logarithm, we get
$$\log r = \log a + \log(1 - \cos\theta)$$
Differentiating w.r. t. θ, we get the differential equation of the family of curves
$$\frac{1}{r}\frac{dr}{d\theta} = \frac{\sin\theta}{1-\cos\theta}$$
Replacing $\frac{dr}{d\theta}$ by $-r^2\frac{d\theta}{dr}$, we have the differential equation of the system of orthogonal trajectories
$$\frac{1}{r}\left(-r^2\frac{d\theta}{dr}\right) = \frac{\sin\theta}{1-\cos\theta}$$
or
$$\frac{dr}{r} + \frac{1-\cos\theta}{\sin\theta}\,d\theta = 0$$
or
$$\frac{dr}{r} + \frac{1-\cos^2\theta}{\sin\theta(1+\cos\theta)}\,d\theta = 0$$

or $$\frac{dr}{r} + \frac{\sin\theta}{1+\cos\theta}\,d\theta = 0$$

Integrating, we obtain

$$\log r - \log(1 + \cos\theta) = \log c$$

or $$\log \frac{r}{1+\cos\theta} = \log c$$

or $$r = c(1 + \cos\theta)$$

which is the required orthogonal trajectories.

EXERCISE 1.14

1. Find the orthogonal trajectories of the family of semi-cubical parabolas $ay^2 = x^3$, where a is a variable parameter.
2. Find the orthogonal trajectories of the following family of curves.
   (i) $y^2 = 4ax$ (ii) $y = ax^n$ (iii) $x^2 + y^2 = 2ay$
   (iv) $x^2 + 2y^2 = a$ (v) $x^{\frac{2}{3}} + y^{\frac{2}{3}} = a^{\frac{2}{3}}$.
3. Find the orthogonal trajectories of :
   (i) $r = a\cos\theta$ (ii) $r^2 = a^2\cos 2\theta$ (iii) $r = c\sin^2\theta$
   (iv) $r(1 + \cos\theta) = 2a$ (v) $r^n \sin n\theta = a^n$ (vi) $r^n = a^n \cos n\theta$.
4. Find the differential equation satisfied by the system of parabolas $y^2 = 4a(x + a)$ and show that the orthogonal trajectories of the system belong to the system itself.
5. Find the orthogonal trajectories of $\frac{x^2}{a^2} + \frac{y^2}{a^2+\lambda} = 1$ where λ is a parameter.
6. Show that the orthogonal trajectories of the system of co-axial circles
   $$x^2 + y^2 + 2\lambda x + c = 0$$
   form another system of co-axial circles
   $$x^2 + y^2 + 2\mu y - c = 0,$$
   where λ and μ are parameters and c is a given constant.

Answers

1. $3y^2 + 2x^2 = c^2$
2. (i) $2x^2 + y^2 = c^2$, (ii) $x^2 + ny^2 = c^2$ (iii) $x^2 + y^2 = 2cx$ (iv) $y = cx^2$
   (v) $x^{\frac{4}{3}} - y^{\frac{4}{3}} = c^{\frac{4}{3}}$
3. (i) $r = c\sin\theta$ (ii) $r^2 = c^2 \sin 2\theta$ (iii) $r^2 = c\cos\theta$ (iv) $r = \frac{2b}{(1-\cos\theta)}$
   (v) $r^n \cos n\theta = c$ (vi) $r^n = c^n \sin n\theta$
5. $x^2 + y^2 + c = 2a^2 \log x$.

1.13 LINEAR DIFFERENTIAL EQUATIONS OF HIGHER ORDERS WITH CONSTANT COEFFICIENTS

A linear differential equation is one, in which the dependent variable and its derivatives occur in the equation only in the first degree and are not multiplied together.

Thus, the general linear differential equation of the nth order is of the form :

$$P_0 \frac{d^n y}{dx^n} + P_1 \frac{d^{n-1} y}{dx^{n-1}} + P_2 \frac{d^{n-2} y}{dx^{n-2}} + \ldots + P_n y = f(x) \qquad \ldots(1)$$

where $P_0, P_1, P_2, \ldots, P_n$ are functions of x alone. When $P_0, P_1, P_2, \ldots, P_n$ are constants (independent of x). Then equation (1) is called *linear differential equation of nth order with constant coefficients.* Therefore, the linear differential equation of nth order with constant coefficient can be written as

$$(a_0 D^n + a_1 D^{n-1} + a_2 D^{n-2} + \ldots + a_n)\, y = f(x) \qquad \ldots(2)$$

or $F(D)\, y = f(x)$, where $F(D) \equiv a_0 D^n + a_1 D^{n-1} + \ldots + a_n$ and $a_0, a_1, a_2, \ldots, a_n$ are constants.

The function $f(x)$ may be equal to zero (*i.e.*, $f(x) = 0$) or may not be zero (*i.e.*, $f(x) \neq 0$)

If $n = 2$, the equation (2) is called linear differential equation of *second order* with constant coefficients which is

$$(a_0 D^2 + a_1 D + a_2)\, y = f(x) \qquad \ldots(3)$$

or $$F(D)\, y = f(x),\ F(D) \equiv (a_0 D^2 + a_1 D + a_2)$$

If $n = 3$, the linear differential equation of *third order* with constant coefficients is

$$(a_0 D^3 + a_1 D^2 + a_2 D + a_3)\, y = f(x) \qquad \ldots(4)$$

or $$F(D)\, y = f(x), \text{ when } D \equiv a_0 D^3 + a_1 D^2 + a_2 D + a_3$$

Similarly, we can write the fourth and higher orders.

Case I. If $f(x) = 0$, then the general linear-equation of second order with constant coefficients is

$$(a_0 D^2 + a_1 D + a_3)\, y = 0 \qquad \ldots(5)$$

or $$F(D)\, y = 0$$

and the general linear equation of third order with constant coefficient is

$$(a_0 D^3 + a_1 D^2 + a_2 D + a_3) = 0 \qquad \ldots(6)$$

or $$f(D)\, y = 0$$

Similarly, we can write the other higher order equations. We now state a theorem, without proof, as follows :

Theorem 1.2. If $y = y_1, y = y_2, y = y_3, \ldots, y = y_n$ are n linearly independent solutions of the differential equation

$$(a_0 D^n + a_1 D^{n-1} + \ldots + a_n)\, y = 0 \qquad \ldots(7)$$

then $y = c_1 y_1 + c_2 y_2 + \ldots + c_n y_n$ is also a solution of the equation (7), where $c_1, c_2, \ldots, c_n$ are arbitrary constants.

Since the solution y contains n arbitrary constants, it is called the *general* or *complete solution* of the equation (7).

1.13.1 Method of solution of linear differential equation of nth order with constant coefficients when f(x) = 0

Let the general linear differential equation of nth order with constant coefficient when $f(x) = 0$ be

$$a_0 \frac{d^n y}{dx^n} + a_1 \frac{d^{n-1} y}{dx^{n-1}} + a_2 \frac{d^{n-2} y}{dx^{n-2}} + \ldots + a_n y = 0 \qquad \ldots(8)$$

Let us consider $y = e^{mx}$ be a *non-zero trial solution* of the equation (8). Then by successive differentiations, we get

$$y_1 = me^{mx}, y_2 = m^2 e^{mx}, \ldots, y_n = m^n e^{mx}.$$

Substituting the values of $y_1, y_2, \ldots, y_n$ in (8), we get

$$(a_0 m^n + a_1 m^{n-1} + a_2 m^{n-2} + \ldots + a_n)\, e^{mx} = 0$$

since $e^{mx} \neq 0$,

$$\therefore \quad a_0 m^n + a_1 m^{n-1} + a_2 m^{n-2} + \ldots + a_n = 0 \qquad \ldots(9)$$

Equation (9) is called *Auxiliary equation (A.E.)* of (8). We then solve the equation (9) to find its roots. Let $m_1, m_2, \ldots, m_n$ be the roots of the A.E. (9). The solution of equation (8) depends upon the nature of the roots of the A.E. (9).

The following cases arise :

(*i*) If all the roots of A.E. are real and distinct , say $m_1, m_2, \ldots, m_n$, then

$$\text{G.S. is } y = c_1 e^{m_1 x} + c_2 e^{m_2 x} + \ldots + c_n e^{m_n x}$$

where $c_1, c_2, \ldots, c_n$ are arbitrary constants.

(*ii*) If two roots are *real* and *equal* and rest are *real unequal,* say the roots are $m_1, m_1, m_2, m_3, \ldots, m_n$, then the G.S. is

$$y = (c_1 + c_2 x)\, e^{m_1 x} + c_3 e^{m_2 x} + \ldots + c_n e^{m_2 x}$$

where $c_1, c_2, \ldots, c_n$ are arbitrary constants.

(*iii*) If three roots are *real equal* and the rest are real and distinct, say the roots are, $m_1, m_1, m_1, m_2, m_3, \ldots, m_n$, then G.S. is

$$y = (c_1 + c_2 x + c_3 x^2)\, e^{m_1 x} + c_4 e^{m_2 x} + \ldots + c_n e^{m_n x}.$$

(*iv*) If two roots are *complex* and other roots are real and distinct, say $m = \alpha \pm i\beta$, $m_3, m_4, \ldots, m_n$

Then G.S. is

$$y = c_1 e^{\alpha x} \cos(\beta x + \varepsilon) + m_3 e^{m_3 x} + c_4 e^{m_4 x} + \ldots + c_n e^{m_n x}$$

or
$$y = e^{\alpha x}(c_1 \cos\beta x + c_2 \sin\beta x) + c_3 e^{m_3 x} + c_4 e^{m_4 x} + \ldots + c_n e^{m_n x}.$$

(*v*) If two roots are *real* and *equal*, two roots are *complex* and the other roots are real and distinct, say the roots are $m_1, m_1, \alpha \pm i\beta, m_5, m_6, \ldots, m_n$, then the G.S. is

$$y = (c_1 + c_2 x)e^{m_1 x} + e^{\alpha x}(c_3 \cos\beta x + c_4 \sin\beta x) + c_5 e^{m_5 x} + \ldots + m_n x + c_n e^{m_n x}.$$

The other possibilities can also be formulated in this way.

We shall illustrate these through the following illustrative examples of second order ($n = 2$).

1.13.2 Illustrative Examples

1: *Find the general solution of the following differential equations :*

(i) $\frac{d^2y}{dx^2} = 0$ (ii) $(D^2 + 5D + 4)\,y = 0$

(iii) $(2D^2 - 3D + 1)\,y = 0$ (iv) $\frac{d^2y}{dx^2} - 4\frac{dy}{dx} + y = 0$

(v) $\frac{d^2x}{dt^2} + 6\frac{dx}{dt} + 9x = 0$ (vi) $(D^2 - 4D + 13)y = 0.$

Sol. (*i*) $\frac{d^2y}{dx^2} = 0$, or $D^2 y = 0$

Here, A.E. is $m^2 = 0$

$\therefore$ $m = 0, 0$

Hence, the G.S. is $y = (c_1 + c_2x)\,e^{0.x} = c_1 + c_2x.$

(*ii*) Given $(D^2 + 5D + 4)\,y = 0.$

The A.E. is $m^2 + 5m + 4 = 0$

or $(m + 1)(m + 4) = 0$

$\therefore$ $m = -1, m = -4$

Hence, the G.S. is $y = c_1 e^{-x} + c_2 e^{-4x}$,

where c_1 and c_2 are arbitrary constants.

(*iii*) $(2D^2 - 3D + 1)\,y = 0$

The A.E. is $2m^2 - 3m + 1 = 0$

or $2m^2 - 2m - m + 1 = 0$

or $2m(m - 1) - (m - 1) = 0$

or $(m - 1)(2m - 1) = 0$

$\therefore m = 1$, and $m = \frac{1}{2}$

Hence, G.S. is $y = c_1 e^x + c_2 e^{\frac{1}{2}x}$,

where c_1 and c_2 are arbitrary constants.

(*iv*) Given $\frac{d^2y}{dx^2} - 4\frac{dy}{dx} + y = 0$ ...(*i*)

or $(D^2 - 4D + 1)\,y = 0$

Here the A.E. is $m^2 - 4m + 1 = 0$

or $$m = \frac{4 \pm \sqrt{16-4}}{2} = \frac{4 \pm 2\sqrt{3}}{2}$$

$$= 2 \pm \sqrt{3}$$

The roots of the equation (*i*) are $2+\sqrt{3}$ and $2-\sqrt{3}$

Hence, G.S. is $$y = c_1 e^{(2+\sqrt{3})x} + c_2 e^{(2-\sqrt{3})x}.$$

(*v*) Given $$\frac{d^2x}{dt^2} + 6\frac{dx}{dt} + 9x = 0$$

or $$(D^2 + 6D + 9)x = 0,$$

where $D \equiv \frac{d}{dt}$

$\therefore$ The A.E. is $$m^2 + 6m + 9 = 0$$

or $$(m+3)^2 = 0$$

that is $$m = -3, -3.$$

Hence, the G.S. is

$$x = (c_1 + c_2 t)\, e^{-3t}$$

(*vi*) Given $(D^2 - 4D + 13)\,y = 0$

Here the A.E. is $m^2 - 4m + 13 = 0$

or $$m = \frac{4 \pm \sqrt{16-52}}{2} = \frac{4 \pm \sqrt{-36}}{2}$$

$$= \frac{4 \pm 6i}{2}, \ (i = \sqrt{-1})$$

$$= 2 \pm 3i$$

Hence, the G.S. is

$$y = e^{2x}\,(c_1 \cos 3x + c_2 \sin 3x)$$

2. *Solve the following differential equations under the given conditions :*

(*i*) $\frac{d^2y}{dx^2} + \frac{dy}{dx} - 2y = 0$, *when* $x = 0$, $y = 3$ *and* $\frac{dy}{dx} = 0$.

(*ii*) $(D^2 + n^2)\,x = 0$, *where* $D \equiv \frac{d}{dt}$, *given* $x = 4$, $\frac{dx}{dt} = 0$, *when* $t = 0$.

(*iii*) $\frac{d^2x}{dt^2} - 3\frac{dx}{dt} + 2x = 0$, *when* $t = 0$, $x = 0$ *and* $\frac{dx}{dt} = 0$.

Sol. (*i*) Given $\frac{d^2y}{dx^2}+\frac{dy}{dx}-2y=0$

or $(D^2+D-2)y=0$

The A.E. is $m^2+m-2=0$

or $m^2+2m-m-2=0$

or $m(m+2)-(m+2)=0$

or $(m+2)(m-1)=0$

The solutions are $m=1, m=-2$

Hence, the G.S. is $y=c_1e^x+c_2e^{-2x}$ ...(*i*)

Given $x=0, y=3$

$\therefore$ $3=c_1+c_2$ ...(*ii*)

Differentiating (*i*) w.r.t. x, we get

$$\frac{dy}{dx}=c_1e^x-2c_2e^{-2x}$$

Using the conditions $x=0, \frac{dy}{dx}=0$, we get

$$0=c_1-2c_2 \qquad ...(iii)$$

Solving (*ii*) and (*iii*) are obtain, $c_1=2, c_2=1$

Hence, the particular solution is

$$y=2e^x+e^{-2x}.$$

(*ii*) Given $(D^2+n^2)x=0, (D\equiv\frac{d}{dt})$

$\therefore$ The A.E. is $m^2+n^2=0$

or $m=\pm ni$

Hence, the G.S. is $x=c_1\cos nt+c_2\sin nt$

Given $t=0, x=4$,

$\therefore$ $4=c_1$

Differentiating w.r.t. t, we get

$$\frac{dx}{dt}=-nc_1\sin nt+nc_2\cos nt$$

given, when $t=0$, $\frac{dx}{dt}=0$

$\therefore 0=nc_2$ or $c_2=0$

Hence, the particular solution of the given equation is $x=4\cos nt$.

(*iii*) Given $\frac{d^2x}{dt^2} - 3\frac{dx}{dt} + 2x = 0$

or $(D^2 - 3D + 2)\,x = 0, D \equiv \frac{d}{dt}$

The A.E. is $m^2 - 3m + 2 = 0$

or $m = \frac{3 \pm \sqrt{9-8}}{2} = \frac{3 \pm 1}{2}$

$\therefore \quad m = 2, 1$

Hence, the G.S. is $x = c_1 e^t + c_2 e^{2t}$ ...(*i*)

Using the condition $t = 0, x = 0$, we get from (*i*),

$$0 = c_1 + c_2 \quad \text{...(ii)}$$

Differentiating (*i*) w.r.t. *t* we get

$$\frac{dx}{dt} = c_1 e^t + 2c_2 e^{2t}$$

Using the condition $t = 0; \frac{dx}{dt} = 0$, we obtain

$$0 = c_1 + 2c_2 \quad \text{...(iii)}$$

Solving (*ii*) and (*iii*), we have

$$c_1 = 0, c_2 = 0$$

Hence, the particular solution of the given equation is $x = 0$.

(*iv*) Given equation is $(D^2 - 3D + 2)\,y = 0$

The A.E. is $m^2 - 3m + 2 = 0$

or $(m-1)(m-2) = 0$

$\therefore$ $m = 1, 2.$

Hence, the G.S. is $y = c_1 e^x + c_2 e^{2x}$.

given $y = 0$, when $x = 0$.

$\therefore$ $0 = c_1 + c_2$

Differentiating the G.S. we get,

$$\frac{dy}{dx} = c_1 e^x + 2c_2 e^{2x}$$

since $\frac{dy}{dx} = 0,$

when $x = 0,$

$\therefore$ $0 = c_1 + 2c_2$

Solving the equations, we get $c_1 = c_2 = 0$.

Therefore, the required solution is $y = 0$.

3. *Find the solution of the equation*

$$l\frac{d^2\theta}{dt^2} + g\theta = 0, \text{ given } \theta = \alpha, \frac{d\theta}{dt} = 0$$

when $t = 0$.

Sol. Given $$l\frac{d^2\theta}{dt^2} + g\theta = 0$$

or $(lD^2 + g)\,\theta = 0, \left(D \equiv \frac{d}{dt}\right)$

The A.E. is $$lm^2 + g = 0$$

or $$m^2 = -\frac{g}{l}$$

or $$m = \pm i\sqrt{\frac{g}{l}}.$$

Therefore, the G.S. is

$$\theta = c_1 \cos\sqrt{\frac{g}{l}}\,t + c_2 \sin\sqrt{\frac{g}{l}}\,t \qquad \text{...(i)}$$

Using the given condition $\theta = \alpha$, when $t = 0$

we get $$\alpha = c_1$$

Differentiating (*i*) w.r.t. t, we get

$$\frac{d\theta}{dt} = -c_1\sqrt{\frac{g}{l}}\sin\sqrt{\frac{g}{l}}\,t + c_2\sqrt{\frac{g}{l}}\cos\sqrt{\frac{g}{l}}\,t$$

using the condition $\frac{d\theta}{dt} = 0$, when $t = 0$

we get $$0 = c_2.\sqrt{\frac{g}{l}} \text{ i.e., } c_2 = 0$$

Hence, the required solution of the given equation is

$$\theta = \alpha \cos\sqrt{\frac{g}{l}}\,t.$$

4. *If* $\frac{d^2V}{dx^2} + p^2 V = 0$, *and* $V = V_o$ *at* $x = 0$ *and* $V = 0$ *at* $x = l$, *show that*

$$V = Vo\frac{\sin p(l - x)}{\sin pl}.$$

Sol. Given $$\frac{d^2V}{dx^2} + p^2V = 0$$

Here the A.E. is $$m^2 + p^2 = 0$$

or $$m = \pm ip \ (i = \sqrt{-1})$$

$\therefore$ G.S. is $$V = c_1 \cos px + c_2 \sin px$$

Given $$V = V_o \text{ at } x = 0$$

$\therefore$ $$V_o = c_1$$

Again, $V = 0$ when $x = l$

$\therefore$ $$0 = c_1 \cos pl + c_2 \sin pl$$

or $$c_2 \sin pl = -c_1 \cos pl = -V_o \cos pl$$

or $$c_2 = -Vo \frac{\cos pl}{\sin pl}$$

Hence, the required solution is

$$V = V_0 \cos p\,x - V_o \frac{\cos pl}{\sin pl} \sin px$$

or $$V = V_o \left(\frac{\sin pl \cos px - \cos pl \sin px}{\sin pl} \right)$$

or $$V = V_o \frac{\sin p(l - x)}{\sin pl}.$$

Remark. General linear differential equations with constant coefficients of order greater than two that is, of order three, four etc. when $f(x) = 0$ are as follows :

$$(D^3 + a_1 D^2 + a_2 D + a_3)\, y = 0,$$
$$(D^4 + a_1 D^3 + a_2 D^2 + a_3 D + a_4)\, y = 0$$

and so on.

The method of solutions are exactly the same as that of second order.

1.13.3 Illustrative Examples

1. *Solve the following differential equations :*

(i) $\frac{d^3y}{dx^3} = y$ (ii) $\frac{d^3y}{dx^3} - 3\frac{dy}{dx} + 2y = 0$

(iii) $\frac{d^3y}{dx^3} - 3\frac{d^2y}{dx^2} + 3\frac{dy}{dx} - y = 0$ (iv) $(D^4 - 5D^2 + 4)\, y = 0,$

(v) $(D^4 - 1)\, y = 0$ (vi) $(D^4 + 8D^2 + 16)\, y = 0$

(vii) $(D^2 + 1)^3 (D^2 + D + 1)\, y = 0.$

Sol. (*i*) Given $$\frac{d^3y}{dx^3} = y$$

or $$(D^3 - 1)\, y = 0 \qquad \ldots(1)$$

$\therefore$ A.E. is $m^3 - 1 = 0$

or $$(m-1)(m^2+m+1) = 0$$

$\therefore$ $m = 1$ and $m^2 + m + 1 = 0$

$$m^2 + m + 1 = 0$$

gives, $$m = \frac{-1 \pm \sqrt{1-4}}{2} = \frac{-1 \pm \sqrt{3}i}{2}$$

Hence, the roots of A.E. are $1, \dfrac{-1+\sqrt{3}i}{2}, \dfrac{-1-\sqrt{3}i}{2}$

$\therefore$ G.S. of (1) is

$$y = c_1 e^x + e^{-\frac{1}{2}x}\left(c_2 \cos \frac{\sqrt{3}}{2}x + c_3 \sin \frac{\sqrt{3}}{2}x\right).$$

(*ii*) Given $$\frac{d^3y}{dx^3} - 3\frac{dy}{dx} + 2y = 0$$

or $$(D^3 - 3D + 2)\, y = 0$$

The A.E. is $$m^3 - 3m + 2 = 0$$

or $$(m-1)(m^2+m-2) = 0$$

or $$(m-1)(m-1)(m+2) = 0$$

Hence the roots of A.E. are $m = 1, 1$, and -2.

$\therefore$ The G.S. of the given equation is

$$y = (c_1 + c_2 x)e^x + c_3 e^{-2x}.$$

(*iii*) Given

$$\frac{d^3y}{dx^3} - 3\frac{d^2y}{dx^2} + 3\frac{dy}{dx} - y = 0$$

or $$(D^3 - 3D^2 + 3D - 1)\, y = 0 \qquad \ldots(1)$$

The A.E. of (1) is

$\therefore$ $$m^3 - 3m^2 + 3m - 1 = 0$$

or $$(m^3 - 1) - 3m(m-1) = 0$$

or $$(m-1)(m^2+m+1) - 3m(m-1) = 0$$

or $$(m-1)(m^2 - 2m + 1) = 0$$

or $$(m-1)(m-1)^2 = 0$$

Hence, the roots of the A.E. are $m = 1, 1, 1.$

∴ The G.S. of equation (1) is

$$y = (c_1 + c_2 x + c_3 x^2)\, e^x,$$

(*iv*) Given $(D^4 - 5D^2 + 4)\,y = 0$ ...(1)

The A.E. is $m^4 - 5m^2 + 4 = 0$

or $(m^2 - 1)(m^2 - 4) = 0$

i.e., $m^2 - 1 = 0$, gives $m = \pm 1$,

and $m^2 - 4 = 0$, gives $m = \pm 2$.

Hence the roots of A.E. are $-1, 1, -2, 2$

∴ The G.S. of equation (1) is

$$y = c_1 e^{-2x} + c_2 e^{-x} + c_3 e^x + c_4 e^{2x}.$$

(*v*) Given $(D^4 - 1)\,y = 0$

∴ A.E. is $m^4 - 1 = 0$

or $(m^2 - 1)(m^2 + 1) = 0$

$m^2 - 1 = 0$

gives $m = \pm 1$

and $m^2 + 1 = 0$,

gives $m = \pm i$

∴ The roots of A.E. are, $-1, 1$, and $\pm i$

∴ The G.S. of the given equation is

$$y = c_1 e^{-x} + c_2 e^x + c_3 \cos x + c_4 \sin x.$$

(*vi*) Given $(D^4 + 8D^2 + 16)\,y = 0$

The A.E. is $m^4 + 8m^2 + 16 = 0$

or $(m^2 + 4)^2 = 0$

which gives $m^2 = -4$,

$m^2 = -4$

i.e., $m = \pm 2i$,

$m = \pm 2i$

∴ The G.S. of the given equation is

$$y = (c_1 + c_2 x) \cos 2x + (c_3 + c_4 x) \sin 2x.$$

(*vii*) Given $(D^2 + 1)^3 (D^2 + D + 1)^2 y = 0$

The A.E. is $(m^2 + 1)^3 (m^2 + m + 1)^2 = 0$

The solutions of the given equation are

$m = \pm i$ (thrice), and $m = -\dfrac{1}{2} \pm i\sqrt{\dfrac{3}{2}}$ (twice).

Hence, the G.S. of the equation is

$$y = \left[\left(c_1 + c_2x + c_3x^2\right)\cos x + \left(c_4 + c_5x + c_6x^2\right)\sin x\right]$$

$$+\left[(c_7 + c_8x)\cos\frac{\sqrt{3}}{2}x + (c_9 + c_{10}x)\sin\frac{\sqrt{3}}{2}x\right]e^{-\frac{1}{2}x}.$$

EXERCISE 1.15

Solve the following differential equations :

1. $\dfrac{d^2y}{dx^2} - 3\dfrac{dy}{dx} - 4y = 0$

2. $\dfrac{d^2y}{dx^2} + \dfrac{dy}{dx} + y = 0$

3. $\dfrac{d^2y}{dx^2} - y\dfrac{dy}{dx} + 12y = 0$

4. $\dfrac{d^2y}{dx^2} + 2\dfrac{dy}{dx} + 2y = 0$ *(W.B.U.T. BCA, 2006)*

5. $(D^2 + D)y = 0$

6. $(D^2 + 6D + 25)y = 0$

7. $(D^2 + 4D + 13)y = 0$

8. $\{D^2 + (a + b)D + ab\}\ y = 0$

9. $(D^2 + 1)y = 0$, $y = 4$, when $x = 0$; $y = 0$, when $x = \dfrac{\pi}{2}$.

10. $(6D^2 - 11D + 3)x = 0$, $D \equiv \dfrac{d}{dt}$, given $x = 7$, $\dfrac{dx}{dt} = 0$, when $t = 0$.

11. $(D + 1)^2\ y = 0$, given $y = 2\log_e 2$, when $x = \log_e 2$ and $y = \dfrac{4}{3}\log_e 3$ when $x = \log_e 3$.

12. $\dfrac{d^2x}{dt^2} = 0$, given $x = 0$ when $t = 0$ and $\dfrac{dx}{dt} = 5$ when $t = 5$.

13. Show that the solution of $\dfrac{d^2x}{dt^2} + k\dfrac{dx}{dt} + \mu x = 0$.

is $x = e^{-\frac{1}{2}kt}(A\cos nt + B\sin nt)$, *if* $k^2 < 4m$, and $n^2 = m - \dfrac{1}{4}k^2$.

14. $\dfrac{d^3y}{dx^3} - 7\dfrac{dy}{dx} - 6y = 0$

15. $\dfrac{d^3y}{dx^3} + 6\dfrac{d^2y}{dx^2} + 11\dfrac{dy}{dx} + 6y = 0$

16. $(D^3 - D^2 - 2D)\ y = 0$

17. $(D^3 - 3D^2 + 4D - 2)\ y = 0$

18. $(D^3 - 3D^2 + 4)\ y = 0$

19. $(D^4 + 2D^2 + 1)\ y = 0$

20. $(D^4 - 4D^2)\ y = 0$

21. $(D^4 + a^4)\ y = 0$

22. $(D^4 + 4D^3 + 8D^2 + 8D + 4)\ y = 0$

23. $(D - 1)^2\ (D^2 + 1)^2\ y = 0$

24. $(D + 1)^3\ (D^2 + 1)\ y = 0$

Answers

1. $y = c_1e^{4x} + c_2e^{-x}$
2. $y = e^{-\frac{x}{2}}\left(c_1 \cos\frac{\sqrt{3}}{2}x + c_2 \sin\frac{\sqrt{3}}{2}x\right)$
3. $y = c_1e^{3x} + c_2e^{4x}$
4. $y = e^{-x}(c_1 \cos x + c_2 \sin x)$
5. $y = c_1 + c_2e^{-x}$
6. $y = e^{-2x}(c_1 \cos 4x + c_2 \sin 4x)$
7. $y = e^{-2x}(c_1 \cos 3x + c_2 \sin 3x)$
8. $y = c_1e^{-ax} + c_2e^{-bx}$
9. $y = 4 \cos x$
10. $x = 9e^{\frac{1}{3}t} - 2e^{\frac{3}{2}t}$
11. $y = 4xe^{-x}$
12. $x = 5t$
14. $y = c_1 + (c_2 x + c_3)\, e^{2x}$
15. $y = c_1e^{-x} + c_2e^{-2x} + c_3e^{-3x}$
16. $y = c_1 + c_2e^{2x} + c_3e^{-x}$
17. $y = e^{x}(c_1 + c_2 \cos x + c_3 \sin x)$
18. $y = e^{2x}(c_1 + c_2x) + c_3e^{-x}$
19. $y = (c_1 + c_2x) \sin x + (c_3 + c_4x) \cos x$
20. $y = c_1 + c_2x + c_3e^{2x} + c_4e^{-2x}$
21. $y = e^{\frac{-a}{\sqrt{2}}x}\left[c_1 \cos\frac{a}{\sqrt{2}}x + c_2 \sin\frac{a}{\sqrt{2}}x\right] + e^{\frac{a}{\sqrt{2}}x}\left[c_3 \cos\frac{a}{\sqrt{2}}x + c_4 \sin\frac{a}{\sqrt{2}}x\right]$
22. $y = e^{-x}\,[(c_1 + c_2x) \cos x + (c_3 + c_4x) \sin x]$
23. $y = (c_1 + c_2x)\, e^{x} + (c_3 + c_4x) \cos x + (c_5 + c_6x) \sin x$
24. $y = e^{-x}(c_1 + c_2x + c_3x^2) + c_4 \cos x + c_5 \sin x.$

1.13.4 Methods of solution of linear differential equations with constant coefficients when f(x) ≠ 0

Before discussing the method of solution of linear differential equations with constant coefficients of second and higher orders, $f(x) \neq 0$, we first state a theorem, without proof, as below:

Theorem 1.3. If $y = u$ is a complete solution of the differential equation $F(D)\,y = 0$ and $y = v$ is a *particular solution* (containing no arbitrary constants) of the equation $F(D)\,y = f(x)$, then the *complete solution* of the equation $F(D)\,y = f(x)$ is $y = u + v$.

Remark In case of a linear differential equation with constant coefficient $F(D) = 0$, $y = u$ is the general solutions (G.S.) of the equation, but for linear differential equation with constant coefficients $F(D)\,y = f(x)$, u is called the *complementary function (C.F.)* and v is called the *particular integral (P.I.)* of the equation. In this case, the general solution (G.S.) is

$$y = u + v$$

i.e.,

$$y = \text{C.F.} + \text{P.I.}$$

Thus to solve a linear differential equation with constant coefficients $F(D)\,y = f(x)$ second and higher orders, first we find G.S. of $F(D)\,y = 0$, which is C.F. of the equation $F(D)\,y = f(x)$, then we find its P.I.

1.13.5 Methods for finding P.I. of linear differential equations of second order with constant coefficients

Let the second order linear differential equation with constant coefficients be

$$F(D)\,y = f(x)$$

where $$F(D) \equiv a_0 D^2 + a_1 D + a_2 \text{ and } D \equiv \frac{d}{dx}.$$

The function $f(x)$ may be of the following forms:

(*i*) $f(x) = x^n$ (function of x) or constant

(*ii*) $f(x) = e^{ax}$

(*iii*) $f(x) = x^n e^{ax}$

(*iv*) $f(x) = \sin ax$ or $\cos ax$

(*v*) $f(x) = \dfrac{x^n \cos ax}{x^n \sin ax}$

(*vi*) $\dfrac{1}{D+a}\phi(x) = e^{-ax}\displaystyle\int e^{ax}\phi(x)\,dx$

we now consider the cases mentioned above to find the P.I.

(*i*) **when f(x) = xⁿ or constant, then**

$$\textbf{P.I.} = \frac{1}{F(D)}x^n = [F(D)]^{-1}x^n$$

In this case we use the Binomial expansion

$$(1 \pm x)^{-1} = 1 \mp nx + \frac{n(n-1)}{2!}x^2 \mp \frac{n(n-1)(n-2)}{3!}x^3 + \ldots\infty$$

As for examples,

(*a*) Find P.I. of $(D^2 + D - 2)\,y = 2x^2$,

Here, $$\text{P.I.} = \frac{1}{D^2+D-2}(2x^2) = \frac{1}{-2\left[1 - \dfrac{D^2+D}{2}\right]}(2x^2)$$

$$= \frac{1}{-2}\left[1 - \frac{D^2+D}{2}\right]^{-1}(2x^2)$$

$$= -1\left[1 + \frac{D^2+D}{2} + \frac{D^4+2D^3+D^2}{4} + \ldots\right]x^2$$

$$= -\left[x^2 + (1+x) + \frac{1}{4}\cdot 2\right]$$

$$= -\left[x^2 + x + \frac{3}{2}\right]$$

$$= -\frac{1}{2}\left[2x^2 + 2x + 3\right].$$

(*b*) Find P.I. of $(D^2 - 3D + 2)\, y = 5$.

Here, $$\text{P.I.} = \frac{1}{D^2 - 3D + 2}(5) = \frac{1}{2\left[1 + \frac{D^2 - 3D}{2}\right]}(5)$$

$$= \frac{1}{2}\left[1 + \frac{D^2 - 3D}{2}\right]^{-1}(5)$$

$$= \frac{1}{2}\left[1 - \frac{D^2 - 3D}{2} + \ldots\right](5)$$

$$= \frac{1}{2}5 = \frac{5}{2}.$$

Alternative method

$$\text{P.I.} = \frac{1}{D^2 - 3D + 2}5 = \frac{1}{D^2 - 3D + 2}5.e^o$$

$$= \frac{1}{0 - 0 + 2}5 = \frac{5}{2}.$$

(*ii*) When $f(x) = e^{ax}$, then

$$\textbf{P.I.} = \frac{1}{F(D)}e^{ax}$$

Rule: Put $D = a$, provided $F(a) \neq 0$, *i.e.*,

$$\text{P.I.} = e^{ax}\frac{1}{F(a)}$$

If $F(a) = 0$, then put $D = D + a$ *i.e.*,

$$\text{P.I.} = e^{ax}\frac{1}{F(D + a)} \qquad \ldots(1)$$

Then use rule (*i*)
or if $F(a) = 0$, then apply,

$$\frac{1}{F(D)}e^{ax} = x\frac{1}{F'(a)}e^{ax}$$

if $f'(a) = 0$, then apply,

$$\frac{1}{F(D)} e^{ax} = x^2 \frac{1}{F''(a)} e^{ax}$$

Let us illustrate these with examples:

Example 1. *Find P.I. of* $(D^2 + 3D - 2)\, y = e^{-2x}$.

Sol. $$\text{P.I.} = \frac{1}{D^2 + 3D - 2} e^{-2x}$$

$$= e^{-2x} \frac{1}{4 - 6 - 2} \qquad [\text{Putting } D = -2, \therefore F(-2) \neq 0]$$

$$= -\frac{1}{4} e^{-2x}$$

Example 2. *Find P.I. of*

$$\frac{d^2y}{dx^2} - 4\frac{dy}{dx} + 3 = 2e^{3x}.$$

Sol. The given equation can be written as

$$(D^2 - 4D + 3)\, y = 2e^{3x}$$

$$\therefore \qquad \text{P.I.} = \frac{1}{D^2 - 4D + 3} 2e^{3x} = e^{3x} \frac{1}{(D+3)^2 - 4(D+3) + 3} 2$$

[Putting D = D + 3, since F(3) = 0]

$$= e^{3x} \frac{1}{D^2 + 9 + 6D - 4D - 12 + 3} (2)$$

$$= e^{3x} \frac{1}{D^2 + 2D} (2) = e^{3x} \frac{1}{2D\left(1 + \frac{D}{2}\right)} (2)$$

$$= e^{3x} \frac{1}{2D}\left(1 + \frac{D}{2}\right)^{-1} (2)$$

$$= e^{3x} \frac{1}{2D}\left(1 - \frac{D}{2} + \ldots\right) (2) \qquad [\text{Using rule } (i)]$$

$$= \frac{1}{2} e^{3x} \frac{1}{D} (2) = \frac{1}{2} e^{3x} . 2x \qquad \left[\because \frac{1}{D}(2) = \int 2dx \equiv 2x\right]$$

$$= xe^{3x}$$

Hence required $\qquad \text{P.I.} = xe^{3x}$.

Alternative method

$$\text{P.I.} = \frac{1}{D^2 - 4D + 3} 2e^{3x} \qquad [\text{Here } F(3) = 0, \because a = 3]$$

$\therefore$ Applying the rule $\frac{1}{F(D)} e^{ax} = x \frac{1}{F'(a)} e^{ax}$

$$= 2x\frac{1}{2D-4}e^{3x} = 2xe^{3x}\frac{1}{2\times 3-4} = xe^{3x}$$

Hence P.I. $= xe^{3x}$.

(iii) **When** $f(x) = x^n e^{ax}$, **then P.I.** $= \dfrac{1}{F(D)} x^n e^{ax}$

Rule: First operates on e^{ax}, putting $D = D + a$ and transferring e^{ax} left of the function $\dfrac{1}{F(D)}$ apply Rule(*i*).

We illustrate this with an example:

Example. *Find P.I. of*

$$(D^2 + 2D + 2)\,y = xe^{-x}.$$

Sol. Here,

$$\text{P.I.} = \frac{1}{D^2 + 2D + 2} xe^{-x}$$

$$= e^{-x}\frac{1}{(D-1)^2 + 2(D-1) + 2}x \qquad \text{[Applying rule (ii)]}$$

$$= e^{-x}\frac{1}{D^2 + 1 - 2D + 2D - 2 + 2}x$$

$$= e^{-x}\frac{1}{1 + D^2}x$$

$$= e^{-x}[1 + D^2]^{-1}x$$

$$= e^{-x}[1 - D^2 + \ldots]x \qquad \text{[Applying rule (i)]}$$

$$= e^{-x}x = xe^{-x}$$

Hence, P.I. $= xe^{-x}$.

(iv) **When** $f(x) = \sin ax$ **(or** $\cos ax$**),**

$$\textbf{P.I.} = \frac{1}{F(D)}\sin ax \text{ (or } \cos ax)$$

Rule: If $F(-a^2) \neq 0$, put $D^2 = -a^2$

i.e.,

$$\text{P.I.} = \frac{1}{F(-a^2)}\sin ax(\text{or } \cos ax)$$

But if $F(-a^2) = 0$, then find P.I. for e^{iax} using rule (*ii*), putting $D = D + ia$. Then replacing $e^{iax} = \cos ax + i\sin ax$, , equate imaginary part for $\sin ax$ or real part for $\cos ax$.

Alternative rule

If $F(-a^2) = 0$, then find P.I. for e^{iax}, using the following rule:

$$\text{P.I.} = \frac{1}{F(D^2)} e^{iax} = x \frac{1}{F'(D^2)} e^{iax}$$

$$= x \frac{1}{F'(D^2)} [\cos ax + i \sin ax]$$

Then equate real part for $\cos ax$

i.e., $$\frac{1}{F(D^2)} \cos ax = x \frac{1}{F'(D^2)} \cos ax,$$

provided $F'(-a^2) \neq 0$

and equate imaginary part for $\sin ax$.

i.e., $$\frac{1}{F(D^2)} \sin ax = x \frac{1}{F'(D^2)} \sin ax, \text{ provided } F'(-a^2) \neq 0.$$

Let us illustrate these with examples:

Example 1. *Find P.I. of $(D^2 - 4)\, y = \sin 2x$.*

Sol. P.I. of the given equation $= \dfrac{1}{D^2 - 4} \sin 2x$

$$= \frac{1}{-2^2 - 4} \sin 2x \qquad \left[\text{Putting } D^2 = -a^2 \ \because \ F(-a^2) \neq 0\right]$$

$$= -\frac{1}{8} \sin 2x.$$

Example 2. *Find the P.I. of $(D^2 + 1)y = \cos(2x + 3)$.*

Sol. $\because$ $(D^2 + 1)\, y = \cos(2x + 3)$

$\therefore$ $$\text{P.I.} = \frac{1}{D^2 + 1} \cos(2x + 3)$$

$$= \frac{1}{-2^2 + 1} \cos(2x + 3),$$

$[\text{Here } a = 2, \text{ putting } D^2 = -2 \because F(-a^2) \neq 0]$

$$= -\frac{1}{3} \cos(2x + 3).$$

Example 3. *Find P.I. of $(D^2 + D + 1)\, y = \cos 2x$.*

Sol. $$\text{P.I.} = \frac{1}{D^2 + D + 1} \cos 2x$$

$$= \frac{1}{-4 + D + 1} \cos 2x \qquad [\text{Putting } D^2 = -a^2 = -4]$$

$$= \frac{1}{D-3}\cos 2x = \frac{(D+3)}{D^2-9}\cos 2x$$

$$= \frac{(D+3)}{-4-9}\cos 2x \qquad \text{[Again putting } D^2 = -a^2 = -4]$$

$$= -\frac{1}{13}(-2\sin 2x + 3\cos 2x)$$

$$= \frac{1}{13}(2\sin 2x - 3\cos 2x).$$

Example 4. *Find P.I. of $(D^2 + 4)\, y = \cos 2x$.*

Sol.

$$\text{P.I.} = \frac{1}{D^2+4}\cos 2x \qquad [\text{Here, } F(-a^2) = -(2)^2 + 4 = -4 + 4 = 0]$$

$$= \text{Real part of } \frac{1}{D^2+4}e^{i2x} \qquad [\because e^{2ix} = \cos 2x + i\sin 2x]$$

$$= \text{Real part of } e^{2ix}\frac{1}{(D+2i)^2+4}(1) \qquad \text{[Using rule } (ii)]$$

$$= \text{Real part of } e^{2ix}\frac{1}{D^2-4+4iD+4}(1)$$

$$= \text{Real part of } e^{2ix}\frac{1}{D^2+4iD}(1)$$

$$= \text{Real part of } e^{2ix}\frac{1}{4iD\left[1+\frac{D}{4i}\right]}(1)$$

$$= \text{Real part of } e^{2ix}\frac{1}{4iD}\left[1+\frac{D}{4i}\right]^{-1}(1)$$

$$= \text{Real part of } e^{2ix}\frac{-i}{4D}\left[1-\frac{D}{4i}+\ldots\right](1)$$

$$= \text{Real part of } -\frac{i}{4}e^{2ix}\frac{1}{D}(1)$$

$$= \text{Real part of } -\frac{i}{4}e^{2ix}\,x = \text{R.P. of } -\frac{i}{4}x(\cos 2x + i\sin 2x)$$

$$= \frac{x}{4}\sin 2x.$$

Hence, $\text{P.I.} = \frac{x}{4} \sin 2x$

Alternative method

$$\text{P.I.} = \frac{1}{D^2 + 4} e^{2ix} \qquad [\therefore \; e^{2ix} = \cos 2x + i \sin 2x]$$

$$= x \cdot \frac{1}{2D} e^{2ix} \text{ [using the rule } \frac{1}{F(D^2)} e^{aix} = x \frac{1}{F'(D^2)} e^{iax}]$$

$$= x \cdot \frac{1}{2D} [\cos 2x + i \sin 2x]$$

$$= x \cdot \frac{1}{2} \left[\frac{\sin 2x}{2} - i \frac{\cos 2x}{2} \right]$$

$$= \frac{x}{4} [\sin 2x - i \cos 2x]$$

$\therefore$ $\text{P.I. of } \frac{1}{D^2 + 4} \cos 2x = \text{Real part of } \frac{x}{4} [\sin 2x - i \cos 2x]$

$$= \frac{x}{4} \sin 2x.$$

(*v*) $$f(x) = \frac{x^n \cos ax}{x^n \sin ax}$$

$$\textbf{P.I.} = \frac{1}{F(D)} \frac{x^n \cos ax}{x^n \sin ax}$$

First we find $\text{P.I.} = \frac{1}{F(D)} x^n e^{iax}$, where $e^{iax} = \cos ax + i \sin ax$.

Rule. First operates on e^{iax} by putting $D = D + a$, then transferring e^{iax} left of the function $\frac{1}{F(D)}$, operates on x^n by rule (*i*).

As an example, we find $\text{P.I.} = \frac{1}{D^2 - 4D + 3} x^2 \cos 3x$

$$= \frac{1}{D^2 - 4D + 3} x^2 e^{3ix}, \qquad (\therefore \; e^{3ix} = \cos 3x + i \sin 3x)$$

$$= e^{3ix} \frac{1}{(D + 3i)^2 - 4(D + 3i) + 3} x^2$$

$$= e^{3ix} \frac{1}{D^2 - 9 + 6iD - 4D - 12i + 3} x^2$$

$$= e^{3ix} \frac{1}{D^2 + 6iD - 4D - 6} x^2$$

$$= e^{3ix} \frac{1}{-6\left[1 - \frac{D^2 + 2(3i-2)D}{6}\right]} x^2$$

$$= -\frac{1}{6} e^{3ix} \left[1 - \frac{D^2 + 2(3i-2)D}{6}\right]^{-1} x^2$$

$$= -\frac{1}{6} e^{3ix} \left[1 + \frac{D^2 + 2(3i-2)D}{6} + \frac{D^4 + 4(3i-2)^2 D^2 + 4(3i-2)D^3}{36} + \ldots\right] x^2$$

$$= -\frac{1}{6} e^{3ix} \left[x^2 + \frac{1}{6}\{2 + 2(3i-2)2x\} + \frac{1}{36}\{4(3i-2)^2 2\}\right]$$

$$= -\frac{1}{6} e^{3ix} \left[x^2 + \frac{1}{3}\{1 + 2(3i-2)x\} + \frac{2}{9}(3i-2)^2\right]$$

$$= -\frac{1}{54} e^{3ix} \left[9x^2 + 3\{1 + 2(3i-2)x\} + 2(-9 + 4 - 12i)\right]$$

$$= -\frac{1}{54} e^{3ix} \left[9x^2 + 6(3i-2)x - 7 - 24i\right]$$

$$= -\frac{1}{54} e^{3ix} \left[9x^2 - 12x - 7 - 12i(x+2)\right]$$

$$= -\frac{1}{54} (\cos 3x + i \sin 3x)\left[9x^2 - 12x - 7 - 12(x+2)i\right].$$

Equating real part, we get

$$\text{P.I.} = \frac{1}{D^2 - 4D + 3} x^2 \cos 3x$$

$$= -\frac{1}{54}\left[\left(9x^2 - 12x - 7\right)\cos 3x + 12(x+2)\sin 3x\right].$$

(vi) $\frac{1}{D+a}\phi(x) = e^{-ax}\int e^{-ax}\phi(x)\,dx$ **and** $\frac{1}{D-a}\phi(x) = e^{ax}\int e^{-ax}\phi(x)\,dx$

Let us illustrate it with examples:

Example 1. *Evaluate* $\frac{1}{D^2+3D+2}e^{e^x}$.

Sol.
$$\frac{1}{D^2+3D+2}e^{e^x} = \frac{1}{(D+1)(D+2)}e^{e^x}$$

$$= \left[\frac{1}{D+1} - \frac{1}{D+2}\right]e^{e^x}$$

$$= \frac{1}{D+1}e^{e^x} - \frac{1}{D+2}e^{e^x}$$

$$= e^{-x}\int e^x e^{e^x}dx - e^{-2x}\int e^{2x}e^{e^x}dx \qquad \ldots(a)$$

Let $I_1 = \int e^x e^{e^x}dx$ and $I_2 = \int e^{2x}e^{e^x}dx$.

Putting $e^x = z$ in both the integrals
and $e^x\,dx = dz$

$\therefore$
$$I_1 = \int e^z dz = e^z = e^{e^x}.$$

$$I_2 = \int z e^z dz = e^z(z-1) = e^{e^x}(e^x - 1).$$

$\therefore$ From (a), $\frac{1}{D^2+3D+2}e^{e^x} = e^{e^x}[e^{-x} - e^{-2x}(e^x - 1)]$.

$$= e^{-2x}e^{e^x}.$$

Example 2. *Find P.I. of* $\frac{d^2y}{dx^2} + y = \text{cosec } x$.

Sol.
$$\text{P.I.} = \frac{1}{(D^2+1)}\text{cosec } x = \frac{1}{(D+i)(D-i)}\text{cosec } x$$

$$= \frac{1}{2i}\left[\frac{1}{D-i} - \frac{1}{D+i}\right]\text{cosec } x$$

$$= \frac{1}{2i}\left[\frac{1}{D-i}\text{cosec } x - \frac{1}{D+i}\text{cosec } x\right].$$

Now
$$\frac{1}{D-i}\text{cosec } x = e^{ix}\int e^{-ix}\text{cosec } x\,dx \qquad \text{[Using rule } (v)]$$

$$= e^{ix}\int \text{cosec } x(\cos x - i\sin x)\,dx$$

$$= e^{ix}\int(\cot x - i)\,dx = e^{ix}(\log\sin x - ix)$$

changing i to $-i$, we get

$$\frac{1}{D+i}\operatorname{cosec} x = e^{-ix}(\log\sin x + ix)$$

$\therefore$
$$\text{P.I.} = \frac{1}{2i}\left[e^{ix}(\log\sin x - ix) - e^{-ix}(\log\sin x + ix)\right]$$

$$= \log\sin x\left(\frac{e^{ix}-e^{-ix}}{2i}\right) - x\left(\frac{e^{ix}+e^{-ix}}{2}\right)$$

$$= \sin x\log\sin x - x\cos x.$$

1.13.6 Illustrative Examples

1. *Solve the equation* $\dfrac{d^2y}{dx^2} - 6\dfrac{dy}{dx} + 9y = 3x + 1$ (W.B.U.T. BCA 2006)

Sol. The given equation is

$$(D^2 - 6D + 9)y = 3x + 1 \qquad \text{...}(i),$$

where $D \equiv \dfrac{d}{dx}$.

The A.E. is $m^2 - 6m + 9 = 0$

or $(m-3)^2 = 0$, *i.e.*, $m = 3, 3$.

$\therefore$ $\text{C.F.} = (c_1 + c_2x)e^{3x}$.

$$\text{P.I.} = \frac{1}{D^2-6D+9}(3x+1) = \frac{1}{(3-D)^2}(3x+1)$$

$$= \frac{1}{3}\left[1-\frac{D}{3}\right]^{-2}(3x+1)$$

$$= \frac{1}{3}\left[1+\frac{2}{3}D+\frac{2.3}{2!}\frac{D^2}{9}+\ldots\right](3x+1)$$

$$= \frac{1}{3}[3x+1+2] = (x+1)$$

$\therefore$ The G.S. is

$$y = (c_1 + c_2x)e^{3x} + (x+1).$$

2. *Solve the equation* $\dfrac{d^2x}{dt^2} + \dfrac{g}{l}x = \dfrac{g}{l}L$,

where g, l, L are constants subject to the conditions $x = a$, $\dfrac{dx}{dt} = 0$ *at* $t = 0$.

Sol. The given equation is

$$\left(D^2+\frac{g}{l}\right)x=\frac{g}{l}L \quad \text{...(}i\text{) where } D\equiv\frac{d}{dt}.$$

The A.E. is $\quad m^2+\frac{g}{l}=0$

or $\quad m=\pm i\sqrt{\frac{g}{l}}$

$$\therefore \quad \text{C.F.}=A\cos\sqrt{\frac{g}{l}}\,t+B\sin\sqrt{\frac{g}{l}}\,t$$

$$\text{P.I.}=\frac{1}{D^2+\frac{g}{l}}\frac{g}{l}L=\frac{1}{\frac{g}{l}\left(1+\frac{l}{g}D^2\right)}\frac{g}{l}L$$

$$=\frac{l}{g}\left(1+\frac{l}{g}D^2\right)^{-1}\frac{g}{l}L$$

$$=\frac{l}{g}\left[1-\frac{l}{g}D^2+\ldots\right]\frac{g}{l}L=\frac{l}{g}\left[\frac{g}{l}L\right]=L$$

$\therefore$ G.S of (*i*) is

$$x=A\cos\sqrt{\frac{g}{l}}\,t+B\sin\sqrt{\frac{g}{l}}\,t+L$$

$$\frac{dx}{dt}=-A\sqrt{\frac{g}{l}}\sin\sqrt{\frac{g}{l}}\,t+\sqrt{\frac{g}{l}}\,B\cos\sqrt{\frac{g}{l}}\,t$$

Using the given conditions *i.e.*, $x=a$ where $t=0$; and $\frac{dx}{dt}=0$ when $t=0$, we get

$$a=A+L\Rightarrow A=(a-L)$$

and $\quad 0=\sqrt{\frac{g}{l}}\,B\Rightarrow B=0$

Hence, the required solution is

$$x=(a-L)\cos\sqrt{\frac{g}{l}}\,t+L.$$

3. *Solve the equation* $(D^2-6D+9)\,y=6e^{3x}+7e^{-2x}-\log 2$.

Sol. A.E. is $m^2-6m+9=0$

$$(m-3)^2=0, \text{ gives } m=3, 3.$$

$$\therefore \quad \text{C.F.}=(c_1+c_2x)e^{3x}.$$

$$\text{P.I.}=\frac{1}{D^2-6D+9}\left(6e^{3x}+7e^{-2x}-\log 2\right)$$

$$=\frac{1}{D^2-6D+9}6e^{3x}+\frac{1}{D^2-6D+9}7e^{-2x}-\frac{1}{D^2-6D+9}(\log 2)e^{\circ}$$

$$=6x\frac{1}{2D-6}e^{3x}+e^{-2x}\frac{1}{4+12+9}7-\frac{1}{9}\log 2$$

$$=6x^2\frac{1}{2}e^{3x}+e^{-2x}\frac{7}{25}-\frac{1}{9}\log 2$$

$$=3x^2e^{3x}+\frac{7}{25}e^{-2x}-\frac{1}{9}\log 2$$

$\therefore$ G.S. is $y=(c_1+c_2x)e^{3x}+3x^2e^{3x}+\frac{7}{25}e^{-2x}-\frac{1}{9}\log 2.$

4. *Solve the equation* $\frac{d^2y}{dx^2}-2\frac{dy}{dx}+y=x^2e^{3x}.$

Sol. **The given equation is**

$$(D^2-2D+1)y=x^2e^{3x}$$

A.E. is $\quad m^2-2m+1=0$

or $(m-1)^2=0$, *i.e.*, $m=1, 1.$

$$\therefore \quad \text{C.F.}=(c_1+c_2x)e^x$$

$$\text{P.I.}=\frac{1}{(D-2D+1)}x^2e^{3x}=\frac{1}{(D-1)^2}x^2e^{3x}$$

$$=e^{3x}\frac{1}{[(D+3)-1]^2}x^2$$

$$=e^{3x}\frac{1}{(D+2)^2}x^2=e^{3x}\frac{1}{4\left(1+\frac{D}{2}\right)^2}x^2$$

$$=\frac{1}{4}e^{3x}\left(1+\frac{D}{2}\right)^{-2}x^2$$

$$=\frac{1}{4}e^{3x}\left[1-2\frac{D}{2}+3\frac{D^2}{4}-4\frac{D^3}{8}+\ldots\right]x^2$$

$$=\frac{1}{4}e^{3x}\left[x^2-2x+\frac{3}{4}.2\right]$$

$$=\frac{1}{8}e^{3x}\left[2x^2-4x+3\right]$$

$\therefore$ G.S. of the given equation is

$$y = (c_1 + c_2 x) e^x + \frac{1}{8} e^{3x} (2x^2 - 4x + 3).$$

5. *Solve the equation $(D^2 - 3D + 5)\, y = \sin 2x$.*

Sol. Given $(D^2 - 3D + 5)\, y = \sin 2x$

or $$(D^2 - 3D + 5)\, y = \frac{1}{2}(1 - \cos 2x) \quad \ldots (i).$$

A.E. is $m^2 - 3m + 5 = 0$

$$\therefore \quad m = \frac{3 \pm \sqrt{9 - 20}}{2} = \frac{3 \pm \sqrt{-11}}{2} = \frac{3 \pm i\sqrt{11}}{2}$$

$$\therefore \quad \text{C.F.} = e^{\frac{3}{2}x}\left[c_1 \cos \frac{\sqrt{11}}{2} x + c_2 \sin \frac{\sqrt{11}}{2} x\right]$$

$$\text{P.I.} = \frac{1}{D^2 - 3D + 5} \frac{1}{2}(1 - \cos 2x)$$

$$= \frac{1}{2}\left[\frac{1}{D^2 - 3D + 5} e^{\circ} - \frac{1}{D^2 - 3D + 5} \cos 2x\right]$$

$$= \frac{1}{2}\left[\frac{1}{0 - 3.0 + 5} 1 - \frac{1}{-4 - 3D + 5} \cos 2x\right]$$

$$= \frac{1}{2}\left[\frac{1}{5} - \frac{1}{1 - 3D} \cos 2x\right]$$

$$= \frac{1}{2}\left[\frac{1}{5} - \frac{1 + 3D}{1 - 9D^2} \cos 2x\right]$$

$$= \frac{1}{2}\left[\frac{1}{5} - \frac{1 + 3D}{1 + 36} \cos 2x\right]$$

$$= \frac{1}{2}\left[\frac{1}{5} - \frac{1}{37}(\cos 2x - 6 \sin 2x)\right]$$

$$= \frac{1}{370}[37 - 5 \cos 2x + 30 \sin 2x]$$

$\therefore$ G.S. of the equation (i) is

$$y = e^{\frac{3}{2}x}\left[c_1 \cos \frac{\sqrt{11}}{2} x + c_2 \sin \frac{\sqrt{11}}{2} x\right] + \frac{1}{370}[37 - 5 \cos 2x + 30 \sin 2x].$$

6. *Solve the equation* $(D^2 + 9)\ y = \sin 3x$.

Sol. A.E. is $m^2 + 9 = 0$,

i.e., $m = \pm 3i$

$\therefore$ $\text{C.F.} = c_1 \cos 3x + c_2 \sin 3x$

$$\text{P.I.} = \frac{1}{D^2 + 9} \sin 3x$$

$$= \text{Imaginary part of } x.\frac{1}{2D} e^{i3x}, (\because e^{i3x} = \cos 3x + i \sin 3x)$$

$$= \text{I.P. of } x.\frac{1}{2D}(\cos 3x + i \sin 3x)$$

$$= \text{I.P. of } x.\frac{1}{2}\left[\frac{\sin 3x}{3} - i\frac{\cos 3x}{3}\right]$$

$$= \text{I.P. of } \frac{1}{6}x[\sin 3x - i \cos 3x]$$

$$= -\frac{x}{6}\cos 3x$$

$\therefore$ G.S. of the given equation is

$$y = c_1 \cos 3x + c_2 \sin 3x - \frac{x}{6}\cos 3x.$$

7. *Solve the equation* $\frac{d^2y}{dx^2} + 3\frac{dy}{dx} + 2y = e^{2x} \sin x$.

Sol. The given equation is

$$(D^2 + 3D + 2)\,y = e^{2x} \sin x \quad \text{... (i)}$$

A.E. is $m^2 + 3m + 2 = 0$

or $(m + 2)(m + 1) = 0$

i.e., $m = -1, m = -2$

$\therefore$ $\text{C.F.} = c_1 e^{-x} + c_2 e^{-2x}$.

$$\text{P.I.} = \frac{1}{(D^2 + 3D + 2)} e^{2x} \sin x$$

$$= e^{2x} \frac{1}{(D+2)^2 + 3(D+2) + 2} \sin x$$

$$= e^{2x} \frac{1}{D^2 + 4 + 4D + 3D + 6 + 2} \sin x$$

$$= e^{2x} \frac{1}{D^2 + 7D + 12} \sin x$$

$$= e^{2x} \frac{1}{-1 + 7D + 12} \sin x \qquad \text{(Putting } D^2 = -1)$$

$$= e^{2x} \frac{(7D - 11)}{(7D + 11)(7D - 11)} \sin x$$

$$= e^{2x} \frac{(7D - 11)}{49D^2 - 121} \sin x$$

$$= e^{2x} \frac{(7D - 11)}{-49 - 121} \sin x = -\frac{1}{125} e^{2x} (7D - 11) \sin x$$

$$= -\frac{1}{170} e^{2x} (7\cos x - 11\sin x)$$

$\therefore$ General solution of (*i*) is

$$y = c_1 e^{-x} + c_2 e^{-2x} - \frac{1}{170} e^{2x} (7\cos x - 11\sin x).$$

8. *Solve the equation* $\frac{d^2y}{dx^2} + 4y = x \sin^2 x.$

Sol. The given equation is

$$(D^2 + 4)\, y = x \sin^2 x = \frac{1}{2} x (1 - \cos 2x) \qquad \dots (i)$$

A.E. is $m^2 + 4 = 0 \;\Rightarrow\; m = \pm 2i$

$\therefore$ C.F. $= c_1 \cos 2x + c_2 \sin 2x$

$$\text{P.I.} = \frac{1}{D^2 + 4} \frac{1}{2} x (1 - \cos 2x)$$

$$= \frac{1}{2} \left[\frac{1}{D^2 + 4} x - \frac{1}{D^2 + 4} x \cos 2x \right]$$

$$= \frac{1}{2} \left[\frac{1}{4} \left(1 + \frac{D^2}{4} \right)^{-1} x - \text{Real part of } \frac{1}{D^2 + 4} x e^{2ix} \right]$$

$$= \frac{1}{2} \left[\frac{1}{4} \left(1 - \frac{D^2}{4} + \dots \right) x - \text{Real part of } e^{2ix} \frac{1}{(D + 2i)^2 + 4} x \right]$$

$$= \frac{1}{2}\left[\frac{1}{4}x - \text{Real part of } e^{2ix}\frac{1}{D^2 + 4iD}x\right]$$

$$= \frac{1}{2}\left[\frac{1}{4}x - \text{Real part of } e^{2ix}\frac{1}{4iD\left(1+\frac{D}{4i}\right)}x\right]$$

$$= \frac{1}{2}\left[\frac{1}{4}x - \text{Real part of } e^{2ix}\frac{-i}{4D}\left(1+\frac{D}{4i}\right)^{-1}x\right] \qquad \because \ (i^2 = -1)$$

$$= \frac{1}{2}\left[\frac{1}{4}x + \text{Real part of } \frac{1}{4}e^{2ix}\frac{i}{D}\left(1 - \frac{D}{4i} - \frac{D^2}{16} + \dots\right)x\right]$$

$$= \frac{1}{2}\left[\frac{1}{4}x + \text{Real part of } \frac{1}{4}e^{2ix}\frac{i}{D}\left(x - \frac{1}{4i}\right)\right]$$

$$= \frac{1}{8}\left[x + \text{Real part of } e^{2ix}\frac{i}{D}\left(x + \frac{i}{4}\right)\right]$$

$$= \frac{1}{8}\left[x + \text{Real part of } e^{2ix}\, i\left(\frac{x^2}{2} + \frac{i}{4}x\right)\right] \qquad \left(\because \ \frac{1}{D} = \int dx\right)$$

$$= \frac{1}{8}\left[x + \text{Real part of } \frac{i}{8}(\cos 2x + i \sin 2x)\left(2x^2 + i\,x\right)\right]$$

$$= \frac{1}{8}\left[x + \text{Real part of } \frac{i}{8}\left\{\left(2x^2 \cos 2x - x \sin 2x\right) + i\left(2x^2 \sin 2x + x \cos 2x\right)\right\}\right]$$

$$= \frac{1}{64}\left[8x - \left(2x^2 \sin 2x + x \cos 2x\right)\right].$$

Hence, the G.S. of (*i*) is

$$y = c_1 \cos 2x + c_2 \sin 2x + \frac{1}{64}\left[8x - 2x^2 \sin 2x - x \cos 2x\right].$$

9. *Solve the equation* $(D^2 - 4D + 4)\, y = 8x^2 e^{2x} \sin 2x$.

Sol. A.E. is $m^2 - 4m + 4 = 0$

or $(m-2)^2 = 0 \Rightarrow m = 2, 2.$

$\therefore$ $\text{C.F.} = (c_1 + c_2 x)\, e^{2x}.$

$$\text{P.I.} = \frac{1}{D^2 - 4D + 4} 8x^2 e^{2x} \sin 2x = \frac{1}{(D-2)^2} 8x^2 e^{2x} \sin 2x$$

$$= 8e^{2x} \frac{1}{(D+2-2)^2} x^2 \sin 2x$$

$$= 8e^{2x} \frac{1}{D^2} \left(x^2 \sin 2x\right)$$

$$= 8e^{2x} \frac{1}{D}\left[-x^2 \frac{\cos 2x}{2} + \int x \cos 2x dx\right] \qquad \text{(Integrating by parts)}$$

$$= 8e^{2x} \frac{1}{D}\left[-\frac{1}{2}x^2 \cos 2x + \frac{1}{2} x \sin 2x + \frac{1}{4} \cos 2x\right]$$

$$= 4e^{2x} \frac{1}{D}\left[-x^2 \cos 2x + x \sin 2x + \frac{1}{2} \cos 2x\right]$$

$$= 4e^{2x} \left[-\int x^2 \cos 2x\, dx + \int x \sin 2x\, dx + \frac{1}{2}\int \cos 2x\, dx\right]$$

$$= 4e^{2x} \left[-\left(\frac{1}{2}x^2 \sin 2x - \int x \sin 2x\, dx\right) + \int x \sin 2x\, dx + \frac{1}{4} \sin 2x\right]$$

$$= 4e^{2x} \left[-\frac{1}{2}x^2 \sin 2x + 2\int x \sin 2x\, dx + \frac{1}{4} \sin 2x\right]$$

$$= 4e^{2x} \left[-\frac{1}{2}x^2 \sin 2x + 2\left(-\frac{x}{2} \cos 2x + \frac{1}{4} \sin 2x\right) + \frac{1}{4} \sin 2x\right]$$

$$= 4e^{2x} \left[-\frac{1}{2}x^2 \sin 2x - x \cos 2x + \frac{1}{2} \sin 2x + \frac{1}{4} \sin 2x\right]$$

$$= 4e^{2x} \left[-\frac{1}{2}x^2 \sin 2x - x \cos 2x + \frac{3}{4} \sin 2x\right]$$

$$= e^{2x} [-2x^2 \sin 2x - 4x \cos 2x + 3 \sin 2x]$$

$$= -e^{2x} [4x \cos 2x + (2x^2 - 3) \sin 2x].$$

Hence, the G.S. of equation (*i*) is

$$y = (c_1 + c_2 x) e^{2x} - e^{2x} [4x \cos 2x + (2x^2 - 3) \sin 2x].$$

10. *Solve the equation* $\dfrac{d^2y}{dx^2} + a^2 y = \tan ax.$

Sol. The given equation is

$$(D^2 + a^2) y = \tan ax$$

A.E. is $m^2 + a^2 = 0 \Rightarrow m = \pm ia$

$$\therefore \qquad \text{C.F.} = c_1 \cos ax + c_2 \sin ax$$

$$\text{P.I.} = \frac{1}{D^2 + a^2} \tan ax = \frac{1}{(D + ia)(D - ia)} \tan ax$$

$$= \frac{1}{2ia}\left[\frac{1}{D - ia} - \frac{1}{D + ia}\right] \tan ax$$

$$= \frac{1}{2ia}\left[\frac{1}{D - ia} \tan ax - \frac{1}{D + ia} \tan ax\right]$$

Now $\frac{1}{D - ia} \tan ax = e^{iax} \int e^{-iax} \tan ax \, dx$

$$= e^{iax} \int (\cos ax - i \sin ax) \frac{\sin ax}{\cos ax} dx$$

$$= e^{iax} \int \left(\sin ax - i \frac{\sin^2 ax}{\cos ax} \right) dx$$

$$= e^{iax} \int \left(\sin ax - i \frac{1 - \cos^2 ax}{\cos ax} \right) dx$$

$$= e^{iax} \int (\sin ax - i (\sec ax - \cos ax)) \, dx$$

$$= e^{iax} \left[-\frac{\cos ax}{a} - \frac{i}{a} \log (\sec ax + \tan ax) + i \frac{\sin ax}{a} \right]$$

$$= -\frac{1}{a} e^{iax} [(\cos ax - i \sin ax) + i \log (\sec ax + \tan ax)]$$

$$= -\frac{1}{a} e^{iax} [e^{-iax} + i \log (\sec ax + \tan ax)]$$

$$= -\frac{1}{a} [1 + ie^{iax} \log (\sec ax + \tan ax)]$$

Changing i to $-i$, we obtain

$$\frac{1}{D + ia} \tan ax = -\frac{1}{a} [1 - ie^{-iax} \log (\sec ax + \tan ax)]$$

$$\therefore \quad \text{P.I.} = \frac{1}{2ia}\left[-\frac{1}{a}\{1 + ie^{iax} \log (\sec ax + \tan ax)\} + \frac{1}{a}\{1 - ie^{-iax} \log (\sec ax + \tan ax)\}\right]$$

$$= -\frac{1}{a^2} \log (\sec ax + \tan ax) \left(\frac{e^{iax} + e^{-iax}}{2} \right)$$

$$= -\frac{1}{a^2}\log(\sec ax + \tan ax)(\cos ax)$$

Hence, G.S. is $y = c_1 \cos ax + c_2 \sin ax - \frac{1}{a^2}\cos ax \log(\sec ax + \tan ax)$.

EXERCISE 1.16

Solve the following differential equations :

1. $(D^2 + 4)\ y = x^2$
2. $(D^2 + 4)\ y = 2x + 3$
3. $(D^2 + 1)\ y = x^3$
4. $(D^2 - 4D + 3)\ y = x^3$
5. $(2D^2 + 3D + 4)\ y = x^2 - 2x$
6. $(D^2 + 9)y = 9e^{3x}$
7. $(D^2 - 4D + 3)\ y = 2e^{3x}$
8. $(D^2 - 6D + 9)\ y = e^{3x}$
9. $(D^2 + 2D + 1)\ y = e^x + e^{-x}$
10. $(D^2 - 4)\ y = (1 + e^x)^2$
11. $(D^2 - 4D + 1)\ y = x^3e^{2x}$
12. $(D^2 - 5D + 6)\ y = e^x \cos 2x$
13. $(D^2 + 4)\ y = \sin 3x$
14. $(D^2 + D + 1)\ y = \cos 2x$
15. $(D^2 + 4)\ y = \cos 2x$
16. $\frac{d^2y}{dx^2} + 4y = x \cos x$
17. $(D^2 + 1)\ y = \sin x \sin 2x$
18. $\frac{d^2y}{dx^2} + y = 4 \cos^3 x$
19. $(D^2 - 4D + 3)\ y = e^{2x} + \cos x$
20. $\frac{d^2y}{dx^2} - 2\frac{dy}{dx} + 2y = 5 \cos x$, given $y = 0$ and $\frac{dy}{dx} = 0$, when $x = 0$.
21. $\frac{d^2y}{dx^2} + 2\frac{dy}{dx} + 5y = \sin 2x$, given $y = 3$, $\frac{dy}{dx} = 0$, when $x = 0$.
22. $\frac{d^2y}{dx^2} - 7\frac{dy}{dx} + 6y = 2 \sin 3x$, given $y = 1$, $\frac{dy}{dx} = 0$, when $x = 0$.
23. $\frac{d^2y}{dx^2} - y = xe^x \sin x$
24. $\frac{d^2y}{dx^2} + 2y = x^2e^{3x} + e^x \cos 2x$
25. $\frac{d^2y}{dx^2} - y = x \sin x + (1 + x^2)\ e^x$
26. $(D^2 + 4)\ y = 4 \tan 2x$
27. $(D^2 + a^2)\ y = \sec ax$
28. $(D^2 - 6D + 13)y = 8e^{3x} \sin 4x + 2^x$.

Answers

1. $y = c_1 \cos 2x + c_2 \sin 2x + \frac{1}{8}(2x^2 - 1)$

2. $y = c_1 \cos 2x + c_2 \sin 2x + \frac{1}{4}(2x + 3)$

3. $y = c_1 \cos x + c_2 \sin x + (x^3 - 6x)$

4. $y = c_1 e^x + c_2 e^{3x} + \frac{1}{81}(27x^3 + 108x^2 + 234x + 240)$

5. $y = e^{-\frac{3}{4}x}\left(c_1 \cos \frac{\sqrt{23}}{4} x + c_2 \sin \frac{\sqrt{23}}{4} x\right) + \frac{1}{32}(8x^2 - 28x + 25)$

6. $y = c_1 \cos 3x + c_2 \sin 3x + \frac{1}{2} e^{3x}$

7. $y = c_1 e^x + c_2 e^{3x} + x e^{3x}$

8. $y = (c_1 + c_2 x) e^{3x} + \frac{1}{2} x^2 e^{3x}$

9. $y = e^{-x}(c_1 + c_2 x + \frac{1}{2} x^2) + \frac{1}{4} e^x$

10. $y = c_1 e^{2x} + c_2 e^{-2x} - \frac{1}{4} - \frac{2}{3} e^x + \frac{1}{4} x e^{2x}$

11. $y = c_1 e^{(2+\sqrt{3})x} + c_2 e^{(2-\sqrt{3})x} - \frac{1}{3}(x^3 + 2x)$

12. $y = c_1 e^{2x} + c_2 e^{3x} - \frac{1}{20} e^x (3 \sin 2x + \cos 2x)$

13. $y = c_1 \cos 2x + c_2 \sin 2x - \frac{1}{5} \sin 3x$

14. $y = e^{-\frac{1}{2}x}\left(c_1 \cos \frac{\sqrt{3}}{2} x + c_2 \sin \frac{\sqrt{3}}{2} x\right) + \frac{1}{13}(2 \sin 2x - 3 \cos 2x)$

15. $y = c_1 \cos 2x + c_2 \sin 2x + \frac{x}{4} \sin 2x$

16. $y = c_1 \cos 2x + c_2 \sin 2x + \frac{1}{3} x \cos x + \frac{2}{9} \sin x$

17. $y = c_1 \cos x + c_2 \sin x + \frac{1}{4} x \sin x + \frac{1}{16} \cos 3x$

18. $y = c_1 \cos x + c_2 \sin x + \frac{3}{2} x \sin x - \frac{1}{8} \cos 3x$

19. $y = c_1 e^x + c_2 e^{3x} + \frac{1}{10}(\cos x - 2 \sin x) - e^{-2x}$

20. $y = e^x (3 \sin x - \cos x) + \cos x - 2 \sin x$

21. $y = \frac{1}{34} e^{-x} (110 \cos 2x + 53 \sin 2x) - \frac{1}{17}(4 \cos 2x - \sin 2x)$

22. $y = \frac{27}{25} e^x - \frac{13}{75} e^{6x} - \frac{1}{75}(\sin 3x - 7 \cos 3x)$

23. $y = c_1e^x + c_2e^{-x} - \frac{1}{25}e^x\,[(10x + 2)\cos x + (5x - 14)\sin x]$

24. $y = c_1 \cos\sqrt{2}\,x + c_2 \sin\sqrt{2}\,x + \frac{1}{11}e^{3x}\left(x^2 - \frac{12}{11}x + \frac{50}{121}\right) + \frac{e^x}{17}(4\sin 2x - \cos 2x)$

25. $y = c_1e^x + c_2e^{-x} - \frac{1}{2}(x\sin x + \cos x) + \frac{1}{12}xe^x(2x^2 - 3x + 9)$

26. $y = c_1 \cos 2x + c_2 \sin 2x - \cos 2x \log(\sec 2x + \tan 2x)$.

27. $y = c_1 \cos ax + c_2 \sin ax + \frac{1}{a}(x \sin ax + \frac{1}{a}\cos ax \log \cos ax)$.

28. $y = e^{3x}(c_1 \cos 2x + c_2 \sin 2x) - \frac{2}{3}e^{3x}\sin 4x + \frac{2^x}{(\log 2)^2 - 6\log 2 + 13}$.

Remark. Third, fourth and higher order linear differential equations can also be solved according to the method described for second order equations.

1.13.7 Illustrative Examples

1. *Solve the equation $(D^3 + 3D^2 + 3D + 1)\,y = e^{-x}$.*

Sol. A.E. of the given equation is

$$m^3 + 3m^2 + 3m + 1 = 0$$

$$\Rightarrow \quad (m + 1)^3 = 0 \quad \Rightarrow \quad m = -1, -1, -1$$

$$\therefore \quad \text{C.F.} = (c_1 + c_2x + c_3x^2)\,e^{-x}$$

$$\text{P.I.} = \frac{1}{D^3 + 3D^2 + 3D + 1}e^{-x} = \frac{1}{(D+1)^3}e^{-x}$$

$$= e^{-x}\frac{1}{(D-1+1)^3}(1)$$

$$= e^{-x}\frac{1}{D^3}(1) = \frac{1}{6}e^{-x}x^3.$$

Hence, the general solution of the given equation is

$$y = (c_1 + c_2x + c_3x^2)e^{-x} + \frac{1}{6}e^{-x}x^3 = (c_1 + c_2x + c_3x^2 + \frac{1}{6}x^3)e^{-x}.$$

2. *Solve the equation $(D^3 + 3D^2 + 2D)\,y = x^2$.*

Sol. The A.E. is $m^3 + 3m^2 + 2m = 0$

or $\quad m(m^2 + 3m + 2) = 0$

or $\quad m(m + 2)(m + 1) = 0$

i.e., $m = 0, m = -1$, and $m = -2$

$$\therefore \quad \text{C.F.} = c_1 + c_2e^{-x} + c_3e^{-2x}$$

$$\text{P.I.} = \frac{1}{D^3 + 3D^2 + 2D} x^2 = \frac{1}{2D\left(1 + \frac{D^2 + 3D}{2}\right)} x^2$$

$$= \frac{1}{2D}\left(1 + \frac{D^2 + 3D}{2}\right)^{-1} x^2$$

$$= \frac{1}{2D}\left[1 - \frac{D^2 + 3D}{2} + \frac{D^4 + 6D^3 + 9D^2}{4} - \ldots\right] x^2$$

$$= \frac{1}{2D}\left[x^2 - \frac{1}{2}(2 + 6x) + \frac{1}{4}.9.2\right]$$

$$= \frac{1}{2D}\left[x^2 - 3x + \frac{7}{2}\right] = \frac{1}{2}\left[\frac{x^3}{3} - \frac{3x^2}{2} + \frac{7}{2}x\right]$$

$$= \frac{1}{12}\left[2x^3 - 9x^2 + 21x\right]$$

$\therefore$ G.S. is $y = c_1 + c_2 e^{-x} + c_3 e^{-2x} + \frac{1}{12}(2x^3 - 9x^2 + 21x)$.

3. *Solve the equation* $\frac{d^3y}{dx^3} - 3\frac{d^2y}{dx^2} + 3\frac{dy}{dx} - y = xe^x + e^x$.

Sol. The given equation is

$$(D^3 - 3D^2 + 3D - 1)\, y = xe^x + e^x$$

or $$(D - 1)^3\, y = xe^x + e^x.$$

A.E. is $(m - 1)^3 = 0$, *i.e.*, $m = 1, 1, 1$

$\therefore$ $$\text{C.F.} = (c_1 + c_2 x + c_3 x^2)e^x$$

$$\text{P.I.} = \frac{1}{(D-1)^3}\left(xe^x + e^x\right) = \frac{1}{(D-1)^3} e^x (x+1)$$

$$= e^x \frac{1}{(D+1-1)^3}(x+1) = e^x \frac{1}{D^3}(x+1)$$

$$= e^x\left(\frac{x^4}{24} + \frac{x^3}{6}\right) = \frac{1}{24}\left(x^4 + 4x^3\right)e^x$$

Hence, the G.S. is

$$y = (c_1 + c_2 x + c_3 x^2)e^x + \frac{1}{24}(x^4 + 4x^3)e^x.$$

4. *Find the general solution of the differential equation* $(D^3 - D^2 + 3D + 5)\, y = x^2 + e^x \cos 2x$.

Sol. The A.E. is $m^3 - m^2 + 3m + 5 = 0$

or $$(m+1)(m^2 - 2m + 5) = 0$$

i.e., $m = -1$ and $$m = \frac{2 \pm \sqrt{4-20}}{2} = \frac{2 \pm 4i}{2} = 1 \pm 2i$$

$\therefore$ $$\text{C.F.} = c_1 e^{-x} + e^x (c_2 \cos 2x + c_3 \sin 2x)$$

$$\text{P.I.} = \frac{1}{(D^2+4)(D^2+1)} (\cos 2x + \sin x)$$

$$= \frac{1}{(D^2+4)(D^2+1)} \cos 2x + \frac{1}{(D^2+4)(D^2+1)} \sin x$$

$$= \frac{\cos 2x}{(D^2+4)(-4+1)} + \frac{\sin x}{(-1+4)(D^2+1)}$$

$$= -\frac{1}{3}\frac{x}{2}\int \cos 2x\, dx + \frac{1}{3}\frac{x}{2}\int \sin x\, dx$$

$$= \frac{1}{6} x \left[-\frac{\sin 2x}{2} - \cos x \right] = -\frac{1}{12} x [\sin 2x + 2\cos x].$$

Hence, general solution of the given equation is

$$y = c_1 e^{-x} + e^x (c_2 \cos 2x + c_3 \sin 2x) - \frac{1}{12} x (\sin 2x + 2\cos x).$$

5. *Solve the equation* $\dfrac{d^4 y}{dx^4} - y = e^x \cos x$.

Sol. The given equation is

$$(D^4 - 1) y = e^x \cos x$$

or $$(D^2+1)(D^2-1) y = e^x \cos x.$$

Here A.E. is $(m^2+1)(m^2-1) = 0$, *i.e.*, $m = \pm 1$ and $m = \pm i$

$\therefore$ $$\text{C.F.} = c_1 e^x + c_2 e^{-x} + (c_3 \cos x + c_4 \sin x)$$

$$\text{P.I.} = \frac{1}{(D^2+1)(D^2-1)} e^x \cos x$$

$$= \frac{1}{2}\left[\frac{1}{D^2-1} - \frac{1}{D^2+1} \right] e^x \cos x$$

$$= \frac{1}{2}\left[\frac{1}{D^2-1} e^x \cos x - \frac{1}{D^2+1} e^x \cos x \right].$$

Now, $$\frac{1}{D^2-1}e^x\cos x = e^x\frac{1}{(D+1)^2-1}\cos x = e^x\frac{1}{D^2+2D+1-1}\cos x$$

$$= e^x\frac{1}{D^2+2D}\cos x = e^x\frac{1}{2D-1}\cos x, \text{(Putting } D^2=-1)$$

$$= e^x\frac{(2D+1)}{4D^2-1}\cos x = e^x\frac{1}{-5}(-2\sin x+\cos x)$$

$$= -\frac{1}{5}e^x(\cos x - 2\sin x)$$

Similarly, we get $\frac{1}{D^2+1}e^x\cos x = \frac{1}{5}e^x(2\sin x+\cos x)$

$$\therefore \qquad \text{P.I.} = \frac{1}{2}\left[-\frac{1}{5}e^x(\cos x-2\sin x)-\frac{1}{5}e^x(2\sin x+\cos x)\right]$$

$$= -\frac{1}{10}e^x(2\cos x) = -\frac{1}{5}e^x\cos x.$$

Hence, the G.S. is $$y = c_1e^x + c_2e^{-x} + (c_3\cos x + c_4\sin x) - \frac{1}{5}e^x\cos x.$$

EXERCISE 1.17

Find the general solutions of the following differential equations:

1. $(D^3+2D^2+D)\,y = e^{2x}+(x^2+x)$
2. $(D^3+4D^2+4D)\,y = 8e^{-2x}+x^2$
3. $(D^3+3D^2+3D+1)\,y = e^{-x}$
4. $(D^3-3D^2-6D)\,y = 1+x^2$
5. $(D^3-7D-6)\,y = e^{2x}(1+x)$
6. $(D^3-1)\,y = xe^x+\cos^2 x$
7. $(D^4-2D^3+D^2)\,y = x^3$
8. $(D^4+D^2+1)\,y = x^2-2e^{-x}\sin 2x$
9. $(D^4+2D^2+1)\,y = x^2\cos x$
10. $(D^2+1)^2\,y = 24x\cos x.$

Answers

1. $y = c_1+(c_2+c_3x)e^{-x}+\frac{x}{6}(2x^2-9x+24)+\frac{1}{18}e^{2x}$

2. $y = c_1+(c_2+c_3x)\,e^{-2x}+2x^2e^{-2x}+\frac{x^3}{12}-\frac{x^2}{4}+\frac{3x}{8}$

3. $y = e^{-x}\left(c_1+c_2x+c_3x^2+\frac{1}{6}x^2\right)$

4. $y = c_1+c_2e^{3x}+c_3e^{-2x}-\frac{1}{6}\left(\frac{25}{18}x-\frac{1}{6}x^2+\frac{1}{3}x^3\right)$

5. $y = c_1e^{-x} + c_2e^{3x} + c_3e^{-2x} - \frac{1}{12}\left(\frac{17}{12} + x\right)e^{2x}$

6. $y = c_1e^{x} + \left(c_2 \cos\frac{\sqrt{3}}{2}x + c_3 \sin\frac{\sqrt{3}}{2}x\right)e^{-\frac{1}{2}x} - \frac{1}{2} + \frac{1}{3}\left(\frac{1}{2}x^2 - x\right)e^{x} - \frac{1}{130}(8 \sin 2x + \cos 2x)$

7. $y = c_1 + c_2x + (c_3 + c_4x)\,e^{x} + \frac{x^5}{20} + \frac{x^4}{2} + 3x^3 + 12x^2$

8. $y = e^{\frac{x}{2}}\left(c_1 \cos\frac{\sqrt{3}}{2}x + c_2 \sin\frac{\sqrt{3}}{2}x\right) + e^{\frac{-x}{2}}\left(c_3 \cos\frac{\sqrt{3}}{2}x + c_4 \sin\frac{\sqrt{3}}{2}x\right)$

$$+ x^2 - 2 - \frac{2e^{-x}}{481}(20 \cos 2x + 9 \sin 2x)$$

9. $y = (c_1 + c_2x) \cos x + (c_3 + c_4x) \sin x - \frac{1}{4}\left(\frac{x^4 - 9x^2}{12} \cos x - \frac{x^3}{3} \sin x\right)$

10. $y = (c_1 + c_2x) \cos x + (c_3 + c_4x) \sin x - x^3 \cos x + 3x^2 \sin x.$

1.14 CAUCHY'S HOMOGENEOUS LINEAR DIFFERENTIAL EQUATIONS

An equation of the form

$$x^n \frac{d^ny}{dx^n} + a_1x^{n-1}\frac{d^{n-1}y}{dx^{n-1}} + a_2x^{n-2}\frac{d^{n-2}y}{dx^{n-2}} + \ldots + a_{n-1}\,x\frac{dy}{dx} + a_ny = f(x)$$

or $(x^nD^n + a_1x^{n-1}D^{n-1} + \ldots + a_{n-1}\,xD + a_n)\,y = f(x)$ ... (1)

where a_i s ($i = 1, 2, \ldots, n$) are constants and $f(x)$ is a function of x alone, is called Chauchy's homogeneous linear equation.

Chauchy's homogeneous linear differential equations can be solved by reducing the equations to linear differential equations with constant coefficients.

These equations can be reduced to linear differential equations with constant coefficients by the substitution

$$x = e^z \text{ or } z = \log x, \therefore \frac{dz}{dx} = \frac{1}{x}.$$

Then $$\frac{dy}{dx} = \frac{dy}{dz}\cdot\frac{dz}{dx} = \frac{dy}{dz}\cdot\frac{1}{x}$$

or $$\boxed{x\frac{dy}{dx} = \frac{dy}{dz}}$$ ... (i)

Differentiating again, with respect to x, we obtain

$$x\frac{d^2y}{dx^2}+\frac{dy}{dx}=\frac{d^2y}{dz^2}\cdot\frac{dz}{dx}=\frac{1}{x}\frac{d^2y}{dz^2}$$

or $$\boxed{x^2\frac{d^2y}{dx^2}+x\frac{dy}{dx}=\frac{d^2y}{dz^2}} \qquad \ldots(ii)$$

or $$(x^2D^2+xD)\,y=\frac{d^2y}{dz^2}$$

Similarly, $$(x^3D^3+3x^2D^2+xD)\,y=\frac{d^3y}{dz^3} \qquad \ldots(iii)$$

and so on.

Using the results (*i*), (*ii*), (*iii*) ... in (1).

We get a linear differential equation with constant coefficients, which can be solved by the methods already discussed in the last article.

We illustrate these with examples.

1.14.1 Illustrative Examples

1. *Solve the differential equation*

$$x^2\frac{d^2y}{dx^2}-x\frac{dy}{dx}-3y=x^2\log x \qquad \ldots(i)$$

(W.B.U.T., BCA, 2006)

Sol. Let $x=e^z$, or $z=\log x$

$$\therefore \qquad \frac{dz}{dx}=\frac{1}{x}$$

$$\frac{dy}{dx}=\frac{dy}{dz}\cdot\frac{dz}{dx}=\frac{1}{x}\frac{dy}{dz}$$

or $$x\frac{dy}{dx}=\frac{dy}{dz} \qquad \ldots(ii)$$

Differentiating again w.r.t. x, we get

$$x\frac{d^2y}{dx^2}+\frac{dy}{dx}=\frac{d^2y}{dz^2}\cdot\frac{dz}{dx}=\frac{1}{x}\frac{d^2y}{dz^2}$$

or $$x^2\frac{d^2y}{dx^2}+x\frac{dy}{dx}=\frac{d^2y}{dz^2} \qquad \ldots(iii)$$

(*i*) Can be written as

$$\left(x^2\frac{d^2y}{dx^2}+x\frac{dy}{dx}\right)-2x\frac{dy}{dx}-3y=x^2\log x$$

Using (*ii*), (*iii*) and $x = e^z$, the above equation reduces to

$$\frac{d^2y}{dz^2} - 2\frac{dy}{dz} - 3y = ze^{2z}$$

or $(D^2 - 2D - 3)\,y = ze^{2z}$, ... (*iv*), where $D \equiv \dfrac{d}{dz}$.

The A.E. of (*iv*) is

$$m^2 - 2m - 3 = 0$$

or $(m-3)(m+1) = 0$

i.e., $m = -1, 3$.

$\therefore$ C.F. of (*iv*) $= c_1 e^{-z} + c_2 e^{3z}$

$$\text{P.I. of } (iv) = \frac{1}{D^2 - 2D - 3} ze^{2z}$$

$$= e^{2z} \frac{1}{(D+2)^2 - 2(D+2) - 3} z = e^{2z} \frac{1}{D^2 + 4D + 4 - 2D - 4 - 3} z$$

$$= e^{2z} \frac{1}{D^2 + 2D - 3} z = e^{2z} \frac{1}{-3\left[1 - \dfrac{D^2 + 2D}{3}\right]} z$$

$$= -\frac{1}{3} e^{2z} \left[1 - \frac{D^2 + 2D}{3}\right]^{-1} z = -\frac{1}{3} e^{2z} \left[1 + \frac{D^2 + 2D}{3} + \ldots\right] z$$

$$= -\frac{1}{3} e^{2z} \left[z + \frac{2}{3}\right] = -\frac{1}{9} e^{2z} (3z + 2)$$

$\therefore$ The G.S. of equation (*iv*) is

$$y = c_1 e^{-z} + c_2 e^{3z} - \frac{1}{9} e^{2z} (3z + 2)$$

Hence, the G.S. of the equation (*i*) is

$$y = c_1 e^{-\log x} + c_2 e^{3 \log x} - \frac{1}{9} e^{2 \log x} (3 \log x + 2)$$

or

$$y = c_1 \frac{1}{x} + c_2 x^3 - \frac{1}{9} x^2 (3 \log x + 2).$$

2. *Solve the equation* $x^2 \dfrac{d^2y}{dx^2} - 4x \dfrac{dy}{dx} + 6y = x.$

Sol. The given equation is

$$x^2 \frac{d^2y}{dx^2} - 4x \frac{dy}{dx} + 6y = x$$

or $$\left(x^2 \frac{d^2y}{dx^2} + x \frac{dy}{dx}\right) - 5x \frac{dy}{dx} + 6y = x \qquad \ldots (i).$$

Putting $x = e^z$ or $z = \log x$ $\therefore$ $\frac{dz}{dx} = \frac{1}{x}$

$\therefore$ $$x \frac{dy}{dx} = \frac{dy}{dz} \qquad \ldots (ii)$$

and $$x^2 \frac{d^2y}{dx^2} + x \frac{dy}{dx} = \frac{d^2y}{dz^2} \qquad \ldots (iii)$$

Using (*ii*) and (*iii*) in (*i*), we get

$$\frac{d^2y}{dz^2} - 5 \frac{dy}{dz} + 6y = e^z$$

or $$(D^2 - 5D + 6)\, y = e^z \qquad \ldots (iv)$$

A.E. of (*iv*) $m^2 - 5m + 6 = 0$ (where $D \equiv \frac{d}{dz}$)

or $(m-3)(m-2) = 0$

i.e., $m = 2, 3$

$\therefore$ C.F. of (*iv*) $= c_1 e^{2z} + c_2 e^{3z}$

$$\text{P.I. of } (iv) = \frac{1}{D^2 - 5D + 6} e^z = e^z \frac{1}{1 - 5 + 6} = \frac{1}{2} e^z$$

$\therefore$ G.S. of the equation (*iv*) is

$$y = c_1 e^{2z} + c_2 e^{3z} + \frac{1}{2} e^z.$$

Hence, the G.S. of the equation (*i*) is

$$y = c_1 e^{2 \log x} + c_2 e^{3 \log x} + \frac{1}{2} x$$

$$= c_1 x^2 + c_2 x^3 + \frac{1}{2} x.$$

3. *Solve the equation* $(x^2 D^2 + xD + 1)\, y = \sin \log x^2$.

Sol. The given equation is

$$x^2 \frac{d^2y}{dx^2} + x \frac{dy}{dx} + y = \sin (2 \log x) \qquad \ldots (i)$$

Putting $x = e^z$, or $z = \log x$ $\quad \therefore \quad \frac{dz}{dx} = \frac{1}{x}$

Hence $$x\frac{dy}{dx} = \frac{dy}{dz}$$

and $$x^2\frac{d^2y}{dx^2} + x\frac{dy}{dx} = \frac{d^2y}{dz^2}.$$

Using these relations in (*i*), we get

$$\frac{d^2y}{dz^2} + y = \sin 2z$$

or $$(D^2 + 1)\,y = \sin 2z, \qquad \ldots (ii), \text{ where } D \equiv \frac{d}{dz}$$

$\therefore$ The A.E. is $m^2 + 1 = 0$

or $m = \pm i$

$\therefore$ C.F. $= c_1 \cos z + c_2 \sin z$

$$\text{P.I.} = \frac{1}{D^2 + 1}\sin 2z = \frac{1}{-4+1}\sin 2z = -\frac{1}{3}\sin 2z$$

$\therefore$ G.S. of equation (*ii*) is

$$y = c_1 \cos z + c_2 \sin z - \frac{1}{3}\sin 2z.$$

Hence, the G.S. of equation (*i*) is

$$y = c_1 \cos(\log x) + c_2 \sin(\log x) - \frac{1}{3}\sin(2\log x).$$

4. *Solve the equation*

$$x^2\frac{d^2y}{dx^2} + 4x\frac{dy}{dx} + 2y = x + \sin x.$$

Sol. Putting $x = e^z$ or $z = \log x$

$$\therefore \quad \frac{dz}{dx} = \frac{1}{x}.$$

Hence, $$x\frac{dy}{dx} = \frac{dy}{dz}$$

and $$x^2\frac{d^2y}{dx^2} + x\frac{dy}{dx} = \frac{d^2y}{dz^2}$$

$\therefore$ The given equation becomes

$$\frac{d^2y}{dz^2} + 3\frac{dy}{dz} + 2y = e^z + \sin(e^z)$$

or $$(D^2 + 3D + 2)y = e^{z} + \sin e^{z} \quad \text{...}(i)$$

A.E. is $m^2 + 3m + 2 = 0$

or $$(m+2)(m+1) = 0$$

i.e., $m = -2, -1$,

$$\therefore \quad \text{C.F.} = c_1 e^{-2z} + c_2 e^{-z}$$

$$\text{P.I. of } (i) = \frac{1}{D^2 + 3D + 2}\left(e^z + \sin e^z\right)$$

$$= \frac{1}{D^2 + 3D + 2} e^z + \frac{1}{D^2 + 3D + 2} \sin e^z$$

$$= e^z \frac{1}{1+3+2} + \frac{1}{(D+2)(D+1)} \sin e^z$$

$$= \frac{1}{6} e^z + \frac{1}{D+2}\left[\frac{1}{D+1} \sin e^z\right] \quad \text{...}(ii)$$

Let $\frac{1}{D+1} \sin e^z = u, \therefore \frac{du}{dz} + u = \sin e^z$ ...(iii)

(iii) is a linear differential equation, whose I.F. $= e^{\int dz} = e^z$, multiplying (iii) by e^z and integrating, we get

$$u.e^z = \int e^z \sin e^z \, dz$$

Putting $e^z = t, e^z dz = dt$

$$\therefore \quad ue^z = \int \sin t \, dt = -\cos t = -\cos e^z$$

$$\therefore \quad u = -e^{-z} \cos e^z$$

$$\therefore \quad \frac{1}{D+1} \sin e^z = -e^{-z} \cos e^z$$

$$\therefore \quad \frac{1}{D+2}\left[\frac{1}{D+1} \sin e^z\right] = \frac{1}{D+2}\left[-e^{-z} \cos e^z\right]$$

Let $$v = \frac{1}{D+2}\left[-e^{-z} \cos e^z\right]$$

$$\therefore \quad \frac{dv}{dz} + 2v = -e^{-z} \cos e^z \quad \text{...}(iv)$$

Here, $$\text{I.F.} = e^{\int 2dz} = e^{2z}$$

Multiplying equation (*iv*) by e^{2z} and integrating, we get

$$v.e^{2z} = \int -e^{-z} \cos e^z . e^{2z}\, dz$$

$$= -\int e^z \cos e^z\, dz = -\sin e^z$$

$$\therefore \quad v = -e^{-2z} \sin e^z$$

$$\therefore \quad \frac{1}{D+2}\left[-e^{-z} \cos e^z\right] = -e^{-2z} \sin e^z.$$

Hence, $$\frac{1}{D+2}\left[\frac{1}{D+1} \sin e^z\right] = -e^{-2z} \sin e^z$$

$\therefore$ from (*ii*), we have

$$\text{P.I. of } (i) = \frac{1}{6} e^z - e^{-2z} \sin e^z$$

$\therefore$ G.S. of (*i*) is

$$y = c_1 e^{-2z} + c_2 e^{-z} + \frac{1}{6} e^z - e^{-2z} \sin e^z.$$

Hence, the G.S. of the given equation is

$$y = c_1 e^{-2 \log x} + c_2 e^{-\log x} + \frac{1}{6} x - e^{-2 \log x} \sin x$$

or $$y = c_1 \frac{1}{x^2} + c_2 \frac{1}{x} + \frac{1}{6} x - \frac{1}{x^2} \sin x.$$

5. *Solve the equation*

$$\frac{d^2y}{dx^2} + \frac{1}{x}\frac{dy}{dx} = \frac{12 \log x}{x^2}.$$

Sol. The given equation is

$$\frac{d^2y}{dx^2} + \frac{1}{x}\frac{dy}{dx} = \frac{12 \log x}{x^2}$$

or $$x^2 \frac{d^2y}{dx^2} + x\frac{dy}{dx} = 12 \log x \qquad \ldots (i)$$

Putting $x = e^z$ or $z = \log x$, we have

$$\frac{dz}{dx} = \frac{1}{x}$$

Then $$x\frac{dy}{dx} = \frac{dy}{dz}$$

and $$x^2 \frac{d^2y}{dx^2} + x\frac{dy}{dx} = \frac{d^2y}{dz^2}$$

$\therefore$ From (*i*), we get

$$\frac{d^2y}{dz^2} = 12z \text{ or } D^2y = 12z \qquad \dots(ii), \left(D \equiv \frac{d}{dz}\right)$$

A.E. is $m^2 = 0$, *i.e.*, $m = 0, 0$.

$\therefore$ $$\text{C.F.} = c_1 + c_2 z$$

$$\text{P.I.} = \frac{1}{D^2}(12z) = 2z^3.$$

Hence, the G.S. of the equation (*ii*) is

$$y = c_1 + c_2 z + 2z^3$$

$\therefore$ The G.S. of equation (*i*) is

$$y = c_1 + c_2 \log x + 2(\log x)^3.$$

Remark. If we write δ for $\frac{d}{dz}$ then we can

write $$x\frac{dy}{dx} = \delta y,\ x^2\frac{d^2y}{dx^2} = \delta(\delta-1)y$$

or $$x^3\frac{d^3y}{dx^3} = \delta(\delta-1)(\delta-2)y$$

....

$\therefore$ $$x^n\frac{d^ny}{dx^n} = \delta(\delta-1)(\delta-2)\dots(\delta-n+1)y.$$

We shall use these relations for third and higher order linear homogeneous equations.

6. *Solve the equation* $x^3\frac{d^3y}{dx^3} - x^2\frac{d^2y}{dx^2} + 2x\frac{dy}{dx} - 2y = x^2$.

Sol. Putting $x = e^z$, *i.e.*, $z = \log x$.

Using the above relations the equation can be transformed to

$[\delta(\delta-1)(\delta-2) - \delta(\delta-1) + 2\delta - 2]y = e^{2z}$

where $$\delta \equiv \frac{d}{dz}.$$

or $$(\delta-1)^2(\delta+2)y = e^{2z} \qquad \dots(i)$$

The A.E. is $(m-1)^2(m+2) = 0$

$\therefore \quad m = 1, 1, -2$

$$\therefore \qquad \text{C.F.} = (c_1 + c_2 z)e^z + c_3 e^{-2z}$$

$$\text{P.I.} = \frac{1}{(\delta - 1)^2 (\delta + 2)} e^{2z} = \frac{1}{4} e^{2z}$$

$\therefore$ The solution of (*i*) is

$$y = (c_1 + c_2 z)e^z + c_3 e^{-2z} + \frac{1}{4} e^{2z}.$$

Hence, the G.S. solution of the given equation is

$$y = (c_1 + c_2 \log x)x + c_3 e^{-2\log x} + \frac{1}{4} e^{2\log x}$$

$$= (c_1 + c_2 \log x)\, x + c_3 \frac{1}{x^2} + \frac{1}{4} x^2.$$

7. *Solve the equation* $(x^3D^3 - xD - 3)\, y = x^2 \log x$.

Sol. Putting $x = e^z$ *i.e.*, $z = \log x$.

The given equation can be written as

$$[\delta(\delta - 1)(\delta - 2) - \delta - 3]y = ze^{2z}$$

or $$[(\delta - 3)(\delta^2 + 1)]\, y = ze^{2z} \qquad \text{... } (i)$$

where $$\delta \equiv \frac{d}{dz}$$

$$\therefore \qquad \text{A.E. is } (m - 3)(m^2 + 1) = 0$$

i.e., $m = 3, m = \pm i$

$$\therefore \qquad \text{C.F.} = c_1 e^{3z} + c_2 \cos z + c_3 \sin z$$

$$\text{P.I.} = \frac{1}{(\delta - 3)(\delta^2 + 1)} ze^{2z} = e^{2z} \frac{1}{(\delta + 2 - 3)\{(\delta + 2)^2 + 1\}} z$$

$$= e^{2z} \frac{1}{(\delta - 1)(\delta^2 + 4\delta + 5)} z = e^{2z} \frac{1}{\delta^3 + 3\delta^2 + \delta - 5} z$$

$$= e^{2z} \frac{1}{-5\left[1 - \dfrac{\delta + 3\delta^2 + \delta^3}{5}\right]} z = e^{2z} - \frac{1}{5}\left[1 - \frac{\delta + 3\delta^2 + \delta^3}{5}\right]^{-1} z$$

$$= -\frac{1}{5} e^{2z}\left[1 + \frac{\delta + 3\delta^2 + \delta^3}{5} + \ldots\right] z = -\frac{1}{5} e^{2z}\left[z + \frac{1}{5}\right] = -\frac{1}{25} e^{2z}\, [5z + 1]$$

$\therefore$ The G.S. of equation (*i*) is

$$y = c_1 e^{3z} + c_2 \cos z + c_3 \sin z - \frac{1}{25} e^{2z} (5z + 1).$$

Hence the general solution of the given equation is

$$y = c_1 x^3 + c_2 \cos(\log x) + c_3 \sin(\log x) - \frac{1}{25} x^2 (5 \log x + 1).$$

8. *Solve the equation*

$$(x^3 D^3 + 2x^2 D^2 + 2)\, y = 10\left(x + \frac{1}{x}\right)$$

where $D \equiv \dfrac{d}{dx}$.

Sol. Given

$$(x^3 D^3 + 2x^2 D^2 + 2)\, y = 10\left(x + \frac{1}{x}\right) \quad \ldots (i)$$

Putting $x = e^z$, *i.e.*, $z = \log x$

so that $xDy = \delta y,\ x^2 D^2 y = \delta(\delta - 1)\, y$

and $x^3 D^3 y = \delta(\delta - 1)(\delta - 2)\, y$, where $\delta \equiv \dfrac{\delta}{dz}$.

Substituting these values in (*i*), the equation reduces to

$$[\delta(\delta - 1)(\delta - 2) + 2\delta(\delta - 1) + 2]\, y = 10(e^z + e^{-z})$$

or

$$(\delta^3 - \delta^2 + 2)\, y = 10(e^z + e^{-z}) \quad \ldots (ii)$$

A.E. is $m^3 - m^2 + 2 = 0$

or $(m + 1)(m^2 - 2m + 2) = 0.$

The roots of the equation are

$$m = -1,\ m = \frac{2 \pm \sqrt{4 - 8}}{2} = 1 \pm i$$

$$\therefore \quad \text{C.F.} = c_1 e^{-z} + e^z (c_2 \cos z + c_3 \sin z)$$

$$\text{P.I.} = 10 \frac{1}{\delta^3 - \delta^2 + 2} (e^z + e^{-z})$$

$$= 10\left[\frac{1}{\delta^3 - \delta^2 + 2} e^z + \frac{1}{\delta^3 - \delta^2 + 2} e^{-z}\right]$$

$$= 10\left[e^z \frac{1}{1 - 1 + 2} + z \cdot \frac{1}{3\delta^2 - 2\delta} e^{-z}\right], \quad \left[\because\ x \cdot \frac{1}{f'(D)} e^x\right]$$

$$= 10\left[\frac{1}{2}e^z + \frac{z}{5}e^{-z}\right]$$

$$= 5e^z + 2ze^{-z}.$$

$\therefore$ The G.S. of equation (*ii*) is

$$y = c_1 e^{-z} + e^z (c_2 \cos z + c_3 \sin z) + 5e^z + 2e^{-z}.$$

Hence, the G.S. of the given equation (*i*) is

$$y = c_1 \frac{1}{x} + x\left[c_2 \cos(\log x) + c_3 \sin(\log x)\right] + 5x + \frac{2}{x}\log x.$$

EXERCISE 1.18

Find the general solution of the following linear homogeneous differential equations:

1. $x^2y'' + 3xy' + 5y = 0$

2. $x^2y'' - xy' - 3y = 0$

3. $xy'' = 2y'$

4. $x^2 \frac{d^2y}{dx^2} + x\frac{dy}{dx} + a^2y = 0$

5. $(x^2D^2 - 2)\, y = x^2 + \frac{1}{x}$

6. $x^2 \frac{d^2y}{dx^2} - x\frac{dy}{dx} + 2y = x \log x$

7. $x^2 \frac{d^2y}{dx^2} + 4x\frac{dy}{dx} + 2y = e^x$

8. $x^2 \frac{d^2y}{dx^2} + 2\frac{dy}{dx} = \log x$

9. $x^2 \frac{d^2y}{dx^2} - x\frac{dy}{dx} + y = \log x$

10. $x^2 \frac{d^2y}{dx^2} - 3x\frac{dy}{dx} + 4y = 2x^2$

11. $x^2 \frac{d^2y}{dx^2} - 2x\frac{dy}{dx} + 2y = x^2 + \sin(5 \log x)$

12. $(x^2D^2 + xD - 1)\, y = \sin(\log x) + x \cos(\log x)$

13. $x^2 \frac{d^2y}{dx^2} + 5x\frac{dy}{dx} + 4y = x^4$

14. $x^2 \frac{d^2y}{dx^2} + 2x\frac{dy}{dx} - 20y = (x + 1)^2$

15. $x^2 \frac{d^2y}{dx^2} + x\frac{dy}{dx} + y = \log x \sin(\log x)$

16. $x^2 \frac{d^3y}{dx^3} - 4x\frac{d^2y}{dx^2} + 6\frac{dy}{dx} = 4,.$

17. $x^3 \frac{d^3y}{dx^3} + 2x^2\frac{d^2y}{dx^2} - 2x\frac{dy}{dx} = 0$

18. $(x^3D^3 + xD - 1)\, y = x^2$

19. $x^3 \frac{d^3y}{dx^3} + 3x^2\frac{d^2y}{dx^2} + x\frac{dy}{dx} + y = x \log x$

20. $(x^4D^3 + 2x^3D^2 - x^2D + x)\, y = 1.$

Answers

1. $y = \frac{1}{x}\,[c_1 \cos(2 \log x) + c_2 \sin(2 \log x)$ 2. $y = c_1 \frac{1}{x} + c_2 x^3$

3. $y = c_1 + c_2 x^3$ 4. $y = c(a \log x + \varepsilon)$

5. $y = c_1 x^{-1} + c_2 x^2 + \frac{1}{3} x^2 \log x - \frac{1}{3x} \log x$

6. $y = \{c_1 \cos(\log x) + c_2 \sin(\log x)\}x + x \log x$

7. $y = \frac{c_1}{x} + \frac{c_2}{x^2} + \frac{e^x}{x^2}$ 8. $y = c_1 + \frac{c_2}{x} + \frac{1}{2}(\log x)^2 - \log x$

9. $y = x(c_1 + c_2 \log x) + \log x + 2$ 10. $y = (c_1 + c_2 \log x)x^2 + x^2(\log x)^3$

11. $y = c_1 x + c_2 x^2 + x^2 \log x + \frac{1}{754}[15 \cos(5 \log x) - 23 \sin(5 \log x)]$

12. $y = c_1 x + \frac{c_2}{x} - \frac{1}{2} \sin(\log x) + \frac{1}{5} x \{2 \sin(\log x) - \cos(\log x)\}$

13. $y = (c_1 + c_2 \log x)x^{-2} + \frac{1}{36} x^4$ 14. $y = c_1 x^4 + c_2 x^{-5} - \frac{x^2}{14} - \frac{x}{9} - \frac{1}{20}$

15. $y = c_1 \cos(\log x) + c_2 \sin(\log x) - \frac{1}{4}(\log x)^2 \cos(\log x) + \frac{1}{4} \log x \sin(\log x)$.

16. $y = c_1 + c_2 x^3 + c_3 x^4 + \frac{2}{3} x$ 17. $y = c_1 + c_2 x^2 + c_3 x^{-1}$

18. $y = \{c_1 + c_2 \log x + c_3 (\log x)^2 x\} + x^2$

19. $y = c_1 x^{-1} + \sqrt{x}\left\{c_2 \cos \frac{\sqrt{3}}{2} \log x + c_3 \sin \frac{\sqrt{3}}{2} \log x\right\} + \frac{x}{2} + \log x$

20. $y = (c_1 + c_2 \log x)x + c_3 x^{-1} + \frac{1}{4} x^{-1} \log x$.

1.15 LEGENDRE'S LINEAR HOMOGENEOUS DIFFERENTIAL EQUATIONS

An equation of the form

$$(a + bx)^n \frac{d^n y}{dx^n} + a_1 (a + bx)^{n-1} \frac{d^{n-1} y}{dx^{n-1}} + \ldots + a_{n-1} (a + bx) \frac{dy}{dx} + a_n y = f(x) \qquad \ldots (i)$$

where a_i's $(i = 1, 2 \ldots, n)$ are constants and $f(x)$ is a function of x alone, is called *Legendre's linear equation.*

The equation (i) can be reduced to linear differential equations with constant coefficients by the substitution.

$$a + bx = e^z, \text{ i.e., } z = \log(a + bx); \text{ and } \frac{dz}{dx} = \frac{b}{a + bx}$$

so that
$$\frac{dy}{dx} = \frac{dy}{dz} \cdot \frac{dz}{dx}$$

$$= \frac{b}{a+bx} \cdot \frac{dy}{dz}$$

or
$$(a+bx)\frac{dy}{dx} = b\,\delta y, \text{ where } \delta \equiv \frac{d}{dz}$$

$$\frac{d^2y}{dx^2} = \frac{d}{dx}\left(\frac{b}{a+bx}\frac{dy}{dz}\right) = \frac{b^2}{(a+bx)^2}\frac{dy}{dz} + \frac{b}{a+bx}\frac{d^2y}{dz^2} \cdot \frac{dy}{dx}$$

$$= -\frac{b^2}{(a+bx)^2}\frac{dy}{dz} + \frac{b}{a+bx}\frac{d^2y}{dz^2} \cdot \frac{b}{a+bx}$$

$$= \frac{b^2}{(a+bx)^2}\left(\frac{d^2y}{dz^2} - \frac{dy}{dz}\right)$$

or
$$(a+bx)^2\frac{d^2y}{dx^2} = b^2(\delta^2 y - \delta y) = b^2\,\delta(\delta - 1)y.$$

Similarly, $(a+bx)^3\dfrac{d^3y}{dx^3} = b^3\,\delta(\delta-1)(\delta-2)y$ and so on

Substituting these values in equation (*i*), we get a linear differential equation with constant coefficients, and the transformed equation can be solved by the methods that discussed earlier. We shall show these by illustrative examples.

1.15.1 Illustrative Examples

1. *Solve the equation*

$$(1+2x)^2\frac{d^2y}{dx^2} - 6(1+2x)\frac{dy}{dx} + 16y = 8(1+2x)^2.$$

Sol. Putting $(1+2x) = e^z$, or $z = \log(1+2x)$

$$\therefore \quad \frac{dz}{dx} = \frac{2}{1+2x}$$

Then
$$(1+2x)\frac{dy}{dx} = 2\,\delta y$$

$$(1+2x)^2\frac{d^2y}{dx^2} = 2^2\,\delta(\delta-1)\,y = 4\,\delta(\delta-1)\,y$$

∴ The given equation reduces to

$$4\delta(\delta-1)y - 6.2\,\delta y + 16y = 8\,e^{2z}$$

or $$[\delta^2 - \delta - 3\delta + 4]\,y = 2e^{2z}$$

or $$[\delta^2 - 4\delta + 4]\,y = 2e^{2z} \qquad \ldots (i)$$

or $$(\delta-2)^2\,y = 2e^{2z}$$

$\therefore$ A.E. is $(m-2)^2 = 0$, given $m = 2, 2$.

$\therefore$ $$\text{C.F.} = (c_1 + c_2 z)\,e^{2z}$$

$$\text{P.I.} = \frac{1}{\delta^2 - 4\delta + 4}\,2e^{2z}$$

$$= 2.z\frac{1}{2\delta - 4}\,e^{2z}$$

$$= 2.z^2\,\frac{1}{\delta}\,e^{2z} = 2z^2 \cdot \frac{e^{2z}}{2}$$

$$= z^2 e^{2z}$$

$\therefore$ G.S. of (*i*) is

$$y = (c_1 + c_2 z)e^{2z} + z^2 e^{2z}$$

Hence, the G.S. of the given equation is

$$y = [c_1 + c_2\log(1+2x)]e^{2\log(1+2x)} + \{\log(1+2x)\}^2 \cdot e^{2\log(1+2x)}$$

$$= [c_1 + c_2\log(1+2x)](1+2x)^2 + \{\log(1+2x)\}^2(1+2x)^2$$

or $$y = (1+2x)^2\,[c_1 + c_2\log(1+2x) + \{\log(1+2x)\}^2].$$

2. *Solve the equation* $(3x+2)^2\dfrac{d^2y}{dx^2} + 3(3x+2)\dfrac{dy}{dx} - 36y = 3x^2 + 4x + 1.$

Sol. Putting $3x + 2 = e^z$, *i.e.*, $z = \log(3x+2)$,

then $$\frac{dz}{dx} = \frac{3}{3x+2}$$

Now $$(3x+2)\frac{dy}{dx} = 3\delta y,$$

where $\delta \equiv \dfrac{d}{dz}$

and $$(3x+2)^2\frac{d^2y}{dx^2} = 3^2\,\delta(\delta-1)\,y = 9\delta(\delta-1)\,y$$

substituting these values in the given equation, is reduces to

$$9\delta(\delta-1)\,y + 3.3\,\delta y - 36y = 3\left(\frac{e^z-2}{3}\right)^2 + 4\left(\frac{e^z-2}{3}\right) + 1$$

or $$9[\delta^2-\delta+\delta-4]\,y=\frac{1}{3}e^{2z}-\frac{1}{3}$$

or $$9[\delta^2-4]\,y=\frac{1}{3}\left(e^{2z}-1\right)$$

or $$(\delta^2-4)\,y=\frac{1}{27}\left(e^{2z}-1\right) \qquad \ldots (i)$$

A.E. is $m^2-4=0$ or $m=\pm 2$

$\therefore$ $$\text{C.F.}=c_1e^{2z}+c_2e^{-2z}$$

$$\text{P.I.}=\frac{1}{27}\frac{1}{\left(\delta^2-4\right)}\left(e^{2z}-1\right)$$

$$=\frac{1}{27}\,.\,z\,.\,\frac{1}{2\delta}\left(e^{2z}-1\right)$$

$$=\frac{1}{27}\,.\,z\,.\,\frac{1}{2}\left(\frac{e^{2z}}{2}-z\right)$$

$$=\frac{1}{108}\,z\left(e^{2z}-2z\right)$$

$\therefore$ G.S. of the equation (*i*) is

$$y=c_1e^{2z}+c_2e^{-2z}+\frac{1}{108}z(e^{2z}-2z).$$

Hence, the G.S. of the given equation is

$$y=c_1(3x+2)^2+c_2(3x+2)^{-2}+\frac{1}{108}[(3x+2)^2\log(3x+2)+1].$$

3. *Solve the equation*

$$(1+x)^2\,y_2+(1+x)\,y_1+y=2\sin\log(1+x).$$

Sol. The given equation is

$$[(1+x)^2D^2+(1+x)\,D+1]\,y=2\sin\log(1+x) \qquad \ldots (i)$$

Putting $(1+x)=e^z$, *i.e.*, $z=\log(1+x)$

Then $$\frac{dz}{dx}=\frac{1}{1+x}$$

$\therefore$ $$(1+x)\,Dy=\delta y,\ \text{where } \delta\equiv\frac{d}{dz}$$

and $$(1+x)^2\,D^2y=\delta(\delta-1)\,y$$

substituting these values in (*i*), we get

$$[\delta(\delta-1)+\delta+1]\,y=2\sin\log e^z=2\sin z, \qquad [\because\ \log_e=1]$$

or, $$[1+\delta^2]\,y=2\sin z \qquad \ldots (ii)$$

$$\text{A.E. is } m^2 + 1 = 0 \Rightarrow m = \pm i$$

$\therefore$ $$\text{C.F.} = c_1 \cos z + c_2 \sin z.$$

Now $$\text{P.I.} = \frac{1}{1+\delta^2} 2\sin z = 2 \cdot \frac{z}{2} \frac{1}{D} \sin z$$

$$= z(-\cos z) = -z\cos z.$$

$\therefore$ The G.S. of equation (*ii*) is

$$y = c_1 \cos z + c_2 \sin z - z\cos z.$$

Hence, the G.S. of (*i*) is

$$y = c_1 \cos \log(1+x) + c_2 \sin \log(1+x) - \log(1+x) \cos \log(1+x).$$

4. *Find the general solution of the equation*

$$(x+1)^2 \frac{d^2y}{dx^2} + (x+1)\frac{dy}{dx} = (2x+3)(2x+4).$$

Sol. The given equation is

$$(x+1)^2 \frac{d^2y}{dx^2} + (x+1)\frac{dy}{dx} = (2x+3)(2x+4)$$

or $$(x+1)^2 D^2y + (x+1)Dy = (2x+3)(2x+4) \quad \Big| \text{ where } D \equiv \frac{d}{dx} \quad \ldots(i).$$

Putting $$x + 1 = e^z \text{ or } z = \log(x+1)$$

then $$\frac{dz}{dx} = \frac{1}{x+1}$$

$\therefore$ $$(x+1)Dy = \delta y, \quad \text{where } \delta \equiv \frac{d}{dz}$$

and $$(x+1)^2 D^2y = \delta(\delta - 1)y$$

Substituting these values in (*i*), we get

$$[\delta(\delta-1) + \delta]\, y = \{2(e^z - 1) + 3\}.\{2(e^z - 1) + 4\}$$

or $$[\delta^2 - \delta + \delta]\, y = \{2e^z + 1\}\{2e^z + 2\}$$

or $$\delta^2 y = 4e^{2z} + 6e^z + 2 \quad \ldots(ii)$$

A.E. is $m^2 = 0$, *i.e.*, $m = 0, 0$

$\therefore$ $$\text{C.F.} = c_1 + c_2 z$$

$$\text{P.I.} = \frac{1}{\delta^2}\left[4e^{2z} + 6e^z + 2\right]$$

$$= \frac{1}{\delta}\left[2e^{2z} + 6e^z + 2z\right]$$

$$= e^{2z} + 6e^z + z^2$$

$\therefore$ G.S. of (*ii*) is

$$y = c_1 + c_2 z + e^{2z} + 6e^z + z^2.$$

Hence, the G.S. of (*i*) is

$$y = c_1 + c_2 \log(x+1) + (x+1)^2 + 6(x+1) + \{\log(x+1)\}^2.$$

5. *Find the general solution of the equation*

$$(2x+3)^2 \frac{d^2y}{dx^2} - 2(2x+3)\frac{dy}{dx} - 12y = 6x.$$

Sol. The given equation is

$$(2x+3)^2 D^2y - 2(2x+3) Dy - 12y = 6x \qquad \dots (i)$$

where $D \equiv \dfrac{d}{dx}$

Putting $(2x+3) = e^z$,

or $z = \log(2x+3)$, then $\dfrac{dz}{dx} = \dfrac{2}{2x+3}$

$$\therefore \qquad (2x+3)\, Dy = 2\delta y, \text{ where } \delta \equiv \frac{d}{dz}$$

and $\qquad (2x+3)^2 D^2y = 2^2\, \delta(\delta-1)y = 4\delta(\delta-1)y$

Substituting these values in (*i*), we get

$$[4\,\delta(\delta-1) - 4\delta - 12]\,y = 6 \cdot \frac{e^z - 3}{2} = 3(e^z - 3)$$

or $\qquad 4[\delta^2 - 2\delta - 3]\,y = 3(e^z - 3) \qquad \dots (ii)$

A.E. is $4(m^2 - 2m - 3) = 0$

or $\qquad m^2 - 2m - 3 = 0$

or $\qquad (m-3)(m+1) = 0$

Hence $m = 3, -1$

$$\therefore \qquad \text{C.F.} = c_1 e^{-z} + c_2 e^{3z}$$

$$\text{P.I.} = \frac{1}{4(\delta^2 - 2\delta - 3)} 3\left[e^z - 3\right]$$

$$= \frac{3}{4}\left[\frac{1}{\delta^2 - 2\delta - 3} e^z - \frac{1}{\delta^2 - 2\delta - 3} 3\right]$$

$$= \frac{3}{4}\left[e^z \frac{1}{1-2-3} - \frac{1}{0 - 2.0 - 3} 3\right] \qquad [\because \; 3 = 3.e^0]$$

$$= \frac{3}{4}\left[-\frac{1}{4}e^{z}+1\right] = \frac{3}{16}\left[4-e^{z}\right]$$

$\therefore$ G.S. of (*ii*) is

$$y = c_1e^{-z} + c_2e^{3z} + \frac{3}{16}[4-e^{z}].$$

Hence, the G.S. of (*i*) is

$$y = c_1\frac{1}{2x+3} + c_2(2x+3)^3 + \frac{3}{16}\left[4-(2x+3)\right]$$

$$= c_1(2x+3)^{-1} + c_2(2x+3)^3 + \frac{3}{16}[1-2x].$$

EXERCISE 1.19

Find the complete solutions (C.S.) of the following Legendre's differential equations:

1. $(5+2x)^2\frac{d^2y}{dx^2} - 6(5+2x)\frac{dy}{dx} + 8y = 0$ **2.** $(x+2)^2\frac{d^2y}{dx^2} - 4(x+2)\frac{dy}{dx} + 6y = 0$

3. $(2x-1)^3\frac{d^3y}{dx^3} + (2x-1)\frac{dy}{dx} - 2y = 0$ **4.** $(1+x)^2\frac{d^2y}{dx^2} + (1+x)\frac{dy}{dx} + y = 4\cos\log(1+x)$

5. $(3x+2)^2\frac{d^2y}{dx^2} + 5(3x+2)\frac{dy}{dx} - 3y = x^2 + x + 1$

6. $(1+x)^2\frac{d^2y}{dx^2} + (1+x)\frac{dy}{dx} + y = \sin 2\{\log(1+x)\}$

7. $x^2D^2y - 3xDy + 5y = x^2\sin(\log x)$.

Answers

1. $y = (5+2x)^2\left[c_1(5+2x)^{\sqrt{2}} + c_2(5+2x)^{-\sqrt{2}}\right]$

2. $y = c_1(x+2)^2 + c_2(x+2)^3$

3. $y = (2x-1)\left\{c_1 + c_2(2x-1)^{\frac{\sqrt{3}}{2}} + c_3(2x-1)^{-\frac{\sqrt{3}}{2}}\right\}$

4. $y = c_1\cos\{\log(1+x)\} + c_2\sin\{\log(1+x)\} + 2\log(1+x)\sin\{\log(1+x)\}$

5. $y = c_1(3x+2)^{\frac{1}{3}} + c_2(3x+2)^{-1} + \frac{(3x+2)^2}{405} - \frac{3x+2}{108} - \frac{7}{27}$

6. $y = c_1\cos\{\log(1+x)\} + c_2\sin\{\log(1+x)\} - \frac{1}{3}\sin 2\{\log(1+x)\}$

7. $y = x^2[c_1\cos(\log x) + c_2\sin(\log x)] - \frac{1}{2}x^2\log x\cos(\log x)$.

EXERCISE 1.20

Multiple Choice Questions

1. The differential equation of all straight lines passing through the origin is
   (a) $y_2 = 0$ (b) $y_1 = 0$
   (c) $xy_1 = y$ (d) $xy_2 = y_1$.
2. The differential equation of parabolas $y^2 = 4ax$ is
   (a) $xy_2 + y_1 = 0$ (b) $2xy_1 = y$
   (c) $xy_1 + 4y = 0$ (d) none of these.
3. The differential equation of a point circle with centre at origin is
   (a) $xy + y_1 = 0$ (b) $x + yy_1 = 0$
   (c) $xy_1 + y = 0$ (d) none of these.
4. The order and degree of the differential equation
   $$\sqrt{y_1^{2/3}} = x^2 + 1 \text{ is}$$
   (a) $1, \frac{1}{2}$ (b) $2, \frac{1}{3}$
   (c) $1, 4$ (d) $1, 1$
5. The differential equation of the curve $x^2 = ay$ is
   (a) $xy_1 = 2y$ (b) $y_1 = xy$
   (c) $2xy = y_1$ (d) none of these.
6. The differential equation of the curve $ax^2 + by = 0$ is
   (a) $y_1 = y$ (b) $2y = xy_1$
   (c) $yy_1 = x^2$ (d) none of these.
7. The order and degree of the differential equation
   $y_2 = 4\{1 + (y_1)^2\}^{3/2}$ is
   (a) $2, 1$ (b) $2, 2$
   (c) $1, 2$ (d) $2, 6$.
8. The order and degree of the differential equation
   $$\frac{d^2y}{dx^2} + x\left(\frac{dy}{dx}\right)^2 = y \text{ is}$$
   (a) $2, 2$ (b) $2, 1$
   (c) $1, 2$ (d) none of these.
9. The differential equation of the curve $y = ax + b$ is
   (a) $y_1 = x$ (b) $y_1 = 0$
   (c) $y_2 = 0$ (d) $y_1 = xy$.
10. The differential equation of all the straight lines that make an angle θ with the positive direction of x-axis is
    (a) $y_1 = 0$ (b) $y_1 = \cos\theta$
    (c) $y_1 = \sin\theta$ (d) $y_1 = \tan\theta$.

11. The equation of all curves whose slope at any point (x, y) is 0, is

(*a*) $y = c$ (*b*) $x = c$

(*c*) $xy = c^2$ (*d*) $x + y = 0$.

12. The curve whose differential equation is $y_1 = 2x + 3$ and passes through the origin is

(*a*) $y = 2$ (*b*) $y = x^2$

(*c*) $y = 3x + x^2$ (*d*) $y = 3x$.

13. The curve which is the solution of the equation $\tan x\, dy - \tan y\, dx = 0$, when $x = \frac{\pi}{6}$, $y = \frac{\pi}{2}$ is

(*a*) $\sin y = 2 \sin x$ (*b*) $\sin y = \sqrt{2} \sin x$

(*c*) $\sec x - \sec y = 2$ (*d*) $2 \sin y = \sin x$.

14. The solution of the equation $x\, dx + y\, dy + (x^2 + y^2)\, dy = 0$ is

(*a*) $x^2 + y^2 = k$ (*b*) $x^2 - y^2 = ke^{2y}$

(*c*) $x^2 + y^2 = ke^{-2y}$ (*d*) none of these.

15. The solution of the equation $y\, dx + x\, dy = xy(dy - dx)$ is

(*a*) $\log(xy) = x + y + c$ (*b*) $\log(xy) = (x - y) + c$

(*c*) $xy = \log(x + y)$ (*d*) $\log(xy) = (y - x) + c$.

16. The integrating factor (I.F.) of $\frac{ds}{dt} - \frac{s}{t} = 2t^2$ is

(*a*) t (*b*) $-\frac{1}{t}$

(*c*) $\frac{1}{t}$ (*d*) $\frac{1}{t^2}$.

17. The integrating factor of $x\frac{dy}{dx} + 2y = \sqrt{1 + x^2}$ is

(*a*) $2 \log x$ (*b*) e^{2x}

(*c*) x^2 (*d*) $\frac{1}{x^2}$.

18. The C.F. of the equation $\frac{d^2y}{dx^2} - 2\frac{dy}{dx} = 3x$ is

(*a*) $c_1x + c_2e^{2x}$ (*b*) $c_1e^x + c_2e^{2x}$

(*c*) $c_1 + c_2e^{2x}$ (*d*) $(c_1 + c_2x)e^x$.

19. Choose the correct answer from the given below:

$y = cx - c^2$ is the general solution of the differential equation

(*a*) $(y')^2 - xy' + y = 0$ (*b*) $y'' = 0$

(*c*) $y' = c$ (*d*) $(y')^2 + xy' + y = 0$.

20. The differential equation governing the solution $y(x) = \sqrt{5x + c}$ where c is an arbitrary constant is

(a) $(y_1)^2 = 5x + c$ (b) $yy' = 5$

(c) $y' = \frac{2.5}{\sqrt{5x+c}}$ (d) $yy' = 2.5$.

21. If the functions 1 and $-e^{-3x}$ form a fundamental system of solutions of the differential equation $y'' + ay' + by = 0$, then

(a) $a = 3, b = 0$ (b) $a = 0, b = +3$

(c) $a = -3, b = 0$ (d) $a = 0, b = -3$.

22. The differential equation $\left(\frac{dx}{dy}\right)^2 + \frac{5}{y^3} = x$ is

(a) linear of degree 2 (b) non-linear of order 1 and degree 2

(c) non-linear of order 1 and degree 6 (d) none of these.

23. The differential equation $(y^2e^{xy^2} + 6x)\,dx + (2xy\,e^{xy^2} - 4y)\,dy = 0$ is

(a) linear, homogeneous and exact

(b) non-linear, homogeneous and exact

(c) non-linear, non-homogeneous and exact

(d) non-linear, non-homogeneous and not exact.

24. The I.F. of $\left(xy^2 - e^{\frac{1}{x^3}}\right) dx - x^2y\,dy = 0$ which makes it exact is

(a) x^4 (b) $\frac{1}{x^4}$

(c) $\frac{1}{y^4}$ (d) $\frac{1}{y^2}$.

25. P.I. of $\frac{1}{D^2 - 4D + 3} 2e^{3x}$ is

(a) $2e^{3x}$ (b) $2x\,e^{3x}$

(c) xe^{3x} (d) $\frac{1}{x}e^{3x}$.

26. P.I. of $\frac{1}{D^2 - D} \cos 2x$ is

(a) $-\frac{1}{10}[\sin 2x + 2\cos 2x]$ (b) $-\frac{1}{10}[2\cos 2x - \sin 2x]$

(c) $\frac{1}{20}[-2\sin 2x + \cos 2x]$ (d) none of these.

27. P.I. of $\frac{1}{D^2 + 4} \sin 2x$ is

(a) $x \sin 2x$ (b) $\frac{1}{4} \sin 2x$

(c) $-\frac{1}{4} \cos 2x$ (d) $\frac{1}{4} \cos 2x$.

28. P.I. of $(D^3 + 3D^2 + 3D + 1)\, y = e^{-x}$ is

(a) xe^{-x} (b) $6e^{-x} x^2$

(c) $\frac{1}{6} e^{-x} x^4$ (d) $\frac{1}{6} x^3 e^{-x}$.

29. The general solution of the equation $y - px = p^2$ is

(a) $y = x^2 + c$ (b) $y = cx + c^2$

(c) $y - x = 1$ (d) none of these.

30. The I.F. of $(x^2y^3 + xy)\, dy = dx$ is

(a) $\log y$ (b) $\frac{1}{y}$

(c) $e^{\frac{y^2}{2}}$ (d) $e^{-\frac{y}{2}}$.

Answers

1. *c*	**2.** *b*	**3.** *b*	**4.** *d*
5. *a*	**6.** *b*	**7.** *b*	**8.** *b*
9. *c*	**10.** *d*	**11.** *a*	**12.** *c*
13. *a*	**14.** *c*	**15.** *d*	**16.** *c*
17. *c*	**18.** *c*	**19.** *c*	**20.** *c*
21. *a*	**22.** *c*	**23.** *c*	**24.** *b*
25. *c*	**26.** *a*	**27.** *c*	**28.** *d*
29. *b*	**30.** *c*.		

Chapter 2

Linear Algebra : Vector Space

2.1 INTRODUCTION

We have ideas of *vectors*, their operations, and two algebraic structures known as *group* and *field*. Group is an algebraic structure (G, *) with one binary operation *, while field is an algebraic structure (F, +, •) with two binary operations +, and •. We now introduce another algebraic structure known as *vector space* (or *linear vector space*) which is a generalisation arising from the ordinary notion of a vector in two or three dimensions (R^2 or R^3), ingeneral in R^n.

2.2 PRELIMINARIES OF VECTOR SPACES

A vector $\boldsymbol{a}$ in two dimension (R^2) or in three dimension (R^3) is represented as $\boldsymbol{a} = (a_1, a_2)$ or $\boldsymbol{a} = (a_1, a_2, a_3)$, where right hand expression is ordered pair of two or ordered triplet of real numbers called the components of the vector. An n-component vector $\boldsymbol{a}$ is an ordered n-tuple of real numbers, written as $\boldsymbol{a} = (a_1, a_2, \ldots, a_n)$.

The addition of two vectors $\boldsymbol{a}$ and $\boldsymbol{b}$ is denoted by $\boldsymbol{a} + \boldsymbol{b}$, which is a vector. The identity element for the operation addition is $\boldsymbol{o}$ (null) vector, as $\boldsymbol{o} + \boldsymbol{a} = \boldsymbol{a} + \boldsymbol{o} = \boldsymbol{a}$.

The scalar multiplication of a vector $\boldsymbol{a}$ by a scalar α is a vector defined by $\alpha\,\boldsymbol{a}$. The scalar product of two vectors $\boldsymbol{a}$ and $\boldsymbol{b}$ is defined by $\boldsymbol{a}\cdot\boldsymbol{b} = ab\cos\theta$, where a and b are the magnitudes (scalars) of the vectors $\boldsymbol{a}$ and $\boldsymbol{b}$ respectively and θ is the included angle between the vectors. If $\boldsymbol{a}\cdot\boldsymbol{b} = 0$, then the vectors $\boldsymbol{a}$ and $\boldsymbol{b}$ are said to be orthogonal, that is the angle between the vectors is $\theta = \frac{\pi}{2}$. The vector $\hat{e}$ is the identity (unit) vector and $|\hat{e}| = 1$.

The addition of vectors is *commutative* and *associative* that is, $\boldsymbol{a} + \boldsymbol{b} = \boldsymbol{b} + \boldsymbol{a}$ and $(\boldsymbol{a} + \boldsymbol{b}) + \boldsymbol{c} = \boldsymbol{a} + (\boldsymbol{b} + \boldsymbol{c})$.

The two algebraic structures, group and field, are defined as follows :

Group. A non-empty set G together with a binary operation $* : G \times G \to G$ is called a group if the following properties hold in G with respect to the operation *.

(*i*) * is a *binary operation* in G, that is, $a * b \in G$ for all $a, b \in G$.

(*ii*) * is *associative* in G, that is, $a * (b * c) = (a * b) * c$ for all $a, b, c \in G$.

(*iii*) There exists an element $e \in G$ such that

$$a * e = e * a = a \text{ for all } a \in G.$$

e is called the *identity* element w.r.t. the operation * in G.

(*iv*) For any element a in G, there exists an element a' in G such that $a * a' = a' * a = e$. a' is called the inverse of a in G w.r.t. the operation *.

In addition to these four properties if $a * b = b * a$ for all $a, b \in G$, that is *commutative* property holds in G w.r.t. the operation *, then G is called an *Abelian group* or *commutative group*.

Field. A non-empty set F together with the two binary operations (+) and (•) defined in F is said to be a field, if

(*i*) F is a commutative group with respect to the operation +.

(*ii*) The set F (excluding the 0 element) is a commutative group with respect to the operation multiplication (•), and

(*iii*) Left and Right distributive properties hold in F that is,

$$a \cdot (b + c) = a{\cdot}b + a{\cdot}c \text{ and}$$

$$(b + c) \cdot a = b \cdot a + c \cdot a.$$

Further, we need to define some more definitions as follows :

Internal composition. Let S be any non-empty set. If for any elements $a, b \in S$, $a * b \in S$ and $a * b$ is unique, then * is said to be an internal composition in the set S. Here a and b both are the elements of S and the map is $S \times S \to S$.

External composition. Let V and F be any two non-empty sets. If $a * x \in V$ for all $a \in F$ and for all $x \in V$ and $a * x$ be unique then * is said to be the *external composition* in V over F.

Here a is an element of the set F and x is an element of the set V and the resulting element $a * x$ is an element of the set V and the map is $F \times V \to V$.

The element of V are called *vectors* and are denoted by ***x, y, z, u, v***, etc. and the elements of F are called *scalars* and are denoted by *a, b, c*, etc.

It is known earlier that the addition of vectors (internal composition) and the multiplication of vectors by scalars (external composition) satisfy the following properties :

(*i*) for any $x, y \in V$, $x + y \in V$, where V is a set of vectors

(*ii*) for any $x, y, z \in V$, $(x + y) + z = x + (y + z)$

(*iii*) for any vector x, there exists a null vector $o \in V$ such that $x + o = o + x = x$

(*iv*) corresponding to the vector $x \in V$, there exists a vector $(-x)$, such that

$$x + (-x) = (-x) + x = o.$$

(*v*) for any vectors $x, y \in V$, $x + y = y + x$.

(*vi*) There exists an external composition in V over the field F, of real numbers, such that

$$a \in F, x \in V \text{ implies } a\,x \in V$$

(*vii*) for any $a, b \in F$ and $x \in V$,

$$(a + b)\,x = ax + bx$$

(*viii*) for any $a \in F$, and $x, y \in V$

$$a\,(x + y) = ax + by$$

(*ix*) for any $a, b \in F$ and $x \in V$

$$(ab)\,x = a\,(bx)$$

(*x*) $1.x = x$, where 1 is the identity element in F and x in any vector of V.

These ideas motivate the following abstract definition of a vector space V over a field F.

2.3 REAL VECTOR SPACE

Definition. A non-empty set V of elements $x, y, z,...$ is called a *real vector space* (or real linear space) and these elements are called *vectors,* if in V there are two defined algebraic operations (called *vector addition* and *scalar multiplication*) as follows :

A. *Vector addition* associates with every pair of vectors x and y of V a unique vector of V, called the *sum* of x and y and denoted by $x + y$, such that the following axioms are satisfied.

A_1. *Commutativity* : For any two vectors x and y of V,

$$x + y = y + x$$

A_2. *Associativity* : For any three vectors x, y, z of V,

$$(x + y) + z = x + (y + z)$$

A_3. There is a unique vector in V, called *zero vector* and is denoted by o, such that for every $x \in V$.

$$x + o = x$$

A_4. For every x in V there is a unique vector in V, namely $-x$ and is such that

$$x + (-x) = o$$

B. *Scalar multiplication.* The real numbers are called *scalars.* Scalar multiplication associated with every x in V and every scalar c in F, a unique vector of V, called the *product* of c and x and denoted by cx (or xc) such that the following axioms are satisfied :

B_1. *Distributivity* for every scalar c and vectors x and y in V,

$$c\,(x + y) = cx + cy$$

B_2. *Distributivity* : For all scalars c and k and every

$$x \in V, (c + k)\,x = cx + kx$$

B_3. *Associativity* : For all scalars c and k and every x in V

$$c\,(kx) = (ck)\,x$$

B_4. For every x in V,

$$1x = x$$

2.3.1 Illustrative Examples

1. *Let F be a set of rational numbers and V be a set of real numbers. The operation (+) on V is the ordinary addition and $a\alpha$ ($a \in F$, $\alpha \in V$) is the ordinary product, prove that V is a vector space.*

   Sol. The real numbers form an abelian group under addition (to prove), $a\alpha$ is a real number belongs to V. Multiplication is associative.

   Also $$1\alpha = \alpha.$$

   Hence, V forms a vector space over the set of rational numbers.

2. *Let F be any field, and V is the set of all n-tuples $x = (x_1, x_2, ..., x_n)$ of scalars x_i in F. If $y = (y_1, y_2, ..., y_n)$ with y_i in F, the sum of x and y is defined by*
   $$x + y = (x_1 + y_1, x_2 + y_2, ..., x_n + y_n),$$
   and the product of a scalar c and a vector x is defined by
   $$cx = (cx_1, cx_2, ..., cx_n), \text{ then prove that V is a vector space.}$$

   Sol. For any $x, y \in V$ and $a \in F$,

   we have

   (*i*)
   $$\begin{aligned} a(x+y) &= a\{(x_1, x_2, ..., x_n) + (y_1, y_2, ..., y_n)\} \\ &= a((x_1 + y_1) + (x_2 + y_2) + ... + (x_n + y_n)) \\ &= a(x_1 + y_1) + a(x_2 + y_2) + ... + a(x_n + y_n) \\ &= (ax_1 + ay_1) + (ax_2 + ay_2) + ... + (ax_n + ay_n) \\ &= (ax_1, ax_2, ..., ax_n) + (ay_1 + ay_2, ..., ay_n) \\ &= a(x_1, x_2, ..., x_n) + a(y_1, y_2, ... y_n) \\ &= ax + ay. \end{aligned}$$

   (*ii*)
   $$\begin{aligned} (a+b)x &= (a+b)(x_1, x_2, ..., x_n) \\ &= ((a+b)x_1, (a+b)x_2, ..., (a+b)x_n) \\ &= (ax_1 + bx_1, ax_2 + bx_2, ..., ax_n + bx_n) \\ &= (ax_1, ax_2, ..., ax_n) + (bx_1, bx_2, ..., bx_n) \\ &= a(x_1, x_2, ..., x_n) + b(x_1, x_2, ..., x_n) \\ &= ax + bx. \end{aligned}$$

   (*iii*)
   $$\begin{aligned} a(bx) &= a(b(x_1, x_2, ..., x_n)) = a(bx_1, bx_2, ..., bx_n) \\ &= (abx_1, abx_2, ..., abx_n) \\ &= (ab)(x_1, x_2, ..., x_n) = (ab)x. \end{aligned}$$

   (*iv*)
   $$\begin{aligned} 1x &= 1(x_1, x_2, ..., x_n) = (1x_1, 1x_2, ... 1x_n) \\ &= (x_1, x_2, ..., x_n) = x. \end{aligned}$$

   Thus, V satisfies all the conditions of vector space, hence V is a vector space.

3. *Prove that R × R is a vector space over R under addition and scalar multiplication defined by*
   $$(x_1, x_2) + (y_1, y_2) = (x_1 + y_1, x_2 + y_2) \text{ and } \alpha(x_1, x_2) = (\alpha x_1, \alpha x_2).$$

Sol. The binary operation (+) is commutative and associative and (0, 0) is the zero element.

The inverse of (x_1, x_2) is $(-x_1, -x_2)$.

Hence $(R \times R, +)$ is an abelian group.

Let $u = (x_1, x_2)$ and $v = (y_1, y_2)$ and $\alpha, \beta \in R$.

Then
$$\begin{aligned}
\alpha(u+v) &= \alpha[(x_1, x_2) + (y_1, y_2)] \\
&= \alpha[(x_1 + y_1, x_2 + y_2)] \\
&= \alpha(x_1 + y_1, x_2 + y_2) \\
&= (\alpha x_1 + \alpha y_1, \alpha x_2 + \alpha y_2) \\
&= (\alpha x_1\ \alpha x_2) + (\alpha y_1 + \alpha y_2) \\
&= \alpha(x_1, x_2) + \alpha(y_1, y_2) = \alpha u + \alpha v
\end{aligned}$$

$$\begin{aligned}
(\alpha + \beta)\, u &= (\alpha + \beta)\,(x_1, x_2) = ((\alpha + \beta)\, x_1, (\alpha + \beta)\, x_2) \\
&= (\alpha x_1 + \beta x_1, \alpha x_2 + \beta x_2) \\
&= (\alpha x_1, \alpha x_2) + (\beta x_1 + \beta x_2) \\
&= \alpha(x_1, x_2) + \beta(x_1, x_2) \\
&= \alpha u + \beta u.
\end{aligned}$$

Also,
$$\begin{aligned}
\alpha(\beta u) &= \alpha(\beta(x_1, x_2)) = \alpha(\beta x_1, \beta x_2) = (\alpha\beta x_1, \alpha\beta x_2) \\
&= (\alpha\beta)\,(x_1, x_2) = (\alpha\beta)\, u
\end{aligned}$$

and
$$1u = 1(x_1, x_2) = (1x_1, 1x_2) = (x_1, x_2) = u.$$

Thus $R \times R$ satisfies all the properties of vector space.

Hence, $R \times R$ is a vector space over R.

4. *Show that the set M_2 (R) of all 2 ×2 matrices is a vector space over R under matrix addition and scalar multiplication defined by*

$$\alpha \begin{bmatrix} a & b \\ c & d \end{bmatrix} = \begin{bmatrix} \alpha a & \alpha b \\ \alpha c & \alpha d \end{bmatrix}.$$

Sol. Addition (+) is a binary operation in the set M_2 (R). Associative and commutative laws hold for the binary operation (+) in M_2 (R). (To be proved)

The matrix $\begin{bmatrix} 0 & 0 \\ 0 & 0 \end{bmatrix} \in M_2$ (R), is the identity element for the operation (+). The inverse of

the matrix $\begin{bmatrix} x_1 & x_2 \\ x_3 & x_4 \end{bmatrix} \in M_2$ (R) is $\begin{bmatrix} -x_1 & -x_2 \\ -x_3 & -x_4 \end{bmatrix}$ which also belongs to M_2 (R).

Hence, M_2 (R) is an abelian group under matrix addition.

Let
$$M_1 = \begin{bmatrix} x_1 & x_2 \\ x_3 & x_4 \end{bmatrix} \text{ and } M_2 = \begin{bmatrix} y_1 & y_2 \\ y_3 & y_4 \end{bmatrix} \in M_2\ (R)$$

and $\alpha, \beta \in R$.

Then $$\alpha(M_1 + M_2) = \alpha\left(\begin{bmatrix} x_1 & x_2 \\ x_3 & x_4 \end{bmatrix} + \begin{bmatrix} y_1 & y_2 \\ y_3 & y_4 \end{bmatrix}\right)$$

$$= \alpha\begin{bmatrix} x_1 + y_1 & x_2 + y_2 \\ x_3 + y_3 & x_4 + y_4 \end{bmatrix} = \begin{bmatrix} \alpha x_1 + \alpha y_1 & \alpha x_2 + \alpha y_2 \\ \alpha x_3 + \alpha y_3 & \alpha x_4 + \alpha y_4 \end{bmatrix}$$

$$= \begin{bmatrix} \alpha x_1 & \alpha x_2 \\ \alpha x_3 & \alpha x_4 \end{bmatrix} + \begin{bmatrix} \alpha y_1 & \alpha y_2 \\ \alpha y_3 & \alpha y_4 \end{bmatrix}$$

$$= \alpha\begin{bmatrix} x_1 & x_2 \\ x_3 & x_4 \end{bmatrix} + \alpha\begin{bmatrix} y_1 & y_2 \\ y_3 & y_4 \end{bmatrix} = \alpha M_1 + \alpha M_2$$

$$(\alpha + \beta) M_1 = (\alpha + \beta)\begin{bmatrix} x_1 & x_2 \\ x_3 & x_4 \end{bmatrix} = \begin{bmatrix} (\alpha + \beta) x_1 & (\alpha + \beta) x_2 \\ (\alpha + \beta) x_3 & (\alpha + \beta) x_4 \end{bmatrix}$$

$$= \begin{bmatrix} \alpha x_1 + \beta x_1 & \alpha x_2 + \beta x_2 \\ \alpha x_3 + \beta x_3 & \alpha x_4 + \beta x_4 \end{bmatrix}$$

$$= \begin{bmatrix} \alpha x_1 & \alpha x_2 \\ \alpha x_3 & \alpha x_4 \end{bmatrix} + \begin{bmatrix} \beta x_1 & \beta x_2 \\ \beta x_3 & \beta x_4 \end{bmatrix}$$

$$= \alpha\begin{bmatrix} x_1 & x_2 \\ x_3 & x_4 \end{bmatrix} + \beta\begin{bmatrix} x_1 & x_2 \\ x_3 & x_4 \end{bmatrix}$$

$$= \alpha M_1 + \beta M_1$$

$$\alpha(\beta M_1) = \alpha\left(\beta\begin{bmatrix} x_1 & x_2 \\ x_3 & x_4 \end{bmatrix}\right) = \alpha\begin{bmatrix} \beta x_1 & \beta x_2 \\ \beta x_3 & \beta x_4 \end{bmatrix} = \begin{bmatrix} (\alpha\beta) x_1 & (\alpha\beta) x_2 \\ (\alpha\beta) x_3 & (\alpha\beta) x_4 \end{bmatrix}$$

$$= (\alpha\beta)\begin{bmatrix} x_1 & x_2 \\ x_3 & x_4 \end{bmatrix} = (\alpha\beta) M_1$$

and $$\begin{bmatrix} 1 & 0 \\ 0 & 1 \end{bmatrix}\begin{bmatrix} x_1 & x_2 \\ x_3 & x_4 \end{bmatrix} = \begin{bmatrix} x_1 & x_2 \\ x_3 & x_4 \end{bmatrix} = M_1.$$

Hence, the set M_2 (R) of all 2×2 matrices is a vector space over R under the operation addition and scalar multiplication.

5. ***Let V be the set of all ordered pairs of real numbers. Addition and multiplication are defined by $(x, y) + (x_1, y_1) = (x + x_1, y + y_1)$***

and $\alpha(x, y) = (x, \alpha y),$

where x, y, x_1, y_1 *and* α *are real numbers. Show that V is not a vector space over R.*

Sol. V is an abelian group under the operation (+) as defined by the problem.

Let $\alpha, \beta \in R$ and $x, y \in V$.

Now $(\alpha+\beta)(x, y) = (x, (\alpha+\beta) y) = (x, \alpha y + \beta y).$

Also
$$\alpha(x, y) + \beta(x, y) = (x, \alpha y) + (x, \beta y)$$
$$= (2x, \alpha y + \beta y).$$

But $(x, \alpha y + \beta y) \neq (2x, \alpha y + \beta y)$

i.e., $(\alpha+\beta)(x, y) \neq \alpha(x, y) + \beta(x, y)$

Hence, V is not a vector space over R.

6. *Show that the set V of all real valued continuous (differentiable or integrable) functions defined in the interval (0, 1) is the vector space over the field R of real numbers with respect to addition and scalar multiplication defined as given below :*

$$(f+g)(x) = f(x) + g(x), \forall f, g \in V \text{ and } \forall x \in [0, 1]$$
$$(\alpha f)(x) = \alpha f(x), \forall f \in V \text{ and } \alpha \in R \text{ and } x \in [0, 1].$$

Sol. (*i*)(V, +) is an abelian group.

(*ii*) V is closed w.r.t. scalar multiplication, since αf, for $\alpha \in R, f \in V$ is also a real valued continuous functions in [0, 1].

(*iii*) If $\alpha \in R$ and $f, g \in V$, then
$$[\alpha(f+g)](x) = \alpha[(f+g)(x)] = \alpha[f(x) + g(x)]$$
$$= \alpha f(x) + \alpha g(x) = [\alpha f + \alpha g)(x)$$

$\therefore$ $\alpha(f+g) = \alpha f + \alpha g.$

(*iv*) If $\alpha, \beta \in R$ and $f \in V$, then
$$[(\alpha+\beta)f](x) = (\alpha+\beta)f(x) = \alpha f(x) + \beta f(x)$$
$$= (\alpha f)(x) + (\beta f)(x)$$

i.e., $(\alpha+\beta)f = \alpha f + \beta f.$

(*v*) If $\alpha, \beta \in R$ and $f \in V$, then
$$[(\alpha\beta)f](x) = (\alpha\beta)f(x) = \alpha[\beta f(x)]$$
$$= \alpha[(\beta f)(x)] = [\alpha(\beta f)](x)$$

i.e., $(\alpha\beta)f = \alpha(\beta f)$

(*vi*) If 1 is the unity of R and $f \in V$, then
$$(1.f)(x) = 1 f(x) = f(x)$$

i.e., $1.f = f$

V (R) satisfies all the properties of vector space, hence V (R) is a vector space.

7. *Prove that the set of all real solutions of the differential equation* $\frac{d^2y}{dx^2} + w^2y = 0$ *forms a linear vector space over a real field.*

Sol. Let V be the set of real solutions of the given equation. Let y_1 and y_2 are two real solutions of the differential equation

$$\frac{d^2y}{dx^2} + w^2y = 0 \qquad ...(i)$$

Then, $$\frac{d^2y_1}{dx^2} + w^2y_1 = 0$$

and $$\frac{d^2y_2}{dx^2} + w^2y_2 = 0$$

adding the above two equation, we get

$$\frac{d^2}{dx^2}(y_1 + y_2) + w^2(y_1 + y_2) = 0.$$

This shows that $(y_1 + y_2)$ satisfies the equation (*i*) that is, $y_1 + y_2$ is a real solution of (*i*). Hence, $y_1 + y_2 \in$ V.

The function $y = 0$ is also a solution of (*i*), which is the null element. If y is a solution of (*i*) so is $-y$.

Also, for every real solution y, αy where α is a real number, is a real solution of the equation (*i*),

since $$\frac{d^2}{dx^2}(\alpha\, y) + w^2(\alpha\, y) = \alpha\frac{d^2y}{dx^2} + \alpha\, w^2y$$

$$= \alpha\left[\frac{d^2y}{dx^2} + w^2y\right] = \alpha.0 = 0$$

$\therefore \quad \alpha y \in$ V

$$(\alpha+\beta)\left\{\frac{d^2y}{dx} + w^2y\right\} = \frac{d^2}{dx^2}\{(\alpha+\beta)\,y\} + w^2\,\{(\alpha+\beta)\,y\}$$

$$= (\alpha+\beta)\frac{d^2y}{dx^2} + (\alpha+\beta)\,w^2y$$

$$= \alpha\frac{d^2y}{dx^2} + \beta\frac{d^2y}{dx^2} + w^2\,\alpha y + w^2\beta y$$

$$= \alpha\left(\frac{d^2y}{dx^2} + w^2\alpha\right)y + \beta\left(\frac{d^2y}{dx^2} + w^2y\right).$$

Thus, all the axioms of vector space are satisfied, hence the set of real solution of the given equation is a vector space.

8. *Given R × R with usual addition and multiplication is defined by $\alpha\,(x, y) = (\alpha x, \alpha^2 y)$. Show that R × R is not a vector space over R.*

Sol. R × R with usual addition is an abelian group.

Now
$$(\alpha + \beta)\,(x, y) = ((\alpha + \beta)\,x, (\alpha + \beta)^2\,y)$$
$$= (\alpha x + \beta x, \alpha^2 y + \beta^2 y + 2\alpha\beta y)$$

and
$$\alpha(x, y) + \beta(x, y) = (\alpha x, \alpha^2 y) + (\beta x, \beta^2 y)$$
$$= (\alpha x + \beta x, \alpha^2 y + \beta^2 y)$$

These show that

$$(\alpha + \beta)\,(x, y) \neq \alpha(x, y) + \beta(x, y).$$

Hence, R × R is not a vector space over R.

9. *Let R^+ be the set of all positive real numbers. Addition and scalar multiplication in R^+ are defined as follows :*

$$a + b = ab \text{ for all } a, b \in R^+;$$

$$\alpha a = a^\alpha \text{ for all } a \in R^+ \text{ and } \alpha \in R.$$

Show that R^+ is a real vector space.

Sol. $(R^+, +)$ is an abelian group with identity 1.

Now,
$$\alpha(a + b) = \alpha(ab) = (ab)^\alpha \text{ [as defined in the problem]}$$
$$= a^\alpha b^\alpha = \alpha a + \alpha b$$
$$(\alpha + \beta)\,a = a^{\alpha + \beta} = a^\alpha a^\beta = \alpha a + \beta a$$
$$\alpha(\beta a) = \alpha a^\beta = (a^\beta)^\alpha = a^{\beta\alpha} = a^{\alpha\beta} = (\alpha\beta)a$$

Also
$$1a = a1 = a.$$

Thus, all the axioms are satisfied in R^+, hence R^+ is a vector space over R.

2.4 PROPERTIES OF VECTOR SPACES

In a vector space V over a field F, if ***o*** is the zero vector of V, then

(*i*) $ao = o\ \forall\ a \in F$

(*ii*) $ox = o\ \forall\ x \in V, o \in F$

(*iii*) $a\,(-x) = -\,(ax)\ \forall\ a \in F$ and $\forall\ x \in V$

(*iv*) $(-a)\,x = -\,(ax), \forall\, a \in F$ and $\forall\, x \in V$

(*v*) $a\,(x - y) = ax - ay, \forall\ a \in F$ and $\forall\ x, y \in V$

(*vi*) $a\,x = o$ implies, $a = 0$ or $x = o$

Proof. (*i*) We have $ao = a(o+o)$ $[\therefore o = o+o]$

$= ao + ao$

Since $a\,o \in V$ and $o + a\,o = ao$

$\therefore$ $o + ao = ao + ao$

Since V is an abelian group under addition, therefore, by right cancellation law in V, we obtain

$$o = ao \;\forall\, a \in F. \qquad [\because ao = o]$$

(*ii*) We have $0x = (0+0)\,x$ $[\because o + o = o]$

$= 0x + 0x \;\forall\, x \in V.$

Hence, $0 + 0x = 0x + 0x$ $[\because 0x \in V \text{ and } 0 + 0x = 0x]$

$\therefore$ $0 = 0x$ [by right cancellation law].

(*iii*) since, $a[x + (-x)] = ax + a(-x)$

i.e., $ao = ax = a\,x + a(-x)$ $[\because x + (x) = o]$

i.e., $o = a\,x + a(-x)$ $[\because ao = o]$

i.e, $a(-x) = -(ax)$ $[\because o + a(-x) = a(-x)]$

(*iv*) We have $[a + (-a)]\,x = ax + (-a)\,x$

i.e., $ox = ax + (-a)\,x$ $[\because a + (-a) = o]$

i.e., $o = ax + (-a)\,x$ $[ox = o]$

Hence, $(-a)\,x$ is the additive inverse of ax

i.e., $(-a)\,x = -(ax)$

(*v*) We have $a(x-y) = a(x + (-y)) = ax + a(-y)$

$= ax + (-(ay))$ $[\because a(-y) = -(ay), \text{by } (iii)]$

$= ax - ay.$

(*vi*) Let us consider $a \neq 0$,

Since a is a non-zero element in F, it has its inverse in F. Let a^{-1} is the inverse of a

$\because$ $ax = o$

$\therefore$ $a^{-1}(ax) = a^{-1}(o)$

or $(a^{-1}a)\,x = o$

or $1x = o$ $(\because a^{-1}a = 1)$

i.e., $x = o.$

Next, let $x \neq 0$. Then it requires to show that $a = 0$.

If possible, let $a \neq 0$, then a^{-1} exists,

$\therefore$ $ax = 0$

$\Rightarrow$ $a^{-1}(ax) = a^{-1}o$

$$\Rightarrow \qquad (a^{-1}a)\,x = 0$$

$$\Rightarrow \qquad 1\,x = 0$$

$$\Rightarrow \qquad x = 0.$$

Thus, we arrive to a contradiction that $x \neq 0$,

Hence, $\qquad a = 0.$

Thus, $x \neq 0$ and $ax = 0 \Rightarrow a = 0$.

2.4.1 Deductions

Let V be a vector space over a field F.

(*a*) If $a, b \in$ F and x be a non-zero vector of V,

then $\qquad ax = bx \quad \Rightarrow \quad a = b.$

(*b*) If $x, y \in$ V and a be a non-zero element of F,

then $\qquad ax = ay \quad \Rightarrow \quad x = y.$

Proof. (*a*) Given $\qquad ax = bx$

i.e., $\qquad ax - bx = 0$

or $\qquad (a - b)\,x = 0$

$\therefore \qquad a - b = 0 \qquad (\because x \neq 0)$

i.e., $\qquad a = b.$

(*b*) Given $\qquad ax = ay$

or $\qquad ax - ay = 0$

or $\qquad a(x - y) = 0$

$\because \qquad a \neq 0,$

$\therefore \qquad x - y = 0,$

i.e., $\qquad x = y.$

EXERCISE 2.1

1. Prove that R^2 is a vector space.
2. Prove that R^3 is a vector space.
3. V is the set of all ordered triples of real numbers of the form (0, y, z); addition and multiplication in V are defined as,

$$(0, y, z) \oplus (0, y', z') = (0, y + y', z + z')$$

and $\qquad c \cdot (0, y, z) = (0, 0, cz),$

prove that V is a vector space.

4. Verify that the set of all real continuous functions in an interval [a, b] forms a linear vector space over a field (real).

5. Verify that the set of all real functions f such that $f(x + 1) = f(x)$ is a vector space over the field of real numbers.
6. Prove that the set V of all functions from R to R in which addition and multiplication are defined as $(f + g)(x) = f(x) + g(x)$ and
$$(\alpha f)(x) = \alpha[f(x)] \text{ for all } f, g \in V, \text{ is a vector space over R.}$$
7. Prove that the set of all ordered pairs of real numbers (x, y) with the operations $(x, y) \oplus (x', y') = (x + x', y + y')$ and $c.(x, y) = (0, 0)$ is not a vector space.
8. Let V be the set of all ordered pairs (x, y) of real numbers and let F be the field of real numbers. Prove that V is not a vector space over the field of real numbers under addition and scalar multiplication defined as,
   (*i*) $(x, y) + (x_1, y_1) = (x + x_1, y + y_1)$
   $$c(x, y) = (|c|\, x, |c|\, y)$$
   (*ii*) $(x, y) + (x_1, y_1) = (x + x_1, y + y_1)$
   $$c(x, y) = (c^2 x, c^2 y).$$
9. Let V be the set of all ordered pairs (x, y) of real numbers and let F be the field of real numbers. Addition and multiplication in V are defined as,
$$(x, y) + (x_1, y_1) = (3y + 3y_1, -x - x_1),$$
$$a(x, y) = (3ay, -ax).$$
   Show that V, with these operations, is not a vector space over F.
10. Let V = $\{a + b\sqrt{2} : a, b \in Q\}$. Show that V is a vector space over Q under addition and multiplication.
11. Show that the following are not vector spaces :
   (*i*) R × R with usual addition and scalar multiplication defined by
   $$\alpha(a, b) = (0, \alpha b)$$
   (*ii*) R × R with addition defined by
   $$(a, b) + (c, d) = (ac, bd)$$
   and usual scalar multiplication.
12. Let V denote the set of all solutions of the differential equation $5\frac{d^2y}{dx^2} - 2\frac{dy}{dx} + 3y = 0$.
   Show that V is a vector space over R.
13. Let V = $\left\{\begin{pmatrix} a & 0 \\ 0 & b \end{pmatrix} : a, b \in R\right\}$ over R. Prove that the set V is a vector space under the usual addition and scalar multiplication in matrices.
14. Let V be the set of pairs (x, y) of real numbers and let F be the field of real numbers. Addition and scalar multiplication are defined as
$$(x, y) + (x_1, y_1) = (x + x_1, 0)$$
$$c(x, y) = (cx, 0).$$
   Is V, with these operations, a vector space ?

2.5 SUBSPACES

In this section, we shall introduce some basic concepts on the structure of a vector space. First, we consider a subset of a given vector space that is itself a vector space with respect to the same operations as those in V.

Definition. Let V be a vector space over a field F and W a nonempty subset of V. If W is a vector space over F with respect to the same operations as those in V, then W is called a *subspace* of V.

Remark. Every vector space has atleast two subspaces, the vector space itself and the subspace $\{o\}$ consisting only of the zero vector. The subspace $\{o\}$ is called the *zero subspace.* As an example, let W be subset of R^3 consisting of all vectors of the form $(a, b, 0)$, where a, b are any real numbers of the field F, with the usual operations of vector addition and scalar multiplication.

To check if W is a subspace of R^3, we have to check whether the axioms of vector spaces hold.

Let $$x = (x_1, x_2, 0)$$

and $$y = (y_1, y_2, 0)$$

be two vectors.

Then $$x + y = (x_1, x_2, 0) + (y_1, y_2, 0) = (x_1 + y_1, x_2 + y_2, 0)$$

is in W, since the third component is zero. Also if c is a scalar, then

$$cx = c(x_1, x_2, 0) = (cx_1, cx_2, 0) \text{ is in W.}$$

Thus vector addition and scalar multiplication are binary operations in W, *i.e.*, W is closed under these two operations. Other properties, of vector space, can also be verified. Hence, W is a subspace of R^3.

We now state and prove two theorems.

Theorem 2.1. *Let V be a vector space over a field F. A nonempty subset W of V is a subspace of V if and only if, W is closed with respect to vector addition and scalar multiplication in V.*

Proof. Let W be a subspace of V. Then W itself is a vector space and hence W is closed w.r.t. vector addition and scalar multiplication.

Conversely, let W be a non-empty subset of V

such that $$u, v \in W \Rightarrow u + v \in W$$

and $$u \in W \text{ and } \alpha \in F \Rightarrow \alpha u \in W.$$

We are to prove that W is a subspace of V. Since W is non-empty, there exists an element $u \in W$

$\therefore$ $$0u = \mathbf{0} \in W$$

Also $$v \in W \Rightarrow (-1)\, v = -v \in W.$$

Thus, W contains **0** and the additive inverse of each of its element.

Hence, W is an additive subgroup of V.

Also, $u \in W$ and $\alpha \in F \Rightarrow \alpha u \in W$

since the elements of W are the element of V, the other axioms of a vector space are true in W.

Hence W is a subspace of V.

Theorem 2.2. *Let V be a vector space over a field F. A nonempty subset W of V is a subspace of V if and only if, $u, v \in W$ and $\alpha, \beta \in F$, implies $\alpha u + \beta v \in W$.*

Proof. Let W be a subspace of V. Let $u, v \in W$ and $\alpha, \beta \in F$.

Then by Theorem 2.1, αu and $\beta v \in W$ and hence $\alpha u + \beta v \in W$.

Conversely, Let $u, v \in W$ and $\alpha, \beta \in F \Rightarrow \alpha u + \beta v \in W$.

Taking $\alpha = \beta = 1$, we get $u, v \in W \Rightarrow u + v \in W$.

Taking $\beta = 0$, we get $\alpha \in F$ and $u \in W \Rightarrow \alpha u \in W$.

Hence by Th. 2.1, W is a subspace of V.

2.5.1 Illustrative Examples

1. *Show that the set W of ordered triad $(a_1, a_2, 0)$ where $a_1, a_2, \in F$, a field, is a sub-space of V_3 over F.*

Sol. Let $u = (a_1, a_2, 0)$ and $v = (b_1, b_2, 0)$ belong to W, where $a_1, a_2, b_1, b_2 \in F$.

If a and b be any two elements of F, we have

$$\begin{aligned} au + bv &= a(a_1, a_2, 0) + b(b_1, b_2, 0) \\ &= (aa_1, aa_2, 0) + (bb_1, bb_2, 0) \\ &= (aa_1 + bb_1, aa_2 + bb_2, 0) \in W \end{aligned}$$

Since $(aa_1 + bb_1), (aa_2 + bb_2) \in F$.

Therefore, W is a vector subspace of V_3 over F.

2. *Let $V = \{(x, y, z) : x, y, z \in R\}$, where R is the field of real numbers. Show that, if $W = \{(x, y, z) : x - 3y + 4z = 0\}$, it is a subspace of V over R.*

Sol. Let $u, v \in W$, then we may write

$$u = (3y_1 - 4z_1, y_1, z_1), \qquad [\because x = 3y - 4z]$$

and $$v = (3y_2 - 4z_2, y_2, z_2).$$

If $a, b \in R$, we have

$$\begin{aligned} au + bv &= a(3y_1 - 4z_1, y_1, z_1) + b(3y_2 - 4z_2, y_2, z_2) \\ &= (3ay_1 - 4az_1, ay_1, az_1) + (3by_2 - 4bz_2, by_2, 3z_2) \\ &= [3(ay_1 + by_2) - 4(az_1 + bz_2), ay_1 + by_2, az_1 + bz_2] \\ &= (3l - 4m, l, m) \in W \end{aligned}$$

where $l = ay_1 + by_2 \in R$ and $m = az_1 + bz_2 \in R$.

Therefore, $a, b \in R$ and $u, v \in W \Rightarrow au + bv \in W$.

Hence, V is a vector sub-space of V in R.

3. *Let W = {f : f ∈ F[x] and f(a) = 0} i.e., W is the set of all polynomials in F[x] having 'a' as a root, where a ∈ F. Prove that W is a subspace over F.*

Sol. If a is a root of $f(x) = 0$ then $x - a \in$ W and hence W is non-empty.

Let $f, g \in F[x]$

and $\alpha, \beta \in F$.

To prove $\alpha f + \beta g \in W$,

we have to show that a is a root of $\alpha f + \beta g = 0$.

Now $(\alpha f + \beta g)(a) = \alpha f(a) + \beta g(a) = \alpha 0 + \beta 0 = 0.$

Hence, a is a root of $\alpha f + \beta g = 0$.

∴ $\alpha f + \beta g \in W$ and hence, W is a subspace of F[x].

4. *Prove that the intersection of two subspaces of a vector space is a subspace.*

Sol. Let A and B be two subspaces of a vector space V over a field F.

To prove, A∩B is a subspace of V.

Now, $0 \in A \cap B$ and hence A∩B is non-empty.

Let $u, v \in A \cap B$

and $\alpha, \beta \in F$.

Then $u, v \in A$

and $u, v \in B$.

∴ $\alpha u + \beta v \in A$ and $\alpha u + \beta v \in B$.

(since A and B are subspaces)

∴ $\alpha u + \beta v \in A \cap B$.

Hence A∩B is a subspace of V.

5. *Prove that the union of two subspaces of a vector space need not be a subspace.* [W.B.U.T., 2007]

Sol. Let $A = \{(a, 0, 0); a \in R\}$

$B = \{(0, b, 0); b \in R\}$

Let $u = (1, 0, 0)$ and $v = (a_2, 0, 0)$,

where $u, v \in A$

and $\alpha, \beta \in R$.

Then $\alpha u + \beta v = \alpha(a_1, 0, 0) + \beta(a_2, 0, 0)$

$= (\alpha a_1, 0, 0) + (\beta a_2, 0, 0)$

$= (\alpha a_1 + \beta a_2, 0, 0) \in A.$

Hence, A is a subspace of R^3.

Similarly, it can be proved that B is a subspace of R^3.

But A∪B is not a subspace of R^3.

for (1, 0, 0) and (0, 1, 0) ∈ A∪B,

but $\quad (1,0,0)+(0,1,0)=(1,1,0) \notin A \cup B.$

Hence, union of two subspaces of a vector space is not a subspace.

6. *Let W be the set of all points in R^3 satisfying the equation $ax + by + cz = 0$. Show that W is a subspace of R^3.*

Sol. Let $\quad u=(a_1, b_1, c_1)$ and $v=(a_2, b_2, c_2) \in W$ and $\alpha, \beta \in R$.

Then, we have

$$aa_1 + bb_1 + cc_1 = 0$$

and $$aa_2 + bb_2 + cc_2 = 0$$

Hence $$\alpha(aa_1 + bb_1 + cc_1) + \beta(aa_2 + bb_2 + cc_2) = 0$$

i.e., $$a(\alpha a_1 + \beta a_2) + b(\alpha b_1 + \beta b_2) + c(\alpha c_1 + \beta c_2) = 0$$

i.e., $\quad \alpha u + \beta v \in W$, so that W is a subspace of R^3.

7. *Show that $S = \{(x, y, z) : x^2 + y^2 = z^2\}$ is not a subspace of R^3.*

Sol. Let $u, v \in S$, then we have

$$u = \left(\sqrt{z_1^2 - y_1^2}, y_1, z_1\right) \text{ and } v = \left(\sqrt{z_2^2 - y_2^2}, y_2, z_2\right).$$

If $\quad \alpha, \beta \in R$,

then $$\alpha u + \beta v = \alpha\left(\sqrt{z_1^2 - y_1^2}, y_1, z_1\right) + \beta\left(\sqrt{z_2^2 - y_2^2}, y_2, z_2\right)$$

$$= \alpha\left(\sqrt{z_1^2 - y_1^2}, \alpha y_1, \alpha z_1\right) + \beta\left(\sqrt{z_2^2 - y_2^2}, \beta y_2, \beta z_2\right)$$

$$= \left(\alpha\sqrt{z_1^2 - y_1^2} + \beta\sqrt{z_2^2 - y_2^2}, \alpha y_1 + \beta y_2, \alpha z_1 + \beta z_2\right) \notin S$$

$\therefore \quad$ S is not a vector subspace of R^3.

8. *Prove that a set of polynomials of degree 2 need not be a subspace.*

Sol. Let V be the set of all polynomials of degree exactly equal to 2; V is a subset of P_2 (polynomials of degree 2), but it is not a subspace of P_2, since the sum of the polynomials $-ax^2 + bx + c$ and $ax^2 + 2kx + d$ is a polynomial of degree 1, is not in V.

9. *If a vector space V is the set of real valued continuous functions over R then show that the set W of solutions of the differential equation*

$$3\frac{d^2y}{dx^2} + 11\frac{dy}{dx} - 4y = 0$$

is a subspace of V.

Sol. Given that the set W contains the solutions of the given equation,

i.e., $$W = \left\{y : 3\frac{d^2y}{dx^2} + 11\frac{dy}{dx} - 4y = 0\right\}$$

where $$y = f(x)$$

$\because$ $y = 0$ satisfies the given equation, hence $0 \in$ W.

If y_1, y_2 be any two elements of W, then we have,

$$3\frac{d^2y_1}{dx^2} + 11\frac{dy_1}{dx} - 4y_1 = 0 \quad ...(1)$$

and $$3\frac{d^2y_2}{dx^2} + 11\frac{dy_2}{dx} - 4y_2 = 0 \quad ...(2)$$

Now for $p, q \in$ R, we obtain.

$$3\frac{d^2}{dx^2}[(py_1) + (qy_2)] + 11\frac{d}{dx}[py_1 + qy_2] - 4(py_1 + qy_2)$$

$$= 3p\frac{d^2y_1}{dx^2} + 3q\frac{d^2y_2}{dx^2} + 11p\frac{dy_1}{dx} + 11q\frac{dy_2}{dx} - 4py_1 - 4qy_2.$$

$$= p\left(3\frac{d^2y}{dx^2} + 11\frac{dy_1}{dx} - 4y_1\right) + q\left(3\frac{d^2y_2}{dx^2} + 11\frac{dy_2}{dx} - 4y_2\right) = p.0 + q.0 = 0$$

[Using (1) and (2)]

Hence, $(py_1 + qy_2)$ satisfies the given equation.

$\therefore$ $(py_1 + qy_2) \in$ W.

Thus the set W is a subspace of V.

EXERCISE 2.2

1. Show that the following subsets of R^3 are subspaces :
   (i) $\{a, 0, c) : a, c \in R\}$ (ii) $\{(a, b, c) : a = b = c\}$
   (iii) $\{(a, b, c) : a = b + c\}$.
2. Show that each of the following subsets of V_3 (R) is not a subspace :
   (i) $S = \{(x, y, z) : x^2 + y^2 + z^2 \le 1\}$
   (ii) $S = \{(x, y, z) : x + y + z = 1\}$
3. Which of the following subsets of R^3 are subspaces of R^3 ?
   The set of all vectors of the form
   (i) $(a, b, 3)$, (ii) (a, b, c) where $c > 0$
   (iii) (a, b, c), where $b = 2a + 1$, (iv) (a, b, c), where $a = c = 0$.
4. Which of the following subsets of P_2 (polynomials of degree 2) are subspaces ?
   The set of all polynomials of the form
   (i) $a_2x^2 + a_1x + a_0$, where $a_0 = 0$
   (ii) $a_2x^2 + a_1x + a_0$, where $a_0 = 2$
   (iii) $a_2x^2 + a_1x + a_0$, where $a_2 + a_1 = a_0$
   (iv) $a_2x^2 + a_1x + a_0$, where $a_2 + a_1 + a_0 = 2$

5. Which of the following subsets of the vector space M_{23} are subspaces ? The set of all matrices of the form

(i) $\begin{bmatrix} l & m & n \\ p & q & r \end{bmatrix}$, where $l = 2n + 1$

(ii) $\begin{bmatrix} 0 & 1 & a \\ b & c & 0 \end{bmatrix}$

(iii) $\begin{bmatrix} l & m & n \\ p & q & r \end{bmatrix}$, where $l + n = 0$ and $m + p + r = 0$.

6. Show that for the set $S \in R_{2\times 2}$, where S is defined by $S = \left\{ \begin{pmatrix} a & b \\ c & d \end{pmatrix} \in R_{2\times 2}; a + b = 0 \right\}$,

S is a subspace of $R_{2\times 2}$.

7. Show that for the subset $W \in R^3$, where
$$W = \{(a, b, c) \in R^3; b = c = 0\}, \text{ W is A subspace of } R^3.$$

8. Show that for the subset $W \in R^3$, where
$$W = \{(x, y, z) \in R^3; xy = z\}, \text{ W is not a subspace of } R^3.$$

9. Let $W = \{(z, 2y, 3z) : x, y, z \in R\}$, where R is the field of real numbers. Prove that W is a subspace of V_3 over R.

10. Let R^3 be the usual three dimensional vector space over the field R of real numbers. Prove that $P = \{(x, y, z) \in R^3, 2x - y + 3z = 0\}$ is a subspace of R^3.

Answers

3. (*iii*) **4.** (*i*), (*iii*) **5.** (*iii*).

We now state a theorem (without proof) as follows :

Theorem 2.3. *Let V be a vector space over a field F. The intersection of any collection of subspaces of V is a subspace V.*

The theorem has been proved for two subspaces in Example 4 (Illustrative Examples 2.5.1)

Example. *Let A and B be subspaces of a vector space V. Then* $A \cap B = \{0\}$ *iff every vector* $v \in A + B$ *can be uniquely expressed in the form* $v = \mathbf{a} + \mathbf{b}$, *where* $\mathbf{a} \in A$ *and* $\mathbf{b} \in B$.

Sol. Let $A \cap B = \{0\}$

and $v \in A + B$.

Let $v = a_1 + b_1 = a_2 + b_2$,

where $a_1, a_2 \in$ A and $b_1, b_2 \in$ B.

Then $a_1 - a_2 = b_2 - b_1$.

But $a_1 - a_2 \in$ A and $b_2 - b_1 \in$ B.

Hence, $a_1 - a_2, b_2 - b_1 \in A \cap B$.

Since $A \cap B = \{0\}$,

$$a_1 - a_2 = 0$$

and $b_2 - b_1 = 0$

so that $a_1 = a_2$

and $b_1 = b_2$.

Hence, the expression of v in the form $a + b$, where $a \in A$ and $b \in B$ is unique.

Conversely, let us suppose that any element in A + B can be uniquely expressed in the form $a + b$, where $a \in A$ and $b \in B$.

Then we claim that $A \cap B = \{0\}$.

If $A \cap B \neq \{0\}$, let $v \in A \cap B$ and $v \neq 0$.

Then $0 = v - v = 0 + 0$.

Thus 0 has been expressed in the form $a + b$ in two different ways which is a contradiction.

Hence, $A \cap B = \{0\}$.

This leads to a definition of *direct sum* of two subspaces of a vector space.

Definition. Let A and B be two subspaces of a vector space V. Then V is called the *direct sum* of A and B

if (*i*) A + B = V, (*ii*) $A \cap B = \{0\}$.

If V is the direct sum of A and B then it is written as

$$V = A \oplus B.$$

Remark. $V = A \oplus B$ iff, every element of V can be uniquely expressed in the form $a + b$, where $a \in A$ and $b \in B$.

Example 1. In V_3 (R), let $A = \{(a, b, 0) : a, b \in R\}$ and

$$B = \{(0, 0, c) : c \in R\}.$$

Then A and B are subspaces of V and $A \cap B = \{0\}$.

Let $v = (a, b, c) \in V_3(R)$.

Then $v = (a, b, 0) + (0, 0, c)$.

So that $A + B = V_3(R)$.

Hence, $V_3(R) = A \oplus B$.

Example 2. In M_3 (R), let A be the set of all matrices of the form $\begin{bmatrix} a & b \\ 0 & 0 \end{bmatrix}$ and B be the set of all matrices of the form $\begin{bmatrix} 0 & 0 \\ c & d \end{bmatrix}$, then A and B are subspaces of M_2 (R) and $A \cap B = \begin{bmatrix} 0 & 0 \\ 0 & 0 \end{bmatrix}$,

$A + B = M_2(R)$.

Hence, $\quad M_2(R) = A \oplus B.$

Remark. If $W_1, W_2, W_3, ..., W_n$ be subspaces of the vector space V over a field F then their sum is called *linear sum*, written as $W_1 + W_2 + W_3 + ... + W_n$, and is defined by

$$W_1 + W_2 + W_3 + ... + W_n = \{x_1 + x_2 + x_3 + ... + x_n : x_i \in W_i, i = 1, 2, ..., n\}.$$

We state and prove a theorem on linear sum as follows :

Theorem 2.4. *The linear sum $W_1 + W_2 + W_3 + ... + W_n$ of subspaces W_i $(i = 1, 2, ..., n)$ of a vector space V is also a subspace of V.*

Proof. Since each W_i is a subspace, so each W_i contains the zero vector.

$$\therefore \quad 0 + 0 + 0 + ... + 0 \in W_1 + W_2 + W_3 + ... + W_n.$$

Hence, the sum $W_1 + W_2 + ... + W_n$ is not empty.

Let x and y be any two elements of the sum

so that $\quad x = x_1 + x_2 + x_3 + ... + x_n$

and $\quad y = y_1 + y_2 + y_3 + ... + y_n,$

where $\quad x_i, y_i \in W_i \ (i = 1, 2, 3, ..., n).$

Then for any scalars λ and $\mu \in F$,

$$\lambda x + \mu y = \lambda(x_1 + x_2 + ... + x_n) + \mu(y_1 + y_2 + ... + y_n)$$

since W_i is a subspace of V, therefore,

$$(\lambda x_i + \mu y_i) \in W_i \text{ for each } i = 1, 2, 3, ..., n.$$

$$\therefore \quad (\lambda x + \mu y) \in (W_1 + W_2 + W_3 + ... + W_n) \text{ and}$$

hence $\quad W_1 + W_2 + ... + W_n \subseteq V.$

Thus, the sum of the subspaces forms a subspace of V.

Remark. From **Theorem 2.4,** it follows that if S is any collection of vectors in V, then there exists a smallest subspace of V which contains S, that is, a subspace which contains S and which is contained in every other subspace containing S.

This leads to the following definition :

Definition. Let S be a set of vectors in a vector space V. The *subspace spanned* by S is defined to be the intersection W of all subspaces of V which contain S. When S is a finite set of vectors, $S = \{a_1, a_2, ..., a_n\}$, we shall simply call W the *subspace spanned by the vectors* $a_1, a_2, ..., a_n$.

Earlier we constructed subspace W of a vector space V in R^2 over a field of scalars F by considering two vectors u and v of V and two scalars $c_1, c_2 \in F$ such that $w_1 = c_1 u + c_2 v$, where $w_1 \in W$,

which is known as linear combination of the vectors u and v. The construction that carried one for two vectors can now easily be performed for more than two vectors. We now give a formal definition.

2.6 LINEAR COMBINATION OF VECTORS

Definition. Let V be a vector space over a field F. Let $v_1, v_2, \ldots v_k$ be the vectors of V. A vector v in V is called a linear combination of $v_1, v_2, \ldots v_k$ if

$$v = c_1v_1 + c_2v_2 + \ldots + c_kv_k$$

for some real scalars $c_1, c_2, \ldots c_k$.

In **figure 2.1**, the linear combination of two vectors v_1 and v_2 in R^2 is shown geometrically.

Figure 2.2 shows v as a linear combination of v_1, v_2 and v_3

such that $$v = v_1 + 2v_2 - v_3$$

when, $$v_1 = (1, 2, 1), v_2 = (1, 0, 2), v_3 = (1, 1, 0)$$

and $$v = (2, 1, 5)$$

i.e., $$(2, 1, 5) = (1, 2, 1) + 2(1, 0, 2) - (1, 1, 0).$$

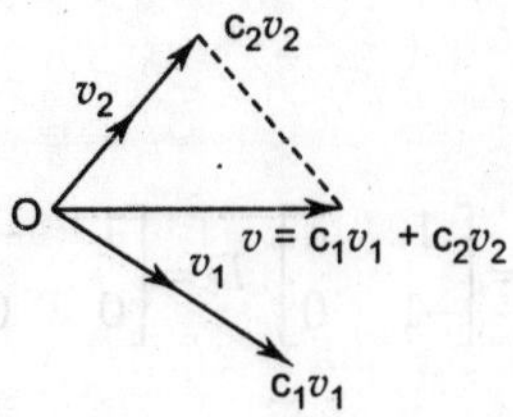

Figure 2.1

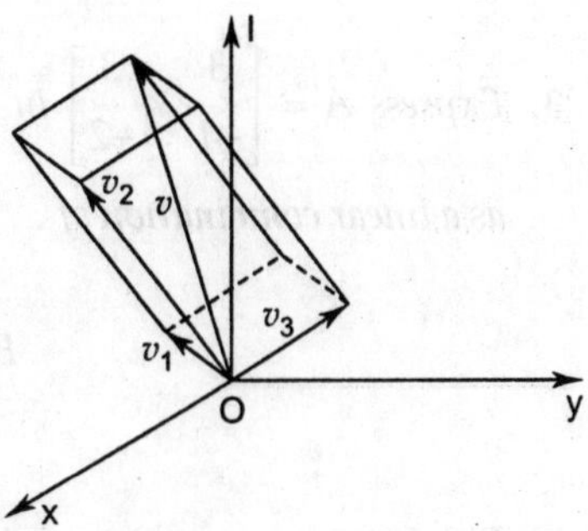

Figure 2.2

2.6.1 Illustrative Examples

1. *Express $v = (5, 9)$ as a linear combination of the vectors*

$$v_1 = (1, 2) \text{ and } v_2 = (2, 3).$$

Sol. Let $$v = c_1v_1 + c_2v_2$$

i.e., $$(5, 9) = c_1(1, 2) + c_2(2, 3), \text{ where } c_1 \text{ and } c_2 \text{ are scalars.}$$

$$= (c_1 + 2c_2, 2c_1 + 3c_2).$$

Equating the components from both sides, we have

$$c_1 + 2c_2 = 5$$

and $$2c_1 + 3c_2 = 9.$$

Solving the equations, we get $c_1 = 3, c_2 = 1$

Hence, $$(5, 9) = 3(1, 2) + 1(2, 3).$$

2. *Express $v = (3, 4, 5)$ as a linear combination of*

$$v_1 = (1, 2, 3), v_2 = (2, 3, 4) \text{ and } v_3 = (4, 3, 2).$$

Sol. Let $$v = c_1 v_1 + c_2 v_2 + c_3 v_3$$

i.e., $$(3, 4, 5) = c_1(1, 2, 3) + c_2(2, 3, 4) + c_3(4, 3, 2)$$
$$= (c_1 + 2c_2 + 4c_3, 2c_1 + 3c_2 + 3c_3, 3c_1 + 4c_2 + 2c_3).$$

Equating the components from both sides, we get

$$c_1 + 2c_2 + 4c_3 = 3$$
$$2c_1 + 3c_2 + 3c_3 = 4$$
and $$3c_1 + 4c_2 + 2c_3 = 5.$$

Solving these equations, we get

$$c_1 = \frac{1}{2}, c_2 = \frac{3}{4} \text{ and } c_3 = \frac{1}{4}$$

Hence, $$(3, 4, 5) = \frac{1}{2}(1, 2, 3) + \frac{3}{4}(2, 3, 4) + \frac{1}{4}(4, 3, 2).$$

3. *Express* $A = \begin{bmatrix} 3 & 3 \\ -1 & -2 \end{bmatrix}$ *in the vector space of* 2×2 *matrices as a linear combination of*

$$B = \begin{bmatrix} 1 & 1 \\ 0 & -1 \end{bmatrix}, C = \begin{bmatrix} 1 & 1 \\ -1 & 0 \end{bmatrix}, D = \begin{bmatrix} 1 & -1 \\ 0 & 0 \end{bmatrix}.$$

Sol. Let $$A = \lambda_1 \begin{bmatrix} 1 & 1 \\ 0 & -1 \end{bmatrix} + \lambda_2 \begin{bmatrix} 1 & 1 \\ -1 & 0 \end{bmatrix} + \lambda_3 \begin{bmatrix} 1 & -1 \\ 0 & 0 \end{bmatrix},$$

where λ_1, λ_2 and λ_3 are scalars.

$$\therefore \quad \begin{bmatrix} 3 & 3 \\ -1 & -2 \end{bmatrix} = \begin{bmatrix} \lambda_1 + \lambda_2 + \lambda_3 & \lambda_1 + \lambda_2 - \lambda_3 \\ -\lambda_2 & -\lambda_1 \end{bmatrix}.$$

Equating the elements of the matrices, we have

$$\lambda_1 + \lambda_2 + \lambda_3 = 3$$
$$\lambda_1 + \lambda_2 - \lambda_3 = 3$$
$$-\lambda_2 = -1$$
$$-\lambda_1 = -2.$$

From these, we get $\lambda_1 = 2, \lambda_2 = 1$

and $$\lambda_3 = 3 - \lambda_1 - \lambda_2 = 3 - 2 - 1 = 0$$

Hence, $$\begin{bmatrix} 3 & 3 \\ -1 & -2 \end{bmatrix} = 2 \begin{bmatrix} 1 & 1 \\ 0 & -1 \end{bmatrix} + 1 \begin{bmatrix} 1 & 1 \\ -1 & 0 \end{bmatrix} + 0 \begin{bmatrix} 1 & -1 \\ 0 & 0 \end{bmatrix}$$

4. *Express (5, 2, 1) as a linear combination of (1, 1, 0) and (3, 0, 1).*

Sol. Let
$$(5, 2, 1) = c_1(1, 1, 0) + c_2(3, 0, 1)$$
$$= (c_1 + 3c_2, c_1, c_2).$$

Equating the components of the vectors from both sides, we get
$$c_1 + 3c_2 = 5$$
$$c_1 = 2, c_2 = 1$$
$$\therefore \quad (5, 2, 1) = 2(1, 1, 0) + 1(3, 0, 1).$$

Remark. If $S = \{v_1, v_2, ..., v_k\}$ is a set of vectors in a vector space V, then the set of all vectors in V that are linear combinations of the vectors in S is denoted by *Span S* or *span* $\{v_1, v_2, ..., v_n\}$.

As an example, let the set S of 2×3 matrices given by

$$S = \left\{\begin{pmatrix}1 & 0 & 0\\0 & 0 & 0\end{pmatrix}, \begin{pmatrix}0 & 1 & 0\\0 & 0 & 0\end{pmatrix}, \begin{pmatrix}0 & 0 & 0\\0 & 1 & 0\end{pmatrix}, \begin{pmatrix}0 & 0 & 0\\0 & 0 & 1\end{pmatrix}\right\}.$$

Then span S is the set in M_{23} consisting of all vectors of the form

$$\lambda_1\begin{pmatrix}1 & 0 & 0\\0 & 0 & 0\end{pmatrix} + \lambda_2\begin{pmatrix}0 & 1 & 0\\0 & 0 & 0\end{pmatrix} + \lambda_3\begin{pmatrix}0 & 0 & 0\\0 & 1 & 0\end{pmatrix} + \lambda_4\begin{pmatrix}0 & 0 & 0\\0 & 0 & 1\end{pmatrix}$$

$$= \begin{pmatrix}\lambda_1 & \lambda_2 & 0\\0 & \lambda_3 & \lambda_4\end{pmatrix}, \text{ where } \lambda_1, \lambda_2, \lambda_3, \lambda_4 \text{ are real scalars.}$$

Now we state and prove a theorem.

Theorem 2.5. *The subspace spanned by a nonempty subset S of a vector space V is the set of all linear combinations of vectors in S.*

Proof. Let W be the subspace spanned by S. Then each linear combination
$$v = \lambda_1 v_1 + \lambda_2 v_2 + ... + \lambda_m v_m$$
of vectors $v_1, v_2, ..., v_m$ in S is also in W. Thus W contains the set L of all linear combinations of vectors in S. On the other hand, the set L contains S and hence the set L is nonempty. If u, v belong to L then u and v can be expressed as a linear combination of the vectors $u_1, u_2, u_3, ... u_n$ and $v_1, v_2, v_3, ..., v_m$ respectively.

i.e., $$u = \mu_1 u_1 + \mu_2 u_2 + ... + \mu_n u_n$$

and $$v = \lambda_1 v_1 + \lambda_2 u_2 + ... + \lambda_m v_m.$$

Now for each scalar k,

$$kv + u = \sum_{i=1}^{m} (k\lambda_i) v_i + \sum_{i=1}^{n} \mu_j u_j.$$

Hence $kv + u$ belongs to L. Thus L is a subspace of V. Thus, we have shown that L is a subspace of V which contains S, and also that any subspace which contains S contains L.

Hence, it follows that L is the intersection of all subspaces containing S, that is, L is the subspace spanned by the set S.

The above theorem leads to the definition of *linear span*.

Definition. Let V be a vector space over the field F and S be any nonempty subset of V. Then the *linear span* of S is defined as the set of all linear combinations of finite sets of elements of S. It is denoted by L (S).

Thus, $$L(S) = \{\lambda_1 u_1 + \lambda_2 u_2 + \dots + \lambda_n u_n,\ u_i \in S, \lambda_i \in F (i = 1, 2, \dots, n\}$$

L(S) is said to be *generated* or *spanned* by the set S and S is said to be the set of *generators* of L(S).

As an example, the set $L(S) = \left\{\begin{pmatrix} \lambda_1 & 0 \\ 0 & \lambda_2 \end{pmatrix}\right\}$ of all 2×2 diagonal matrices is the linear span of $\begin{pmatrix} 1 & 0 \\ 0 & 0 \end{pmatrix}$ and $\begin{pmatrix} 0 & 0 \\ 0 & 1 \end{pmatrix}$.

2.7 BASES AND DIMENSION

We now move to the task of assigning a *dimension* to certain vector spaces. Although the term 'dimension' generally associated with something geometrical, we try to find a suitable algebraic definition of the dimension of a vector space. It will be done through the concept of a *basis* for the space. The concepts of dimension and basis depend on linear combination of vectors, *linearly dependent* and *linearly independent* of vectors. In the above lines we discussed the linear combination of vectors. We now define linearly dependent and linearly independent of vectors.

Definition. Let V be a vector space over F. A subset S of V is said to be *linearly dependent* (*or simply, dependent*) if there exist distinct vectors $a_1, a_2, \dots, a_n$ in S and scalars $\lambda_1, \lambda_2, \dots, \lambda_n$ in F, not all of which are 0, such that

$$\lambda_1 a_1 + \lambda_2 a_2 + \dots \lambda_n a_n = 0.$$

On the other hand, if

$$\lambda_1 a_1 + \lambda_2 a_2 + \dots + \lambda_n a_n = 0$$

provided all $\lambda_i = 0, i = 1, 2, \dots, n$, then the vectors of the subset S are called *linearly independent*.

The following are the consequences of the definition :

(*i*) Any set which contains a linearly dependent set is linearly dependent.

(*ii*) Any subset of a linearly independent set is linearly independent.

(*iii*) Any set which contains the **0** vector is linearly dependent.

(*iv*) A set S of vectors is linearly independent if and only if each finite subset of S is linearly independent, that is, iff, for any distinct vectors $a_1, a_2, \dots, a_n$ of S, $\lambda_1 a_1 + \lambda_2 a_2 + \dots \lambda_n a_n = 0$ implies $c_i = 0$.

(v) If a vector can be expressed as a scalar multiple of another vector then the vectors are linearly dependent *i.e.*, if $b = \lambda a$, then the vectors a and b are linearly dependent.

2.7.1 Illustrative Examples

1. *Show that in V_n (F), the set of n unit vectors $\{e_1, e_2, \ldots e_n\}$ is a linearly independent set of vectors.*

Sol. The linear combination of the n unit vectors,

$$\lambda_1 e_1 + \lambda_2 e_2 + \ldots + \lambda_n e_n = 0$$

i.e., $\lambda_1(1, 0, 0, \ldots, 0) + \lambda_2(0, 1, 0, \ldots, 0) + \ldots + \lambda_n(0, 0, 0, \ldots, 1) = (0, 0, \ldots, 0)$

is possible only when

$$\lambda_1 = \lambda_2 = \ldots = \lambda_n = 0.$$

Hence, the set of n unit vectors are linearly independent.

2. *Prove that in V_3 (R) the vectors (1, 2, 1), (2, 1, 0) and (1, –1, 2) are linearly independent.*

Sol. Let $\lambda_1(1, 2, 1) + \lambda_2(2, 1, 0) + \lambda_3(1, -1, 2) = (0, 0, 0)$

or $(\lambda_1 + 2\lambda_2 + \lambda_3, 2\lambda_1 + \lambda_2 - \lambda_3, \lambda_1 + 2\lambda_3) = (0, 0, 0)$

$$\therefore \quad \lambda_1 + 2\lambda_2 + \lambda_3 = 0 \quad \ldots(i)$$

$$2\lambda_1 + \lambda_2 - \lambda_3 = 0 \quad \ldots(ii)$$

$$\lambda_1 + 2\lambda_3 = 0 \quad \ldots(iii)$$

Solving equations (*i*), (*ii*) and (*iii*), we get

$$\lambda_1 = \lambda_2 = \lambda_3 = 0.$$

Hence, the given vectors are linearly independent.

3. *How the following vectors are related ?*

(i) $a = (2, 1, -3), b = (4, 2, -6)$

(ii) $a = (3, 2, 4), b = (5, 3, 1)$.

Sol. (*i*) The vectors a and b are linearly dependent,

since $\quad b = (4, 2, -6) = 2(2, 1, -3) = 2a$

i.e., $\quad b = 2a$.

(*ii*) Vectors a and b are linearly independent, since a and b cannot be expressed one as a scalar multiple of the other.

4. *In V_3 (R), prove that the vectors (1, 4, –2), (–2, 1, 3) and (–4, 11, 5) are linearly dependent.*

Sol. Let $\lambda_1(1, 4, -2) + \lambda_2(-2, 1, 3) + \lambda_3(-4, 11, 5) = 0$

$$\therefore \quad \lambda_1 - 2\lambda_2 - 4\lambda_3 = 0 \quad \ldots(i)$$

$$4\lambda_1 + \lambda_2 + 11\lambda_3 = 0 \quad \ldots(ii)$$

$$-2\lambda_1 + 3\lambda_2 + 5\lambda_3 = 0 \quad \ldots(iii)$$

From (*i*) and (*ii*), $\quad \dfrac{\lambda_1}{-18} = \dfrac{\lambda_2}{-27} = \dfrac{\lambda_3}{9}$

or $$\frac{\lambda_1}{2} = \frac{\lambda_2}{3} = \frac{\lambda_3}{-1} = k \text{ (say)}$$

$\therefore$ $$\lambda_1 = 2k, \lambda_2 = 3k, \lambda_3 = -k.$$

Taking $$k = 1, \lambda_1 = 2, \lambda_2 = 3, \lambda_3 = -1$$

which satisfy eqn (*iii*).

$\because$ λ_1, λ_2 and λ_3 are non-zero scalars, therefore, the vectors are linearly dependent.

Remark. Some times we use the concept of the rank of a matrix to test the linearly dependent and independent of vectors.

If the matrix formed by the given vectors is nonsingular then the vectors are linearly independent but if the matrix is singular then the given vectors are linearly dependent.

5. *Show that the vectors (3, 0, 2), (7, 0, 9) and (4, 1, 2) are linearly independent.*

Sol. Let A be the matrix represented by the given vectors.

i.e., $$A = \begin{pmatrix} 3 & 0 & 2 \\ 7 & 0 & 9 \\ 4 & 1 & 2 \end{pmatrix}$$

$\therefore$ The determinant formed by the elements of the matrix A, is

$$|A| = \begin{vmatrix} 3 & 0 & 2 \\ 7 & 0 & 9 \\ 4 & 1 & 2 \end{vmatrix} = 3(-9) + 2(7) = -27 + 14 = -13 \neq 0.$$

Thus the matrix formed by the elements of the given vectors is non-singular and its rank = 3.

Hence, the given vectors are linearly independent.

6. *Show that the vectors (2, 2, 8), (1, 0, 4), (1, 2, 4) are linearly dependent.*

Sol. Let A be the matrix represented by the given vectors,

i.e., $$A = \begin{pmatrix} 2 & 2 & 8 \\ 1 & 0 & 4 \\ 1 & 2 & 4 \end{pmatrix}.$$

The determinant formed by the given vectors is

$$|A| = \begin{vmatrix} 2 & 2 & 8 \\ 1 & 0 & 4 \\ 1 & 2 & 4 \end{vmatrix} = 2(0-8) - 2(4-4) + 8(2-0) = -16 - 0 + 16 = 0.$$

Since $|A| = 0$, therefore the matrix formed by the elements of the given vectors is singular and its rank is < 3.

Hence, the given vectors are linearly dependent.

We are now in a position to define *basis* and *dimension* of a vector space.

Definition. Let V be a vector space. A *basis* for V is a linearly independent set of vectors in V which spans the space V.

The number of elements (vectors) present in the basis of a vector space V is known as the *dimension* of V.

The dimension of V is denoted by ***dim* V**.

The space V is *finite dimensional* if it has a finite basis. The vector space V is called *n*-dimensional, if in its basis contains *n* elements (vectors).

As an example, if S = {(1, 0, 0), (0, 1, 0), (0, 0, 1)} is the basis of the vector space R^3, then dimension of R^3 is 3, since its basis contains three linearly independent vectors.

Remarks. 1. A vector space which is not finitely generated is called an infinite dimensional space.

2. The vector space {0} is of zero dimension.

3. A vector space may have more than one basis, that is, basis of a vector space is not unique.

4. The set S = $\{e_1, e_2, \ldots e_n\}$, of unit vectors, is a basis for E^n which is known as *'standard basis'* of E^n.

We need to state and prove few Theorems involving basis and dimension.

Theorem 2.6. *(Existence of basis)*

There exists a basis for each finite dimensional vector space.

Proof. Let V be a finite dimensional vector space over the field F. Then by definition V is linear span of the set S having finite number of vectors belonging to V.

Let $$S = \{a_1, a_2, \ldots, a_n\} \subseteq V \text{ and } V = L(S).$$

If we assume that no member of S is zero, that is $\mathbf{0} \notin S$, then it will not affect the generality of the theme, because the contribution of **0** in the linear combination of the elements of S is zero.

Thus $S \subseteq V \Rightarrow$ S is linearly independent or S is linearly dependent.

Let S be linearly independent, then S will be a basis of V, implies basis of a finite dimensional vector space exists.

Next, let S be linearly dependent.

We know that a set S of non-zero vectors $\{a_1, a_2, \ldots, a_n\}$ form a vector space V over F is linearly dependent, if and only if, some vectors $a_k (k > 1)$ belonging to S is a linear combination of the preceding vectors

$a_1, a_2, \ldots, a_{k-1}$, in S.

Let $$a_k = \lambda_1 a_1 + \lambda_2 a_2 + \ldots + \lambda_{k-1} a_{k-1} \qquad \ldots(i)$$

Let $$S_1 = \{a_1, a_2, \ldots, a_{k-1}, a_{k+1}, \ldots, a_n\}.$$

But $\quad S = \{a_1, a_2, ..., a_{k-1}, a_k, a_{k+1}, ..., a_n\}$.

Hence, $\quad S_1 \subseteq S \Rightarrow L(S_1) \subseteq L(S)$.

But $\quad L(S) = V$,

$\therefore \quad L(S_1) \subseteq V$ ...(*ii*).

Now, we prove $V \subseteq L(S_1)$.

Let $\quad a \in V$ and

a = a linear combination of the vectors $a_1, a_2, ..., a_k, ..., a_n$

= linear combination of $(a_1, a_2, ..., a_{k-1}, a_{k+1}, ... a_n)$

= linear combination of the elements of $S_1 + \lambda_k$(*l.c.* of $a_1, a_2, ... a_{k-1}$)

= *l.c.* of the elements of S_1

and hence $a \in L(S_1)$

$\therefore \quad a \in V \Rightarrow a \in L(S_1)$, hence $V \subseteq L(S_1)$.

Thus $\quad L(S_1) \subseteq V$ and $V \subseteq L(S_1) \Rightarrow V = L(S_1)$.

Let S_1 be linearly independent, then S_1 will be basis of V, implies basis of a finite dimensional vector space exists.

If S_1 be linearly dependent, then as before we get a set of $(n-2)$ vectors which generate the vector space V(F).

Continuing this process we get a linearly independent subset of S which generates V(F) and is therefore a basis of V(F).

Theorem 2.7. *Any vector of a given vector space can be expressed as a linear combination of the vectors belonging to a given basis in one and only one way.*

Proof. Let us consider a set of vectors say, $a_1, a_2, ... a_n$ which form a basis of the vector space V_n. Let ϑ be a vector of V_n and ϑ can be expressed as a linear combination with the set of vectors $a_1, a_2, ..., a_n$, that is

$$\vartheta = \lambda_1 a_1 + \lambda_2 a_2 + ... + \lambda_n a_n \qquad ...(1)$$

where λ_i's are scalars. Since the sum of any two vectors, and the scalar multiplication of any vector of V_n is a vector of V_n, so ϑ is a vector of V_n.

If possible, let us now assume another representation of ϑ as,

$$\vartheta = \mu_1 a_1 + \mu_2 a_2 + ... + \mu_n a_n \qquad ...(2)$$

where μ_i's are scalars.

Therefore, from (1) and (2) we have

$$\lambda_1 a_1 + \lambda_2 a_2 + ... + \lambda_n a_n = \mu_1 a_1 + \mu_2 a_2 + ... + \mu_n a_n$$

or, $(\lambda_1 - \mu_1)\, a_1 + (\lambda_2 - \mu_2)\, a_2 + ... + (\lambda_n - \mu_n)\, a_n = 0$.

Since $a_1, a_2, ..., a_n$ form the basis of V_n, so $a_1, a_2, ..., a_n$ are linearly independent.

$\therefore \quad \lambda_1 - \mu_1 = 0, \lambda_2 - \mu_2 = 0, ..., \lambda_n - \mu_n = 0$

i.e., $\lambda_1 = \mu_1, \lambda_2 = \mu_2, ..., \lambda_n = \mu_n.$

Hence, the representation of ϑ is unique.

Theorem 2.8. *Let V be a vector space which is spanned by a finite set of vectors $a_1, a_2, ..., a_m$. Then any independent set of vectors in V is finite and contains no more than m vectors.*

Proof. If we can show that every subset S of V which contains more than m vectors is linearly dependent then it suffices to prove the Theorem.

Let S be such a set. In S, let us consider there are n distinct vectors $\alpha_1, \alpha_2, ..., \alpha_n$, where $n > m$. Since, $a_1, a_2, ..., a_m$ span V, there exist scalars λ_{ij} in F such that

$$\alpha_j = \sum_{i=1}^{m} \lambda_{ij}\, a_i$$

For any n scalars $c_1, c_2, ..., c_n$, we have

$$c_1\alpha_1 + c_2\alpha_2 + ... + c_n\alpha_n = \sum_{j=1}^{n} c_j\, \alpha_j = \sum_{j=1}^{n} c_j \sum_{i=1}^{m} \lambda_{ij}\, a_i$$

$$= \sum_{j=1}^{n}\sum_{i=1}^{m} (\lambda_{ij}\, c_j)\, a_i = \sum_{i=1}^{m} \left(\sum_{j=1}^{n} \lambda_{ij}\, c_j \right) a_i$$

Since $n > m$, there exist scalars $c_1, c_2, ..., c_n$, not all 0, such that

$$\sum_{j=1}^{n} \lambda_{ij}\, c_j = 0, 1 \le i \le m.$$

Hence, $c_1\alpha_1 + c_2\alpha_2 + ... + c_n\alpha_n = 0$.

This proves that S is a linearly dependent set.

Corollary. 2.1. *If V is a finite-dimensional vector space, then any two bases of V have the same (finite) number of elements.*

Proof. Since V in finite-dimensional, it has a finite basis $\{a_1, a_2, ..., a_m\}$.

By Theorem 2.8 every basis of V is finite and contains, no more than m elements. Thus if $\{\alpha_1, \alpha_2, ..., \alpha_n\}$ is a basis, then $n \le m$. By the same argument, $m \le n$.

Hence, $m = n.$

In the above lines, we have shown that the number of elements in a basis is the same. But basis of a vector space may not be unique. We prove it by the following theorem.

Theorem 2.9. *The basis of a vector space V is not unique.*

Proof. Let $\{a_1, a_2, ..., a_m\}$ be a basis of V.

Let $\alpha = c_1 a_1 + c_2 a_2 + ... + c_m\, a_m$, and $c_m \ne 0$, (c_i's are scalars).

Let us take another set of vectors $\{a_1, a_2, ..., a_{m-1}, \alpha\}$.

Now $d_1a_1 + d_2a_2 + ... + d_{m-1}a_{m-1} + d_m\alpha$

$$= (d_1a_1 + d_2a_2 + ... + d_{m-1}a_{m-1}) + d_m(c_1a_1 + c_2a_2 + ... + c_m a_m)$$

(where d_i's are scalars)

$$= (d_1 + d_m c_1)a_1 + (d_2 + d_m c_2)a_2 + ... + (d_{m-1} + d_m c_{m-1})a_{m-1} + d_m c_m a_m$$

since $a_1, a_2, ..., a_{m-1}, a_m$ are linearly independent,

therefore, $d_1a_1 + d_2a_2 + ... + d_{m-1}a_{m-1} + d_m\alpha = 0$

which is possible, when

$$d_1 + d_m c_1 = 0, d_2 + d_m c_2 = 0, ..., d_m c_m = 0$$

i.e., when $d_m = 0, d_{m-1} = 0, ..., d_2 = 0, d_1 = 0,$ $(\because c_m \neq 0)$.

Hence, $a_1, a_2, ..., a_{m-1}, \alpha$ are linearly independent.

Next, let the vector space generated by the above mentioned vectors be denoted by V′.

Let us take a vector ϑ of V′.

Then
$$\vartheta = \lambda_1a_1 + \lambda_2a_2 + ... + \lambda_{m-1}a_{m-1} + \lambda_m\alpha$$
$$= \lambda_1a_1 + \lambda_2a_2 + ... + \lambda_{m-1}a_{m-1} + \lambda_m(c_1a_1 + c_2a_2 + ... + c_m a_m)$$
$$= \mu_1a_1 + \mu_2a_2 + ... + \mu_{m-1}a_{m-1} + \lambda_m a_m.$$

where
$$\mu_1 = \lambda_1 + c_1\lambda_m, \mu_2 = \lambda_2 + c_2\lambda_m, ...,$$
$$\mu_{m-1} = \lambda_{m-1} + c_{m-1}\lambda_m \text{ and } \mu_m = c_m\lambda_m.$$

Hence $\vartheta \in$ V.

Thus every vector of V′ is also a vector V.

Similarly, it can be shown that every vector of V is also a vector of V′.

Hence, it follows that V = V′.

Thus, $a_1, a_2, ..., a_{m-1}, a_m$ and $a_1, a_2, ..., a_{m-1}, \alpha$ form two bases of the same vector space V. This proves the theorem.

We shall state now some useful theorems without proof.

Theorem 2.10. *If W is a subspace of a finite-dimensional vector space V, every linearly independent subset of W is finite and is part of a (finite) basis for W.*

Theorem 2.11. *If W is a proper subspace of a finite-dimensional vector space V, then W is finite-dimensional and dim W < dim V.*

Theorem 2.12. *If W_1 and W_2 are finite-dimensional subspaces of a vector space V, then $W_1 + W_2$ is finite-dimensional and*

$$\dim W_1 + \dim W_2 = \dim(W_1 \cap W_2) + \dim(W_1 + W_2).$$

Corollary 2.2. Let S be a linearly independent subset of a vector space V. If ϑ is a vector in V which is not in the subspace spanned by S, then the set obtained by joining ϑ to S is linearly independ.

Corollary 2.3. In a finite-dimension vector space V every nonempty linearly independent set of vectors is a part of a basis.

2.7.2 Illustrative Examples

1. *Show that the vectors (3, 0, 2), (7, 0, 9) and (4, 1, 2) form a basis for E^3.*

Sol. We have already proved that the given vectors are linearly independent, (see Example 5 of 2.7.1 Illustrative Examples).

Now we have to show that the vectors span (E^3). To show it we require to prove that for any vector (a_1, a_2, a_3) in E^3, it is possible to find real scalars $\lambda_1, \lambda_2, \lambda_3$, such that

$$(a_1, a_2, a_3) = \lambda_1(3, 0, 2) + \lambda_2(7, 0, 9) + \lambda_3(4, 1, 2)$$

Now, $$(a_1, a_2, a_3) = (3\lambda_1 + 7\lambda_2 + 4\lambda_3, \lambda_2, 2\lambda_1 + 9\lambda_2 + 2\lambda_3)$$

Equating the like components from both sides, we get

$$3\lambda_1 + 7\lambda_2 + 4\lambda_3 = a_1$$

$$\lambda_2 = a_2$$

and $$2\lambda_1 + 9\lambda_2 + 2\lambda_3 = a_3.$$

From these equations, real values of $\lambda_1, \lambda_2, \lambda_3$ can be found. Hence the given vectors form a basis for E^3.

2. *Find the basis of all real solutions of the differential equation*

$$\frac{d^2y}{dx^2} + w^2y = 0.$$

Sol. All real solutions of the differential equation can be written as

$$y = a \sin wx + b \cos wx,$$

where a and b are real numbers. The functions $\sin wx$ and $\cos wx$ are two linearly independent solutions of the equation.

Hence $\{\sin wx, \cos wx\}$ is the basis.

3. *Show that the vectors (2, 2, 8), (1, 0, 4) and (1, 2, 4) do not form a basis for E^3.*

Sol. To form a basis the vectors must be linearly independent. We have shown that the given vectors are linearly dependent (see Example 6 of 2.7.1 Illustrative Examples). Hence, the vectors do not form a basis for E^3.

4. *Find a basis for E^3 that contains the vectors (1, 2, 0) and (2, 0, 1).*

Sol. Let the given vectors be

$$a_1 = (1, 2, 0) \text{ and } a_2 = (2, 0, 1).$$

The standard basis for E^3 is $\{e_1, e_2, e_3\}$,

where $$e_1 = (1, 0, 0), e_2 = (0, 1, 0) \text{ and } e_3 = (0, 0, 1)$$

$$a_1 = (1, 2, 0) = 1e_1 + 2e_2 + 0.e_3$$

$\therefore$ a_1 can replace any one vector from $\{e_1, e_2, e_3\}$ to form a new basis. Let us consider a new basis $\{a_1, e_2, e_3\}$.

Let $$a_2 = (2, 0, 1) = \lambda_1(1, 2, 0) + \lambda_2(0, 1, 0) + \lambda_3(0, 0, 1).$$

Equating the components from both sides, we get

$$\lambda_1 = 2,\ 2\lambda_1 + \lambda_2 = 0 \text{ and } \lambda_3 = 1$$

i.e., $$\lambda_1 = 2,\ \lambda_2 = -4,\ \lambda_3 = 1$$

$\therefore$ $$(2, 0, 1) = 2(1, 2, 0) - 4(0, 1, 0) + 1(0, 0, 1)$$
$$= 2a_1 - 4e_2 + e_3$$

$\therefore$ a_2 can replace e_1 or e_3 from $\{a_1, e_2, e_3\}$ to form a new basis.

Hence, the new basis is $\{a_1, a_2, e_3\}$.

5. *Prove that the vectors (1, 1, 1), (1, 1, 0) and (1, 0, 0) form a basis in E^3. Also prove that the vector (1, 3, 1) can replace any one of the three vectors of the basis to form a new basis.*

Sol. Let A be the matrix represented by the given vectors,

i.e., $$A = \begin{bmatrix} 1 & 1 & 1 \\ 1 & 1 & 0 \\ 1 & 0 & 0 \end{bmatrix}. \text{ Then } |A| = \begin{vmatrix} 1 & 1 & 1 \\ 1 & 1 & 0 \\ 1 & 0 & 0 \end{vmatrix} = -1.$$

Hence, the rank of A is 3 (since the matrix is non-singular).

Therefore, the vectors forming A are linearly independent. Now, we have to show that the vectors span E^3.

Let (a_1, a_2, a_3) be a vector of E^3 and let

$$(a_1, a_2, a_3) = \lambda_1(1, 1, 1) + \lambda_2(1, 1, 0) + \lambda_3(1, 0, 0)$$
$$= (\lambda_1 + \lambda_2 + \lambda_3, \lambda_1 + \lambda_2, \lambda_1).$$

Equating the corresponding components from both sides, we get

$$\lambda_1 + \lambda_2 + \lambda_3 = a_1$$
$$\lambda_1 + \lambda_2 = a_2$$
and $$\lambda_1 = a_3.$$

From these equations real values of $\lambda_1, \lambda_2, \lambda_3$ can be found.

Hence, the vector span E^3.

$\therefore$ The given vectors form a basis for E^3.

Again, let $$(1, 3, 1) = \mu_1(1, 1, 1) + \mu_2(1, 1, 0) + \mu_3(1, 0, 0)$$

where μ_1, μ_2, μ_3 are scalars.

Equating like components from both sides, we have

$$\mu_1 + \mu_2 + \mu_3 = 1$$
$$\mu_1 + \mu_2 = 3$$
$$\mu_1 = 1$$

Solving these equations, we obtain

$$\mu_1 = 1, \mu_2 = 2, \mu_3 = -2.$$

Since μ_1, μ_2 and μ_3 are non-zero scalars and (1, 3, 1) is a non-null vector, hence (1, 3, 1) together with any two vectors of the given basis can form a new basis.

6. *Find the dimension of the subspace spanned by the vectors (1, 0, 2), (2, 0, 1), (1, 0, 1) in V_3 (R).*

Sol. Let A be the matrix represented by the given vectors,

i.e.,
$$A = \begin{pmatrix} 1 & 0 & 2 \\ 2 & 0 & 1 \\ 1 & 0 & 1 \end{pmatrix}.$$

Then
$$|A| = \begin{vmatrix} 1 & 0 & 2 \\ 2 & 0 & 1 \\ 1 & 0 & 1 \end{vmatrix} = 0$$

$\therefore$ A is singular. Hence the given three vectors forming A are linearly dependent and so the given three vectors do not form the basis of the subspace in V_3 (R).

Now, let $\lambda_1(1, 0, 2) + \lambda_2(2, 0, 1) = \mathbf{0}$

i.e., $(\lambda_1 + 2\lambda_2, 0, 2\lambda_1 + \lambda_2) = \mathbf{0} = (0, 0, 0).$

Equating like components from both sides, we get

$$\lambda_1 + 2\lambda_2 = 0$$

$$2\lambda_1 + \lambda_2 = 0.$$

These two equations have no other solutions except $\lambda_1 = 0$ and $\lambda_2 = 0$.

Since λ_1 and λ_2 are zero, so the vectors (1, 0, 2) and (2, 0, 1) are linearly independent.

So (1, 0, 2) and (2, 0, 1) form the basis of the subspace of V_3 (R).

Hence, the dimension of the basis is 2.

7. *Show that the vectors (1, 0, 0), (0, 1, 0), (0, 0, 1) and (1, 2, 3) generate the same space as generated by the vectors (1, 0, 0), (0, 1, 0) and (0, 0, 1).*

Sol. Let us consider the linear combination of the set of vectors (1, 0, 0), (0, 1, 0), (0, 0, 1), (1, 2, 3).

$$\lambda_1(1, 0, 0) + \lambda_2(0, 1, 0) + \lambda_3(0, 0, 1) + \lambda_4(1, 2, 3)$$
$$= (\lambda_1 + \lambda_4, \lambda_2 + 2\lambda_4, \lambda_3 + 3\lambda_4)$$
$$= (\lambda_1 + \lambda_4)(1, 0, 0) + (\lambda_2 + 2\lambda_4)(0, 1, 0) + (\lambda_3 + 3\lambda_4)(0, 0, 1).$$

Thus the two sets of vectors generate the same space.

8. *If u, v, w be linearly independent vectors of V (F), where F is the field of real numbers, show that (u + v), (v + w), (w + u) are also linearly independent.*

Sol. Since *u*, *v* and *w* are linearly independent, therefore, their linear combination

$$\lambda u + \mu v + \vartheta w = 0$$

implies, $\lambda = \mu = \vartheta = 0.$

Now $\lambda_1(u+v) + \lambda_2(v+w) + \lambda_3(w+u)$

$$= (\lambda_1 u + \lambda_2 v + \lambda_3 w) + (\lambda_3 u + \lambda_1 v + \lambda_2 w)$$

Since u, v, w are linearly independent,

therefore

$$\lambda_1 u + \lambda_2 v + \lambda_3 w = 0$$

and $\lambda_3 u + \lambda_1 v + \lambda_2 w = 0$

implies, $\lambda_1 = \lambda_2 = \lambda_3 = 0.$

Hence, $\lambda_1(u+v) + \lambda_2(v+w) + \lambda_3(w+u) = 0$

where $\lambda_1 = \lambda_2 = \lambda_3 = 0.$

$\therefore$ The vectors $u+v, v+w$ and $w+u$ are linearly independent.

9. *Show that* $S = \left\{\begin{pmatrix}1 & 0\\0 & 0\end{pmatrix}, \begin{pmatrix}0 & 1\\1 & 0\end{pmatrix}, \begin{pmatrix}0 & 0\\0 & 1\end{pmatrix}\right\}$ *spans the subspace of* M_{22} *consisting of all symmetric matrices.*

Sol. An arbitrary symmetric matrix has the form $A = \begin{bmatrix}a & b\\b & c\end{bmatrix}$, where a, b, c are any real numbers.

Let $$\begin{bmatrix}a & b\\b & c\end{bmatrix} = \lambda_1\begin{bmatrix}1 & 0\\0 & 0\end{bmatrix} + \lambda_2\begin{bmatrix}0 & 1\\1 & 0\end{bmatrix} + \lambda_3\begin{bmatrix}0 & 0\\0 & 1\end{bmatrix},$$

where $\lambda_1, \lambda_2, \lambda_3$ are scalars.

$$= \begin{bmatrix}\lambda_1 & \lambda_2\\\lambda_2 & \lambda_3\end{bmatrix}.$$

The solution is $\lambda_1 = a, \lambda_2 = b, \lambda_3 = c$

Since we have obtained a solution for every choice of a, b and c, we conclude that S spans the given subspace.

EXERCISE 2.3

1. Express (3, –5) as a linear combination of a = (1, 2), b = (–3, 1) in V_2 (R).
2. Express the vector (1, 7, –4) as a linear combination of the vectors (1, –3, 2), (2, –1, 1) in V_3 (R).
3. Express (1, 1, 2) as a linear combination of (0, 2, 1), (2, 2, 4) in V_3 (R).

4. Show that the vector (3, –4, 2) cannot be expressed as a linear combination of the vectors, (2, 1, 5) and (–1, 3, 2).
5. Show that the vector (0, 2, 1) cannot be expressed as a linear combination of the vectors (1, 1, 2) and (2, 2, 4).
6. Can the matrix $M = \begin{bmatrix} 2 & 0 \\ 4 & -5 \end{bmatrix}$ be expressed as the linear combination of the matrices

$$P = \begin{bmatrix} 0 & -3 \\ 2 & 0 \end{bmatrix}, Q = \begin{bmatrix} 0 & 0 \\ 2 & 1 \end{bmatrix} \text{ and } R = \begin{bmatrix} 2 & 3 \\ 0 & 5 \end{bmatrix}?$$

7. Show that the following vectors are linearly dependent.
   (i) (4, 3), (8, 6)
   (ii) $\left(\frac{3}{2}, \frac{1}{3}, 1\right), (9, 2, 6)$
   (iii) (1, 1, 2), (2, 1, 0), (0, 4, 1)
   (iv) (1, 2, 3), (4, 1, 5), (–4, 6, 2).
8. Show that the set of matrix $\left\{\begin{pmatrix} 2 & 0 \\ 0 & 0 \end{pmatrix}, \begin{pmatrix} 0 & 0 \\ 2 & 0 \end{pmatrix}, \begin{pmatrix} 0 & -1 \\ 3 & 0 \end{pmatrix}\right\}$ in $M_2(R)$ is linearly independent.
9. Are the following sets of vectors linearly independent or dependent ?
   (i) {2, –4), (1, 9), (3, 5)},
   (ii) {(–1, 5, 0), (16, 8, –3), (–64, 56, 9)}
   (iii) {(1, –1, 1), (1, 1, –1), (–1, 1, 1) (0, 1, 0)}
   (iv) $\left\{\left(\frac{1}{4}, 0, -\frac{1}{4}\right), \left(0, \frac{1}{2}, -\frac{1}{2}\right), \left(\frac{1}{3}, -\frac{1}{3}, 0\right)\right\}$
   (v) {(1, 9, 9, 8), (2, 0, 0, 3), (2, 0, 0, 8)}.
10. Show that the vectors (1, 2, 1), (2, 1, 0), (1, –1, 2) form a basis of the vector space E^3 over the field of real numbers.
11. Is the vector (2, –5, 3) in the subspace of R^3 spanned by the vectors (1, –3, 2), (2, –4, –1) and (1, –5, 7) ?
12. Show that the following sets of vectors form a basis of the vector space R^3.
   (i) {(2, –1, 0), (3, 5, 1), (1, 1, 2)}
   (ii) {(1, –3, 2), (2, 4, 1), (1, 1, 1)}
   (iii) {(1, 2, –3), (2, 5, 1), (–1, 1, 4)}
   (iv) {(2, –3, 1), (0, 1, 2), (1, 1, 2)}
   (v) {(1, 2, 2), (1, –1, 2), (1, 0, 1)}.
13. Show that the following sets of vectors do not form a basis of the vector space R^3.
   (i) {(3, 2, 1), (3, 1, 5), (3, 4, –7)}
   (ii) {(1, 2, 3), (2, 3, 4), (3, 4, 5)}
   (iii) {(1, 2, 1), (1, 3, 5), (–1, 0, 1), (1, –1, 2)}
   (iv) {(1, 2, –3), (2, –3, 1), (–3, 1, 1)}.

14. If α, β, γ be linearly independent vectors of V(F), where F is the field of real numbers, show that $(\alpha + \beta)$, $(\alpha - \beta)$, $(\alpha - 2\beta + \gamma)$ are also linearly independent.
15. If $\{\alpha, \beta, \gamma\}$ be a basis of a vector space R^3, then show that $\{\alpha + \beta, \beta + \gamma, \gamma + \alpha\}$ is also a basis of R^3.
16. In the vector space V^3 of real numbers,

$$\alpha = (1, 2, 1), \beta = (3, 1, 5), \gamma = (3, -4, 7),$$

show that the subspaces spanned by

$$S = \{\alpha, \beta\} \text{ and } T = \{\alpha, \beta, \gamma\} \text{ are the same.}$$

17. In the vector space R^3, $\alpha = (1, 2, 1)$, $\beta = (3, 1, 5)$, $\gamma = (3, -4, 7)$, show that there exist more than one basis for the subspace spanned by the set $S = \{\alpha, \beta, \gamma\}$.
18. Find the dimension of the subspace spanned by the following vectors in V_3 (R) :
    (*i*) $(1, 1, 1), (-1, -1, -1)$ (*ii*) $(1, 2, -3), (0, 0, 1), (-1, 2, 1)$.
19. Show that the set of matrices $\left\{\begin{pmatrix} 1 & 1 \\ 0 & 0 \end{pmatrix}, \begin{pmatrix} 0 & 0 \\ 1 & 1 \end{pmatrix}, \begin{pmatrix} 1 & 0 \\ 0 & 1 \end{pmatrix}, \begin{pmatrix} 0 & 1 \\ 1 & 1 \end{pmatrix}\right\}$ form a basis for the vector space M_{22}.
20. If the vectors $(0, 1, c)$, $(1, c, 1)$, and $(c, 1, 0)$ of V_3 (R) are linearly dependent, find the value of c.

Answers

1. $(3, -5) = -\frac{12}{7}(1, 2) - \frac{11}{7}(-3, 1)$
2. $(1, 7, -4) = -3(1, -3, 2) + 2(2, -1, 1)$.
3. $(1, 1, 2) = 0\,(0, 2, 1) - \frac{1}{2}(2, 2, 4)$
6. No.
9. (*i*) linearly dependent
   (*ii*) linearly dependent
   (*iii*) linearly independent
   (*iv*) linearly dependent
   (*v*) linearly independent
11. No.
18. (*i*) 1,
    (*ii*) 3
20. $0, \pm\sqrt{2}$.

EXERCISE 2.3A

(Short and objective problems)

1. Are the vectors $\{e_1, e_2, e_3\}$ linearly dependent ?
2. Which of the following vectors span R^2 ?
   (*i*) $(1, 2), (-1, 1)$ (*ii*) $(0, 0), (1, 1), (-2, -2)$
   (*iii*) $(2, 4), (-1, 2)$ (*iv*) $(1, 3), (2, -3), (0, 2)$.
3. Which of the following sets of vectors span R^3 ?
   (*i*) $\{(1, -1, 2), (0, 1, 1)\}$, (*ii*) $\{(2, 2, 3), (-1, -2, 1), (0, 1, 0)\}$
   (*iii*) $\{(1, 0, 0), (0, 1, 0), (0, 0, 1), (1, 1, 1)\}$.
4. For what values of c are the vectors $(-1, 0, -1)$, $(2, 1, 2)$ and $(1, 1, c)$ in R^3 linearly dependent ?
5. Is the set of vectors $\{e_1, e_2, e_3\}$ form a basis for E^3 ?
6. Which of the following sets of vectors are bases for R^2 ?
   (*i*) $\{(1, 3), (1, -1)\}$ (*ii*) $\{(1, 3), (-2, 6)\}$
   (*iii*) $\{(0, 0), (1, 2), (2, 4)\}$ (*iv*) $\{(1, 2), (2, -3), (3, 2)\}$.
7. Find a basis for R^3 that includes the vectors
   (*a*) $(1, 0, 2)$, (*b*) $(1, 0, 2)$ and $(0, 1, 3)$.
8. Find all values of a for which $\{(a^2, 0, 1), (0, a, 2), (1, 0, 1)\}$ is a basis for R^3.
9. Dimension of M_{23} is
   (*a*) 2 (*b*) 3
   (*c*) 5 (*d*) 6.

Answers

1. No. (independent)
2. (*i*), (*iii*), (*iv*) 3. (*i*), (*ii*), (*iii*)
4. $c = 1$ 5. yes 6. (*i*), (*ii*)
7. (*a*) $\{e_1, (1, 0, 2)\}$ (*b*) $\{e_1, (1, 0, 2), (0, 1, 3)\}$
8. $a = 0, \pm 2$ 9. (*d*).

2.8 COORDINATES AND ORDERED BASES

So far, on discussions of basis for a vector space, we did not pay much attention to the *coordinates* and *order* of the vectors in the basis. One of the useful features of a basis B in an n-dimensional vector space V is that it essentially enables one to introduce coordinates in V analogous to the 'natural coordinates' x_i of a vector $x = (x_1, x_2, ..., x_n)$ in the space.

If V is an n-dimensional vector space, then V has a basis S with n vectors in it. Let $S = \{v_1, v_2, ..., v_n\}$ be a basis for V, then $S_1 = \{v_2, v_1, ..., v_n\}$ is a different ordered basis for V. Since S is an ordered basis for the vector space, so every vector v in V can be uniquely expressed in the form

$$v = c_1v_1 + c_2v_2 + \dots + c_nv_n,$$

where c_i's, $i = 1, 2, \dots, n$, are scalars.

Then $[v]_s = \begin{bmatrix} c_1 \\ c_2 \\ \vdots \\ c_n \end{bmatrix}$ is referred to as the coordinate vector of v with respect to the ordered basis S. The entries of $[v]_s$ are called the coordinates of v with respect to S.

Remark. The coordinate vector $[v]_s$ depends on the order in which the vectors in S are listed; a change in the order of this listing may change the coordinates of v with respect to S. All bases considered in this chapter are ordered bases.

2.8.1 Illustrative Examples

1. *Let S = {e_1, e_2, e_3} be the natural (standard) basis for R^3 and let v = (3, –2, 4). Compute the coordinate vector of v with respect to the basis S.*

   Sol. Since S is the natural basis, the linear combination of the vector v is

   $$v = 3e_1 - 2e_2 + 4e_3$$

   Hence the coordinate vector of v w.r.t. the basis S

   is $$[v]_S = \begin{bmatrix} 3 \\ -2 \\ 4 \end{bmatrix}.$$

2. *Compute the coordinate vector of V with respect to the given basis S for V where*

   (i) V is R^2, $S = \left\{\begin{bmatrix} 1 \\ 0 \end{bmatrix}, \begin{bmatrix} 0 \\ 1 \end{bmatrix}\right\}, v = \begin{bmatrix} 3 \\ -2 \end{bmatrix}.$

   (ii) V is R^3, $S = \{(1, -1, 0), (0, 1, 0), (1, 0, 2)\}, v = (2, -1, -2).$

   Sol. *(i)* $S = \{e_1, e_2\}$

   $\therefore$ $v = 3e_1 - 2e_2$

   Hence $[v]_S = \begin{bmatrix} 3 \\ -2 \end{bmatrix}.$

   (ii) Let $v = c_1(1, -1, 0) + c_2(0, 1, 0) + c_3(1, 0, 2)$

   or $(2, -1, -2) = (c_1 + c_3, -c_1 + c_2, 2c_3).$

   $\therefore$ $c_1 + c_3 = 2, -c_1 + c_2 = -1, 2c_3 = -2.$

   Solving these equations, we get

$$c_3 = -1, c_1 = 3, c_2 = 2$$

Hence $$[v]_S = \begin{bmatrix} 3 \\ 2 \\ -1 \end{bmatrix}.$$

3. *Compute the vector v if the coordinate vector $[v]_S$ is given with respect to the basis S for V*

(i) *V is* R^2, $$S = \left\{\begin{bmatrix} 2 \\ 1 \end{bmatrix}, \begin{bmatrix} -1 \\ 1 \end{bmatrix}\right\}, [v]_S = \begin{bmatrix} 1 \\ 2 \end{bmatrix}$$

(ii) *V is* M_{22}, $$S = \left\{\begin{bmatrix} -1 & 0 \\ 1 & 0 \end{bmatrix}, \begin{bmatrix} 2 & 2 \\ 0 & 1 \end{bmatrix}, \begin{bmatrix} 1 & 2 \\ -1 & 3 \end{bmatrix}, \begin{bmatrix} 0 & 0 \\ 2 & 3 \end{bmatrix}\right\}$$

$$[v]_S = \begin{bmatrix} 2 \\ 1 \\ -1 \\ 3 \end{bmatrix}.$$

Sol. (*i*) $$v = c_1(2, 1) + c_2(-1, 1) = 1(2, 1) + 2(-1, 1)$$

or $$v = (2 - 2, 1 + 2) = (0, 3) \qquad (\because c_1 = 1, c_2 = 2)$$

(*ii*) $$v = 2\begin{bmatrix} -1 & 0 \\ 1 & 0 \end{bmatrix} + 1\begin{bmatrix} 2 & 2 \\ 0 & 1 \end{bmatrix} - \begin{bmatrix} 1 & 2 \\ -1 & 3 \end{bmatrix} + 3\begin{bmatrix} 0 & 0 \\ 2 & 3 \end{bmatrix}$$

$$= \begin{bmatrix} -2 & 0 \\ 2 & 0 \end{bmatrix} + \begin{bmatrix} 2 & 2 \\ 0 & 1 \end{bmatrix} + \begin{bmatrix} -1 & -2 \\ 1 & -3 \end{bmatrix} + \begin{bmatrix} 0 & 0 \\ 6 & 9 \end{bmatrix}$$

$$= \begin{bmatrix} -2+2-1+0 & 0+2-2+0 \\ 2+0+1+6 & 0+1-3+9 \end{bmatrix} = \begin{bmatrix} -1 & 0 \\ 9 & 7 \end{bmatrix}$$

EXERCISE 2.4

1. Let S = $\{e_1, e_2\}$ be the natural basis for R^2 and let $v = (4, -5)$. Compute the coordinate vector of v with respect to the basis S.
2. Let S = $\{e_1, e_2, e_3\}$ be the natural basis for R^3 and let $v = (-1, 3, -2)$. Compute $[v]_S$.
3. Let S = $\{(2, 3), (3, 5)\}$ be the basis for R^2 and let $v = (7, 11)$. Compute $[v]_S$.
4. Compute the coordinate vector of V with respect to the given basis S for V, where V is R^3, $$S = \{(-1, 2, 0), (0, -1, 1), (3, -4, 2)\}$$

and $v = (-1, 2, 4)$,

5. Compute the coordinate vector v with respect to the given basis S for V.

(i) V is M_{22}, $S = \left\{\begin{bmatrix}1 & 0\\0 & 0\end{bmatrix}, \begin{bmatrix}0 & 0\\1 & 0\end{bmatrix}, \begin{bmatrix}0 & 1\\0 & 0\end{bmatrix}, \begin{bmatrix}0 & 0\\0 & 1\end{bmatrix}\right\}$,

$$v = \begin{bmatrix}1 & 0\\-1 & 2\end{bmatrix}$$

(ii) V is M_{22}, $S = \left\{\begin{bmatrix}1 & -1\\0 & 0\end{bmatrix}, \begin{bmatrix}0 & 1\\1 & 0\end{bmatrix}, \begin{bmatrix}1 & 0\\0 & -1\end{bmatrix}, \begin{bmatrix}1 & 0\\-1 & 0\end{bmatrix}\right\}, v = \begin{bmatrix}1 & 3\\-2 & 2\end{bmatrix}$.

6. Compute the vector v if the coordinate vector $[v]_S$ is given with respect to the basis S for V.

(i) V is R^3, $S = \{(0, 1, -1), (1, 0, 0), (1, 1, 1)\}, [v]_S = \begin{bmatrix}-1\\1\\2\end{bmatrix}$

(ii) V is M_{22}, $S = \left\{\begin{bmatrix}1 & -2\\0 & 0\end{bmatrix}, \begin{bmatrix}-1 & 3\\0 & 1\end{bmatrix}, \begin{bmatrix}1 & 0\\0 & 0\end{bmatrix}, \begin{bmatrix}0 & -1\\1 & 0\end{bmatrix}\right\}, [v]_S = \begin{bmatrix}0\\1\\0\\2\end{bmatrix}$.

Answers

1. $\begin{bmatrix}4\\-5\end{bmatrix}$

2. $\begin{bmatrix}-1\\3\\-2\end{bmatrix}$

3. $\begin{bmatrix}2\\1\end{bmatrix}$

4. $\begin{bmatrix}4\\2\\1\end{bmatrix}$

5. (i) $\begin{bmatrix}1\\-1\\0\\2\end{bmatrix}$ (ii) $\begin{bmatrix}-1\\2\\-2\\4\end{bmatrix}$

6. (i) (3, 1, 3) (ii) $\begin{bmatrix}-1 & 1\\2 & 1\end{bmatrix}$.

2.9 LINEAR TRANSFORMATIONS

Definition. Let V and W be two vector spaces over the field F. A *linear transformation* T from V into W is a function assigning a unique vector T(u) in W to each u in V such that

(*i*) $T(u + v) = T(u) + T(v)$, for every u, v in V.

(*ii*) $T(cu) = c\,T(u)$, for every u in V and every scalar $c \in F$.

(*i*) and (*ii*) can be combined into a single statement as,

$$T(cu + v) = cT(u) + T(v).$$

Remark 1. Instead of T(u) it is convenient to write simply Tu.

Remark 2. The linear transformation T also represents a *mapping* V into W written as $T : V \to W$.

A map $T : V \to W$ is also called a *homomorphism* of V into W if T satisfies the above two properties.

Remark 3. If V = W, the linear transformation $T : V \to V$ is also called a linear operator on V.

2.9.1 Illustrative Examples

1. *If V be any vector space, the identity transformation I, defined by* $Iu = u$*, is a linear transformation from V to V.*

   Sol. The *zero transformation* O defined by $Ou = O$, is a linear transformation from V to V.

2. *Let R be the field of real numbers and let V be the space of all functions from R into R which are continuous.*

   Sol. Let us define T by

   $$(Tf)(x) = \int_0^x f(t)\,dt.$$

   Then T is a linear transformation from V into V. The function Tf is not only continuous but has a continuous first derivative.

3. *The following are the geometrical examples of transformations :*

   (*i*) *Projection* $T: R^3 \to R^2$ *defined by* $T(x, y, z) = (x, y)$.

   (*ii*) *Dilation* $T: R^3 \to R^3$ *defined by* $Tu = ru, r > 1$.

   (*iii*) *Contraction* $T: R^3 \to R^3$ *defined by* $Tu = ru, 0 < r < 1$.

   (*iv*) *Reflection* $T: R^2 \to R^2$ *defined by* $T(x, y) = (x, -y)$.

   (*v*) *Rotation* $T: R^2 \to R^2$ *defined by* $Tu = \begin{bmatrix} \sin\theta & -\sin\theta \\ \sin\theta & \cos\theta \end{bmatrix} u.$

4. *Let* $T : P \to Q$ *be a linear transformation be defined by* $T(ax + b) = x(ax + b)$*. Prove that T is a linear transformation.*

   Sol. Let $(ax + b)$ and $cx + d \in P$

and let k be a scalar.

Then $\quad T[(ax+b)+(cx+d)] = x[(ax+b)+(cx+d)]$

$$= x(ax+b)+x(cx+d)$$
$$= T(ax+b)+T(cx+d)$$

and $\quad T[k(ax+b)] = x[k(ax+b)] = k[x(ax+b)]$

$$= k\,T(ax+b).$$

Hence, T is a linear transformation.

Theorem 2.13. *The set $\alpha(U, V)$ of all linear transformations from a vector space U to a vector space V is a vector space for suitably defined addition and scalar multiplication.*

Proof. If T_1, T_2 are two linear transformations of $\alpha(U, V)$, we define a linear transformation (*map*) $T_1 + T_2 : U \to V$ as follows:

$$(T_1 + T_2)\,x = T_1x + T_2x$$

where $\quad x \in U$

It can be verified that $T_1 + T_2$ is a linear transformation.

$T_1 + T_2$ is called the *sum* of the linear transformations T_1 and T_2.

Again, if $T \in \alpha(U, V)$ and k is a scalar, we define a map $kT : U \to V$ as follows:

$$(kT)(x) = k(Tx), \text{ where } x \in U.$$

It can be verified that kT is a linear transformation. The set $\alpha(U, V)$ is an abelian group with respect to addition, the zero transformation $O : U \to V$ for which $O(x) = 0$, for every $x \in U$, being the zero and the negative of a linear transformation T being the linear transformation

$$-T : U \to V \text{ given by } (-T)\,x = -(Tx),$$

for every $x \in U$.

Deductions. Verify the following identities:

(*i*) $\alpha(T_1 + T_2) = \alpha T_1 + \alpha T_2$

(*ii*) $(\alpha + \beta)\,T = \alpha T + \beta T$

(*iii*) $(\alpha\beta)\,T = \alpha(\beta T)$

(*iv*) $1T = T$

where $T_1, T_2 \in t(U, V)$, and α, β are scalars.

Proof. (*i*) $\quad \alpha(T_1 + T_2)\,x = \alpha(T_1x + T_2x) = \alpha(T_1x) + \alpha(T_2x)$

$$= (\alpha T_1)x + (\alpha T_2)x$$

$\therefore \quad \alpha(T_1 + T_2) = \alpha T_1 + \alpha T_2$

(*ii*) $\quad ((\alpha + \beta)\,T)\,x = (\alpha + \beta)(Tx) = \alpha(Tx) + \beta(Tx)$

$$= (\alpha T)x + (\beta T)x$$
$$= (\alpha T + \beta T)\,x$$

$\therefore \quad (\alpha + \beta)\,T = \alpha T + \beta T$

(*iii*) $\quad (\alpha\beta)T)\,x = (\alpha\beta)\,(Tx) = (\alpha\beta(Tx)) = (\alpha(\beta T)\,x)$

$$= (\alpha(\beta T))x$$

$\therefore$ $$(\alpha\beta)T = \alpha(\beta T)$$

(*iv*) $$(1T)x = (1(Tx)) = (1T)x = Tx$$

$\therefore$ $$1T = T.$$

2.10 ALGEBRA OF LINEAR TRANSFORMATIONS

Theorem 2.14. *Let V and W be two vector spaces over the field F. Let T and L be linear transformations from V into W. Then (T + L) defined by*

$$(T+L)\alpha = T\alpha + L\alpha$$

is a linear transformation from V into W.

If c is any element of F, the function (cT) defined

by $$(cT)(\alpha) = c(T\alpha)$$

is a linear transformation from V into W. The set of all linear transformations from V into W, together with the addition and scalar multiplication defined above, is a vector space over the field F.

Proof. Since for the linear transformations T and L from V into W, the sum T + L is defined as

$$(T+L)(\alpha) = T\alpha + L\alpha,$$

Then
$$\begin{aligned}(T+L)(c\alpha+\beta) &= T(c\alpha+\beta) + L(c\alpha+\beta)\\ &= c(T\alpha) + T\beta + c(L\alpha) + L\beta\\ &= c(T\alpha + L\alpha) + (T\beta + L\beta)\\ &= c(T+L)(\alpha) + (T+L)\beta\end{aligned}$$

which shows that (T + L) is a linear transformation.

Also,
$$\begin{aligned}(cT)(k\alpha+\beta) &= c\,[T(k\alpha+\beta)]\\ &= c\,[k(T\alpha) + T\beta]\\ &= ck\,[(T\alpha)] + c(T\beta)\\ &= k\,[c(T\alpha)] + c(T\beta)\\ &= k\,[(cT)\alpha] + (cT)\beta\end{aligned}$$

which shows that (cT) is a linear transformation.

Theorem 2.15. *Let U, V, W be vector spaces over the field F. Let T be a linear transformation from U into V and L is a linear transformation from V into W. Then the composition of functions LT defined by*

(LT) α = L(Tα) is a linear transformation from V into W.

Proof.
$$\begin{aligned}(LT)(c\alpha+\beta) &= L[T(c\alpha+\beta)] = L[cT\alpha + T\beta]\\ &= c\,[L(T\alpha)] + L(T\beta)\\ &= c(LT)(\alpha) + (LT)(\beta).\end{aligned}$$

Hence, LT is a linear transformation from V into W.

Theorem 2.16. *Let U and V be finite dimensional vector spaces over a field F. Let $\{u_1, u_2, \ldots, u_n\}$ be a basis for U and let $v_1, v_2, \ldots, v_n$ be any n vectors in V (not necessarily distinct). Then there exists a unique linear transformation $T : U \to V$ such that*

$$T(u_i) = v_i, \; i = 1, 2, \ldots, n.$$

Proof. Let $u = \alpha_1 u_1 + \alpha_2 u_2 + \ldots + \alpha_n u_n$ be a vector of U.

Let us define $T(u) = \alpha_1 v_1 + \alpha_2 v_2 + \ldots + \alpha_n v_n$.

Let x, y be any two vectors of U.

Let $x = \alpha_1 u_1 + \alpha_2 u_2 + \ldots + \alpha_n u_n$

and $y = \beta_1 u_1 + \beta_2 u_2 + \ldots + \beta_n u_n$.

Then $x + y = (\alpha_1 + \beta_1)\, u_1 + (\alpha_2 + \beta_2)\, u_2 + \ldots + (\alpha_n + \beta_n)\, u_n$

and
$$\begin{aligned} T(x + y) &= (\alpha_1 + \beta_1)\, v_1 + (\alpha_2 + \beta_2)\, v_2 + \ldots + (\alpha_n + \beta_n)\, v_n \\ &= (\alpha_1 v_1 + \alpha_2 v_2 + \ldots + \alpha_n v_n) + (\beta_1 v_1 + \beta_2 v_2 + \ldots + \beta_n v_n) \\ &= Tx + Ty \end{aligned}$$

Similarly, $T(\alpha x) = \alpha T x$

Hence, T is a linear transformation.

Also $u_1 = 1u_1 + 0u_2 + \ldots + 0u_n$

$\therefore$ $T(u_1) = 1v_1 + 0v_2 + \ldots + 0v_n = v_1$

Thus $T(u_i) = v_i, i = 1, 2, \ldots, n.$

Now, it requires to prove the uniqueness.

Let $T' : U \to V$ be any other linear transformation

such that $T'(u_i) = v_i, i = 1, 2, \ldots, n.$

Let $u = \alpha_1 u_1 + \alpha_2 u_2 + \ldots + \alpha_n u_n \in U$

Then
$$\begin{aligned} T'(u) &= \alpha_1 T'(u_1) + \alpha_2 T'(u_2) + \ldots + \alpha_n T'(u_n) \\ &= \alpha_1 v_1 + \alpha_2 v_2 + \ldots + \alpha_n v_n = T(u) \end{aligned}$$

Hence $T' = T.$

These complete the proof.

Example. The vectors $u_1 = (1, 2)$, $u_2 = (3, 4)$ are linearly independent and they form a basis for R^2. According to the theorem 2.16, there is a unique transformation from R^2 into R^3 such that

$$T(u_1) = (3, 2, 1)$$
$$T(u_2) = (6, 5, 4).$$

If so, it will be possible to find $T(e_1)$.

We find scalars c_1 and c_2 such that

$$Te_1 = c_1 u_1 + c_2 u_2$$

we know that $Te_1 = c_1\, Tu_1 + c_2\, Tu_2$.

If $(1, 0) = c_1(1, 2) + c_2(3, 4)$,

then we get $c_1 = -2$ and $c_2 = 1$.

Thus $T(1, 0) = -2(3, 2, 1) + (6, 5, 4) = (0, 1, 2).$

2.11 SINGULAR AND NON-SINGULAR TRANSFORMATIONS

Definition. A linear transformation T on a finite dimensional vector space V is said to be *non-singular* if there exists a linear transformation T *on V such that T*T = I, the identity transformation on V. A linear transformation T is said to be *singular* if it is non-singular.

2.12 ISOMORPHISM OF VECTOR SPACES

If V and W are vector spaces over the field F, any linear transformation T of V onto W is called an *isomorphism* of V onto W provided

(*i*) T is one-one and onto mapping

(*ii*) $T(au + bv) = aTu + bTv \; \forall \; a, b \in F$ and $\forall \; u, v \in V$.

If the two vector spaces V(F) and W(F) are isomorphic, then it is denoted by $V(F) \cong W(F)$.

We state and prove a theorem as follows :

Theorem 2.17 *Every n-dimensional vector space over the field F is isomorphic to the space* $V_n(F)$.

Proof. Let V be a vector space of dimension n.

Let $\{v_1, v_2, ..., v_n\}$ be a basis for V.

We know that if $v \in V$, v can be written uniquely as

$$v = \alpha_1 v_1 + \alpha_2 v_2 + ... + \alpha_n v_n,$$

where $\alpha_i \in F$.

Let us consider the mapping

$$T : V \to V_n(F)$$

given by $T(\alpha_1 v_1 + \alpha_2 v_2 + ... + \alpha_n v_n) = (\alpha_1, \alpha_2, ..., \alpha_n)$.

$\therefore$ The mapping T is one-one and onto.

Let $u, w, \in V$.

Then $v = \alpha_1 v_1 + \alpha_2 v_2 + ... + \alpha_n v_n$

and $w = \beta_1 v_1 + \beta_2 v_2 + ... + \beta_n v_n$

$$\begin{aligned} T(v + w) &= T[(\alpha_1 + \beta_1)v_1 + (\alpha_2 + \beta_2) v_2 + ... + (\alpha_n + \beta_n) v_n] \\ &= ((\alpha_1 + \beta_1), (\alpha_2 + \beta_2), ... (\alpha_n + \beta_n)) \\ &= (\alpha_1, \alpha_2, ..., \alpha_n) + (\beta_1, \beta_2, ..., \beta_n) \\ &= T(v) + T(w) \end{aligned}$$

Also,
$$\begin{aligned} T(\alpha v) &= T(\alpha\alpha_1 v_1 + \alpha\alpha_2 v_2 + ... + \alpha\alpha_n v_n) \\ &= (\alpha\alpha_1, \alpha\alpha_2, ..., \alpha\alpha_n) \\ &= \alpha(\alpha_1, \alpha_2, ..., \alpha_n) = \alpha T(v). \end{aligned}$$

Hence, T is an isomorphism of V to V_n (F).

Corollary 2.4. Any two vector spaces of the same dimension over a field F are isomorphic, for, if the vector spaces are of dimension n, each is isomorphic to $V_n(F)$ and hence they are isomorphic.

Theorem 2.18. *Let U and V be two vector spaces over a field F. Let T : U → V be an isomorphism. Then T maps a basis of U on to a basis of V.*

Proof. Let $\{u_1, u_2, ..., u_n\}$ be a basis for U. We shall prove that $T(u_1), T(u_2), ..., T(u_n)$ are linearly independent and that they span V.

$$\text{Now,} \quad \alpha_1 T(u_1) + \alpha_2 T(u_2) + ... + \alpha_n T(u_n) = 0$$

$$\Rightarrow \quad T(\alpha_1 u_1) + T(\alpha_2 u_2) + ... + T(\alpha_n u_n) = 0$$

$$\Rightarrow \quad T(\alpha_1 u_1 + \alpha_2 u_2 + ... + \alpha_n u_n) = 0$$

$$\Rightarrow \quad \alpha_1 u_1 + \alpha_2 u_2 + ... + \alpha_n u_n = 0 \qquad (\because \text{T is one-one})$$

$$\Rightarrow \quad \alpha_1 = \alpha_2 = ... = \alpha_n = 0$$

($\because u_1, u_2, ..., u_n$ are linearly independent)

$\therefore$ $T(u_1), T(u_2), ..., T(u_n)$ are linearly independent.

Again, let $v \in V$. Then since T is onto, there exists a vector $u \in V$ such that $T(u) = v$.

$$\text{Let} \quad u = \alpha_1 v_1 + \alpha_2 v_2 + ... + \alpha_n v_n$$

$$\begin{aligned} \text{Then} \quad v &= T(u) \\ &= T(\alpha_1 v_1 + \alpha_2 v_2 + ... + \alpha_n v_n) \\ &= \alpha_1 T(v_1) + \alpha_2 T(v_2) + ... + \alpha_n T(v_n). \end{aligned}$$

Thus v is a linear combination of the vectors $T(v_1), T(v_2), ..., T(v_n)$

$\therefore$ $T(v_1), T(v_2), ..., T(v_n)$ span V and hence is a basis for V.

Corollary 2.5. Two finite dimensional vector spaces U and V over a field F are isomorphic iff they have the same dimensions.

2.12.1 Illustrative Examples

1. *Let $T : R^3 \to R^2$ be defined by*

$$T(x, y, z) = (x, y).$$

Prove that T is a linear transformation.

Sol. Let u and v be two vectors such that

$$u = (a_1, a_2, a_3), v = (b_1, b_2, b_3).$$

$$\begin{aligned} \text{Then} \quad T(u + v) &= T((a_1, a_2, a_3) + (b_1, b_2, b_3)) \\ &= T(a_1 + b_1, a_2 + b_2, a_3 + b_3) \\ &= (a_1 + b_1, a_2 + b_2) \qquad \text{[by given definition]} \\ &= (a_1, a_2) + (b_1, b_2) = T(u) + T(v). \end{aligned}$$

If c is a real number, then

$$\begin{aligned} T(cu) &= T(ca_1, ca_2, ca_3) = (ca_1, ca_2) \\ &= c(a_1, a_2) = cT(u). \end{aligned}$$

Hence, L is a linear transformation.

2. *Let $T : R^3 \to R^2$ be a linear transformation for which it is defined as*

$$T(1, 0, 0) = (2, -1), T(0, 1, 0) = (3, 1) \text{ and } T(0, 0, 1) = (-1, 2).$$

Determine $T(-3, 4, 2)$.

Sol. Since $(-3, 4, 2) = 3i + 4j + 2k,$

$$\therefore \quad T(-3, 4, 2) = T(-3i + 4j + 2k)$$
$$= -3T(i) + 4T(j) + 2T(k)$$
$$= -3(2, -1) + 4(3, 1) + 2(-1, 2)$$
$$= (-6 + 12 - 2, 3 + 4 + 4) = (4, 11).$$

3. *Let $T : R^3 \to R^2$ be defined by*

$$T\left(\begin{bmatrix} u_1 \\ u_2 \\ u_3 \end{bmatrix}\right) = \begin{bmatrix} u_1 + 1 \\ u_2 - u_3 \end{bmatrix}.$$

Determine whether T is a linear transformation.

Sol. Let $$u = \begin{bmatrix} u_1 \\ u_2 \\ u_3 \end{bmatrix} \text{ and } v = \begin{bmatrix} v_1 \\ v_2 \\ v_3 \end{bmatrix}.$$

Then $$T(u+v) = T\left(\begin{bmatrix} u_1 \\ u_2 \\ u_3 \end{bmatrix} + \begin{bmatrix} v_1 \\ v_2 \\ v_3 \end{bmatrix}\right) = T\begin{pmatrix} u_1 + v_1 \\ u_2 + v_2 \\ u_3 + v_3 \end{pmatrix} = \begin{pmatrix} (u_1 + v_1) + 1 \\ (u_2 + v_2) - (u_3 + v_3) \end{pmatrix}$$

Again, $$T(u) + T(v) = T\begin{bmatrix} u_1 \\ u_2 \\ u_3 \end{bmatrix} + T\begin{bmatrix} v_1 \\ v_2 \\ v_3 \end{bmatrix} = \begin{bmatrix} u_1 + 1 \\ u_2 - u_3 \end{bmatrix} + \begin{bmatrix} v_1 + 1 \\ v_2 - v_3 \end{bmatrix}$$

$$= \begin{bmatrix} u_1 + v_1 + 2 \\ (u_2 - u_3) + (v_2 - v_3) \end{bmatrix}.$$

These show that $T(u+v) \neq T(u) + T(v)$.

Hence T is not a linear transformation.

4. *Let $T : R^3 \to R^3$ be defined by*

$$T(x, y, z) = (x + y, 0, 2x - z).$$

Determine whether T is a linear transformation.

Sol. Let $u = (u_1, u_2, u_3)$ and $v = (v_1, v_2, v_3)$.

Then $$T(u+v) = T((u_1, u_2, u_3) + (v_1, v_2, v_3))$$
$$= T(u_1 + v_1, u_2 + v_2, u_3 + v_3)$$
$$= (u_1 + u_2 + v_1 + v_2, 0, 2u_1 + 2v_1 - u_3 - v_3) \quad ...(i)$$

Again, $$T(u) + T(v) = T(u_1, u_2, u_3) + T(v_1, v_2, v_3)$$
$$= (u_1 + u_2, 0, 2u_1 - u_3) + (v_1 + v_2, 0, 2v_1 - v_3)$$
$$= (u_1 + u_2 + v_1 + v_2, 0, 2u_1 + 2v_1 - u_3 - v_3) \quad ...(ii)$$

$\therefore$ From (*i*) and (*ii*), we have

$$T(u+v) = T(u) + T(v)$$

Also
$$\begin{aligned} T(cu) &= T(cu_1, cu_2, cu_3) = (cu_1 + cu_2, 0, 2cu_1 - cu_3) \\ &= (c(u_1 + u_2), 0, c(2u_1 - u_3)) \\ &= c(u_1 + u_2, 0, 2u_1 - u_3) = cT(\boldsymbol{u}). \end{aligned}$$

Hence, T is a linear transformation.

EXERCISE 2.5

1. Let $T : R^2 \to R^3$, be defined by
$$T(x, y) = (x + 1, y, x + y).$$
Is T a linear transformation ?
2. Let $L : R^2 \to R^2$ be defined by
$$L(x, y) = (x^2 + x, y - y^2).$$
Is L a linear transformation ?
3. Which of the following are linear transformations ?
   (*i*) $T(a, b, c) = (a + b, 0, 2a - c)$
   (*ii*) $T(a, b) = (a - b, 0, 2a + 1)$
   (*iii*) $T\left(\begin{bmatrix} a \\ b \end{bmatrix}\right) = \begin{bmatrix} a^2 - b^2 \\ a^2 + b^2 \end{bmatrix}$ (*iv*) $T(x, y) = (x^2, y^2)$ for $(x, y) \in R^2$.
4. A linear transformation T is defined as
$$T\left(\begin{bmatrix} a \\ b \\ c \end{bmatrix}\right) = \begin{bmatrix} a + b \\ b \\ a - c \end{bmatrix};$$
is T a linear transformation ?
5. Given $T : R^3 \to R^3$, a linear transformation defined by
$$T\left(\begin{bmatrix} a \\ b \\ c \end{bmatrix}\right) = \begin{bmatrix} a + b \\ a - c \\ a + c \end{bmatrix}.$$
Find the standard matrix representing T.
[**Hint :** Find $T(e_1)$, $T(e_2)$, $T(e_3)$ which are the columns of the matrix].
6. Prove that $R^2 \to R^2$ defined by $T(a, b) = (2a - 3b, a + 4b)$ is a linear transformation.
7. Show that $T : R^3 \to R^3$ defined by
$$T(x, y, z) = (3x + 3y, x + y + z, 0)$$
is a linear transformation.
8. If $T : R^3 \to R^3$ be a linear transformation such that
$$T(1, 0, 0) = (1, 3, 2)$$
$$T(1, 1, 0) = (3, 4, 0)$$
$$T(1, 1, 1) = (2, 1, 3),$$
show that $T(1, 2, 1) = (0, 2, 1)$.

9. Let V be the set of all polynomials of degree $\leq n$ in R $[x]$ including the zero polynomial. Let $T : V \to V$ be defined by $T(f) = \dfrac{df}{dx}$.

Prove that T is a linear transformation.

10. Let $T : M_{22} \to M_{22}$ is defined by

$$T\left(\begin{bmatrix} a & b \\ c & d \end{bmatrix}\right) = \begin{bmatrix} b & c-d \\ c+d & 2a \end{bmatrix}.$$ Is T a linear transformation ?

11. Let $T : M_{22} \to R'$ be defined by $T\left(\begin{bmatrix} a & b \\ c & d \end{bmatrix}\right) = a + d$. Is T a linear transformation ?

12. Let $T : R^2 \to R^2$ be a linear transformation such that

$$T(1, 1) = (1, -2), T(-1, 1) = (2, 3).$$

(*i*) What is $T(-1, 5)$? (*ii*) What is $T(x, y)$?

13. Obtain the linear transformation determined by the matrix

$T : V_2(R) \to R_2(R)$ given by $\begin{bmatrix} \cos\theta & -\sin\theta \\ \sin\theta & \cos\theta \end{bmatrix}$ w.r.t. standard basis.

14. Obtain the linear transformation determined by the matrix

$T : V_2(R) \to V_3(R)$ given by $\begin{bmatrix} 2 & 1 & -1 \\ 1 & 1 & -1 \end{bmatrix}$ w.r.t. the standard basis.

Answers

1. No. 2. No. 3. (*i*) 4. yes

5. $\begin{bmatrix} 1 & 1 & 0 \\ 0 & 1 & -1 \\ 1 & 0 & 1 \end{bmatrix}$ 10. yes 11. yes 12. (*i*) {1}, (*ii*) {*a*, 1}.

13. $T(x, y) = (x\cos\theta + y\sin\theta, -x\sin\theta + y\cos\theta)$,

14. $T(a, b) = (2a + b, a + b, -a - b)$.

2.13 THE KERNEL AND RANGE OF A LINEAR TRANSFORMATION

In this section, we formulate the notions of one-to-one linear transformations and also onto linear transformations. We also develop methods for determining when a linear transformation is one-to-one or onto.

Let us first define a one-to-one linear transformation as follows :

Definition. A linear transformation

$$T : V \to W$$

is said to be *one-to-one* if for all v_1, v_2 $(v_1 \neq v_2)$ in V, $T(v_1) = T(v_2)$ implies that $v_1 = v_2$.

Remark. It follows from the definition that T is one-to-one if T (v_1) and T (v_2) are distinct when ever v_1 and v_2 are distinct. Geometrically, one-to-one and not one-to-one are shown in Figure 2.3 and Figure 2.4 respectively.

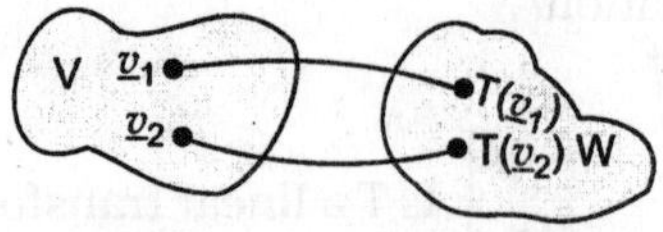

Figure 2.3. *T is one-to-one*

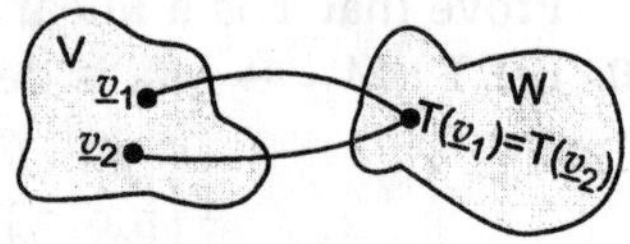

Figure 2.4. *T is not one-to-one*

1. *Given a linear transformation* $T : R^2 \to R^2$ *defined by*

$$T(x, y) = (x + y, x - y).$$

Determine whether T is one-to-one.

Sol. Let $u = (a_1, a_2)$, and $v = (b_1, b_2)$.

If T is one-to-one, then

$$T(u) = T(v)$$

i.e., $T(a_1, a_2) = T(b_1, b_2)$

i.e., $(a_1 + a_2, a_1 - a_2) = (b_1 + b_2, b_1 - b_2)$

i.e., $a_1 + a_2 = b_1 + b_2$

and $a_1 - a_2 = b_1 - b_2$

from these two equations,

we get $a_1 = b_1$ and $a_2 = b_2$.

Hence, $u = v$ and thus T is one-to-one.

2. *Let* $L : R^2 \to R^3$ *be defined by*

$$L(x, y) = (x, x + y, y).$$

Is L one-to-one ?

Sol. Let $u = (a_1, a_2)$ and $v = (b_1, b_2)$.

Then $L(u) = L(a_1, a_2) = (a_1, a_1 + a_2, a_2)$

$L(v) = L(b_1, b_2) = (b_1, b_1 + b_2, b_2)$.

If T is one-to-one, then

$$L(u) = L(v)$$

i.e., $(a_1, a_1 + a_2, a_2) = (b_1, b_1 + b_2, b_2)$

i.e., $a_1 = b_1$

$a_2 = b_2$.

Hence $u = v$. Therefore L is one-to-one.

Definition of Kernel

Let $T : V \to W$ be a linear transformation. The *Kernel* of T, denoted by Ker T, is the subset of V consisting of all vectors v such that T (v) = o_w.

Remark 1. Kernel T is not an empty set, since o_v is in Ker T.

Example 1.

Let $T : R^3 \to R^2$ be defined by $T(x, y, z) = (x, y)$.

Then the vector (0, 0, 3) is in Ker T,

since T (0, 0, 3) = (0, 0). But the vector (2, 0, –1) is not in Ker T, since T (2, 0, –1) = (2, 0).

Remark 2. Ker T consists of all vectors x in R^3 of the form $(0, 0, r)$, where r is any real number. It may be noted that Ker T consists of the z-axis in three dimensional space R^3.

Example 2. *Let* $T : R^2 \to R^2$ *be defined by*

$$T(x, y) = (x + y, x - y).$$

Then show that Ker T consists of all vectors x *in* R^2 *such that* $T(x) = o$.

Sol. Since $T(x) = 0$,

Therefore, $x + y = 0$ and $x - y = 0$

Solving we get $x = 0, y = 0$,

Hence, Ker T = {o}.

Theorem 2.19. *If* $T : V \to W$ *is a linear transformation, then Ker T is a subspace of V.*

Proof. Ker T is not empty set, since o_v is in Ker T. Let u and v be in Ker T. Then, since T is a linear transformation, therefore

$$T(u + v) = T(u) + T(v) = o_w + o_w = o_w.$$

Hence $u + v \in \text{Ker T}$.

Next, since T is a linear transformation, so for any scalar c,

we have $T(cu) = cT(u) = co_w = o_w,$

Hence, cu is in Ker T.

Therefore, Ker T is a subspace of V.

3. *Let* $T : R^4 \to R^2$ *is defined by*

$$T\left(\begin{bmatrix} x \\ y \\ z \\ t \end{bmatrix}\right) = \begin{bmatrix} x + y \\ z + t \end{bmatrix}.$$

Find a basis for Ker T and dim (Ker T).

Sol. Ker T consists of all vectors u in R^4 such that $Tu = o$.

This leads to the linear system

$$x + y = 0$$

and $z + t = 0$

which has a pair of solution, (1, –1, 0, 0), (0, 0, 1, –1)

Hence, a basis of Ker T consists of the vectors

$$\begin{bmatrix} 1 \\ -1 \\ 0 \\ 0 \end{bmatrix} \text{ and } \begin{bmatrix} 0 \\ 0 \\ 1 \\ -1 \end{bmatrix};$$

since the basis of Ker T consists of two vectors, hence dim (Ker T) = 2.

Remark. A linear transformation $T : V \rightarrow W$ is one-to-one if and only if Ker T = $\{o_v\}$. Also T is one-to-one, if and only if, dim (Ker T) = 0.

Now we define *range* of a linear transformation.

Definition. If $T : V \rightarrow W$ is a linear transformation, then the *range* of T, denoted by *range* T, is the set of all vectors in W that are images, under T, of vectors in V. Thus a vector w is in range T, if there exists some vector $v \in V$ such that $T(v) = w$. If range T = W, then T is onto.

Theorem 2.20. *If $T : V \rightarrow W$ is a linear transformation, then range of T is a subspace of W.*

Proof. Range T is not empty set, since $o_w = T(o_v)$, so o_w is in range T.

Let w_1 and $w_2 \in$ range T.

Then $w_1 = T(v_1)$ and $w_2 = T(v_2)$

for some v_1 and $v_2 \in V$.

$$\therefore \qquad w_1 + w_2 = T(v_1) + T(v_2) = T(v_1 + v_2),$$

since T is a linear transformation.

This implies that $w_1 + w_2 \in$ range T.

Also, if c is a scalar, then

$$cw_1 = cT(v_1) = T(cv_1),$$

so cw_1 is in range T.

Hence, range T is a subspace of V.

4. *Let T be a linear transformation from R^3 to R^2 defined by $T(u, v, w) = (u, v)$. Is T onto ?*

Sol. Let us choose any vector $a = (a_1, a_2)$ in R^2 and we shall find a vector $b = (b_1, b_2, b_3)$ in R^3 such that $T(b) = a$.

Since $\qquad T(b) = (b_1, b_2),$

we find that if $\qquad b_1 = a_1$ and $b_2 = a_2,$

Then $T(b) = a$.

Therefore, T is onto and the dimension of range T is 2.

5. *Let $\qquad T : R^3 \rightarrow R^3$ be defined by*

$$T\left(\begin{pmatrix} x \\ y \\ z \end{pmatrix}\right) = \begin{bmatrix} 1 & 0 & 1 \\ 1 & 1 & 2 \\ 2 & 1 & 3 \end{bmatrix} \begin{bmatrix} x \\ y \\ z \end{bmatrix}.$$

(i) Is T onto ? *(ii) Find a basis for range T.*

(iii) Find Ker T *(iv) Is T one-to-one ?*

Sol. (*i*) For any $u = \begin{bmatrix} u_1 \\ u_2 \\ u_3 \end{bmatrix}$ in R^3,

where u_1, u_2, u_3 are any real scalars,

we can find $v = \begin{bmatrix} x \\ y \\ z \end{bmatrix}$

so that $T(v) = w.$

Let us find a solution to the linear system

$$\begin{bmatrix} 1 & 0 & 1 \\ 1 & 1 & 2 \\ 2 & 1 & 3 \end{bmatrix} \begin{bmatrix} x \\ y \\ z \end{bmatrix} = \begin{bmatrix} u_1 \\ u_2 \\ u_3 \end{bmatrix}.$$

The augmented matrix is

$$\begin{bmatrix} 1 & 0 & 1 & \vdots & u_1 \\ 1 & 1 & 2 & \vdots & u_2 \\ 2 & 1 & 3 & \vdots & u_3 \end{bmatrix} \sim \begin{bmatrix} 1 & 0 & 1 & \vdots & u_1 \\ 0 & 1 & 1 & \vdots & u_2 - u_1 \\ 0 & 0 & 0 & \vdots & u_3 - u_2 - u_1 \end{bmatrix},$$

[applying

$$R_2' = R_2 - R_1$$
$$R_3' = R_3 - R_2 - R_1]$$

Thus a solution exists only for $u_3 - u_2 - u_1 = 0$,

so T is not onto.

(*ii*) Again, $$T\left(\begin{bmatrix} x \\ y \\ z \end{bmatrix}\right) = \begin{bmatrix} 1 & 0 & 1 \\ 1 & 1 & 2 \\ 2 & 1 & 3 \end{bmatrix} \begin{bmatrix} x \\ y \\ z \end{bmatrix} = \begin{bmatrix} x+z \\ x+y+2z \\ 2x+y+3z \end{bmatrix}$$

$$= x\begin{bmatrix} 1 \\ 1 \\ 2 \end{bmatrix} + y\begin{bmatrix} 0 \\ 1 \\ 1 \end{bmatrix} + z\begin{bmatrix} 1 \\ 2 \\ 3 \end{bmatrix}$$

This shows that the set of vectors

$$\left\{\begin{bmatrix} 1 \\ 1 \\ 2 \end{bmatrix}, \begin{bmatrix} 0 \\ 1 \\ 1 \end{bmatrix}, \begin{bmatrix} 1 \\ 2 \\ 3 \end{bmatrix}\right\}$$

span range T. That is, range T is the subspace of R^3 spanned by the columns of the matrix defining T. The first two vectors in this set are linearly independent, since they are not constant multiples of each other. The third vector is the sum of the first two vectors. Therefore, $\begin{bmatrix}1\\1\\2\end{bmatrix}, \begin{bmatrix}0\\1\\1\end{bmatrix}$ form a basis for range T and dim (range T) = 2

[$\because$ in the basis there are two vectors]

(*iii*) To find ker L, we have to find all v in R^3 so that

$$T(v) = O_{R^3}.$$

The homogeneous system of equations are

$$x + z = 0, x + y + 2z = 0, \text{ and } 2x + y + 3z = 0.$$

Solving these equations, we get $x = -z$, and $y = -z$.

Thus ker T consists of all vectors of the form $\begin{bmatrix}-r\\-r\\r\end{bmatrix} = r\begin{bmatrix}-1\\-1\\1\end{bmatrix}$, where r is any real number.

dim (ker T) = 1

(*iv*) Since ker T $\neq \{O_{R^3}\}$, so T is not one-to-one.

6. *Let* $T : R^4 \to R^3$ *be defined by*

$$T(x, y, z, t) = (x + y, z + t, x + z).$$

Find a basis for range T.

Sol. We have $T(x, y, z, t) = x\,(1, 0, 1) + y\,(1, 0, 0) + z\,(0, 1, 1) + t\,(0, 1, 0)$.

Thus the set $S = \{(1, 0, 1), (1, 0, 0), (0, 1, 1), (0, 1, 0)\}$ spans range T.

To find a subset of S that is, a basis for range T, we proceed as follows :

$$x(1, 0, 1) + y\,(1, 0, 0) + z\,(0, 1, 1) + t\,(0, 1, 0) = (0, 0, 0).$$

The matrix whose rows are the vectors of the set S is

$$\begin{bmatrix}1 & 0 & 1\\1 & 0 & 0\\0 & 1 & 1\\0 & 1 & 0\end{bmatrix}$$

Now, $\begin{bmatrix}1 & 0 & 1\\1 & 0 & 0\\0 & 1 & 1\\0 & 1 & 0\end{bmatrix} \underset{\sim}{R_1 \leftrightarrow R_1} \begin{bmatrix}1 & 0 & 0\\1 & 0 & 1\\0 & 1 & 1\\0 & 1 & 0\end{bmatrix} \underset{\sim}{R_2 - R_1} \begin{bmatrix}1 & 0 & 0\\0 & 0 & 1\\0 & 1 & 1\\0 & 1 & 0\end{bmatrix} \underset{\sim}{R_3 - R_2} \begin{bmatrix}1 & 0 & 0\\0 & 0 & 1\\0 & 1 & 0\\0 & 1 & 0\end{bmatrix}$

$\underset{\sim}{R_4 - R_3} \begin{bmatrix}1 & 0 & 0\\0 & 0 & 1\\0 & 1 & 0\\0 & 0 & 0\end{bmatrix} \underset{\sim}{R_2 \leftrightarrow R_3} \begin{bmatrix}1 & 0 & 0\\0 & 1 & 0\\0 & 0 & 1\\0 & 0 & 0\end{bmatrix}$. [reduces to row echelon form]

Hence, the set $\{(1, 0, 0), (0, 1, 0), (0, 0, 1)\}$ is a basis for range T.

Here dim (ker T) = 1, dim (range T) = 2 and dim (domainT) = 3

From the problem we can write

dim (ker T) + dim (range T) = dim (domain T).

Now we state a theorem without proof as follows :

Theorem 2.21. *If $T : V \to W$ is a linear transformation of an n-dimensional vector space V into a vector space W, then dim (ker T) + dim (range T) = dim V.*

Remark. The dimension of ker T is also called *nullity* of T, and the dimension of range T is called the *rank* of T. We define formally, the rank and nullity as follows :

Definition. Let V and W be vector spaces over the field F and let T be a linear transformation from V into W. The *null space* of T is the set of all vectors v in V such $Tv = 0$.

If V is finite-dimensional, the rank of T is the dimension of the range of T and the nullity of T is the dimension of the null space of T.

The following is the most important relation between rank, nullity and dimension :

Theorem 2.22. *Let V and W be vector spaces over the field F and let T be a linear transformation from V into W. Let us assume that V is finite-dimensional. Then*

$$rank(T) + nullity(T) = dim\ V.$$

Proof. Let N be the *null space* of T and $\{u_1, u_2, ..., u_m\}$ be a basis for N. There exist vectors $u_{m+1}, u_{m+2}, ..., u_n$ in V such that $\{u_1, u_2, ..., u_n\}$ is a basis for V. It is required to prove that $\{Tu_{m+1}, Tu_{m+2}, ..., Tu_n\}$ is a basis for the range of T. Since $u_1, u_2, ..., u_n$ are the elements of the basis of V, therefore $Tu_1, Tu_2, ..., Tu_n$ span the range of T, and since $Tu_j = O$, for $j \le m$, then $Tu_{m+1}, ..., Tu_n$ span the range. Next to prove that these vectors are linearly independent. Let there exists scalars $c_{m+1}, c_{m+2}, ..., c_n$ such that

$$\sum_{i=m+1}^{n} c_i (Tu_i) = o$$

This implies $T\left(\sum_{i=m+1}^{n} c_i\, u_i\right) = o$ and hence the vector

$$u = \sum_{i=m+1}^{n} c_i\, u_i$$

is in the null space of T.

Since the vectors $u_1, u_2, ..., u_m$ form a basis for N, there must exist scalars $b_1, b_2, ... b_m$ such that

$$u = \sum_{i=1}^{m} b_i\, u_i.$$

Thus, $\sum_{i=1}^{m} b_i\, u_i - \sum_{j=m+1}^{n} c_i\, u_j = o$

Since $u_1, u_2, ..., u_n$ are linearly independent, therefore,

$$b_1 = b_2 = ... = b_m = c_{m+1} = c_{m+2} = ... = c_n = 0.$$

If r is the rank of T, and since $Tu_{m+1}, ... Tu_n$ form a basis for the range T, hence $r = n - m$. Since m is nullity of T and n is the dimension of V, hence

$$\text{rank (T)} + \text{nullity (T)} = \dim \text{V}.$$

7. *Let* $T : R^3 \to R^3$ *be a linear transformation defined by*

$$T\left(\begin{bmatrix} x \\ y \\ z \end{bmatrix}\right) = \begin{bmatrix} x+z \\ x+y \\ y-z \end{bmatrix}$$

Find rank (T), nullity (T).

Sol. A vector $\begin{bmatrix} x \\ y \\ z \end{bmatrix}$ is in ker T if

$$T\left(\begin{bmatrix} x \\ y \\ z \end{bmatrix}\right) = \begin{bmatrix} 0 \\ 0 \\ 0 \end{bmatrix}.$$

We now find a basis for solution space of homogeneous system

$$x + z = 0$$

$$x + y = 0$$

$$y - z = 0.$$

These give $\quad y = -x, y = z.$

Thus ker T consists of all vectors of the form $\begin{bmatrix} -r \\ r \\ r \end{bmatrix} = r\begin{bmatrix} -1 \\ 1 \\ 1 \end{bmatrix}$ where r is any real number.

$\therefore$ A basis of ker T is $\left\{\begin{bmatrix} -1 \\ 1 \\ 1 \end{bmatrix}\right\}$

$\therefore$ $\dim$ (ker T) (or nullity of T) = 1

Also, every vector in range T is of the form $\begin{bmatrix} x+z \\ x+y \\ y-z \end{bmatrix}$,

which can be expressed as

$$x\begin{bmatrix}1\\1\\0\end{bmatrix}+y\begin{bmatrix}0\\1\\1\end{bmatrix}+z\begin{bmatrix}1\\0\\-1\end{bmatrix}$$

Here
$$A=\begin{bmatrix}1&0&1\\1&1&0\\0&1&-1\end{bmatrix}$$

$\therefore$
$$|A|=\begin{vmatrix}1&0&1\\1&1&0\\0&1&-1\end{vmatrix}=1(-1)+(1)=0$$

$\therefore$ The three vectors $\begin{bmatrix}1\\1\\0\end{bmatrix},\begin{bmatrix}0\\1\\1\end{bmatrix},\begin{bmatrix}1\\0\\-1\end{bmatrix}$ are linearly dependent.

But any two vectors are linearly independent, because they are not constant multiples of each other.

$\therefore$ The basis for range T (or rank of T) is

$$\left\{\begin{bmatrix}1\\1\\0\end{bmatrix},\begin{bmatrix}0\\1\\0\end{bmatrix}\right\}\text{ or, }\left\{\begin{bmatrix}1\\1\\0\end{bmatrix},\begin{bmatrix}1\\0\\-1\end{bmatrix}\right\}\text{ or }\left\{\begin{bmatrix}0\\1\\1\end{bmatrix},\begin{bmatrix}1\\0\\-1\end{bmatrix}\right\}$$

Hence rank of T is 2. Also, dimension of $R^3 = 3$.

Thus, rank (T) + nullity (T) = 2 + 1 = 3 = $dim\ (R^3)$.

8. *If $T : R^3 \to R^3$ is given by $T(x, y, z) = (x - y, y - z, z - x)$, show that T is a linear transformation and obtain Kernel T.* [W.B.U.T, BCA, 2006]

Sol. Let u and v be two vectors in R^3, such that

$$u=(a_1,a_2,a_3),\ v=(b_1,b_2,b_3).$$

Then
$$\begin{aligned}T(u+v)&=T((a_1,a_2,a_3)+(b_1,b_2,b_3))\\&=T(a_1+b_1,a_2+b_2,a_3+b_3)\\&=(a_1+b_1-a_2-b_2,a_2+b_2-a_3-b_3,a_3+b_3-a_1-b_1)\quad\text{...(i)}\end{aligned}$$

Again
$$Tu=T(a_1,a_2,a_3)=(a_1-a_2,a_2-a_3,a_3-a_1)$$

and
$$Tv=T(b_1,b_2,b_3)=(b_1-b_2,b_2-b_3,b_3-b_1)$$

$\therefore$
$$\begin{aligned}Tu+Tv&=(a_1-a_2,a_2-a_3,a_3-a_1)+(b_1-b_2,b_2-b_3,b_3-b_1)\\&=(a_1-a_2+b_1-b_2,a_2-a_3+b_2-b_3,a_3-a_1+b_3-b_1)\\&=(a_1+b_1-a_2-b_2,a_2+b_2-a_3-b_3,a_3+b_3-a_1-b_1)\quad\text{...(ii)}\end{aligned}$$

$\therefore$ From (*i*) and (*ii*), we have

$$T(u+v)=Tu+Tv.$$

If c be any real number, then

$$\begin{aligned} T(cu) &= T\,(c(a_1-a_2), c\,(a_2-a_3), c\,(a_3-a_1)) \\ &= T\,(ca_1-ca_2, ca_2-ca_3, ca_3-ca_1) \\ &= (ca_1-ca_2-ca_2+ca_3, ca_2-ca_3-ca_3+ca_1, ca_3-ca_1-ca_1+ca_2) \\ &= (c(a_1-a_2-a_2+a_3, c\,(a_2-a_3-a_3+a_1), c\,(a_3-a_1-a_1+a_2) \\ &= c\,((a_1-a_2)-(a_2-a_3), (a_2-a_3)-(a_3-a_1), (a_3-a_1)-(a_1-a_2)) \\ &= cTu. \end{aligned}$$

Hence, T is a linear transformation.

A vector (x, y, z) is in Ker T if

$$T\,(x, y, z) = (0, 0, 0)$$

i.e., $(x-y, y-z, z-x) = (0, 0, 0)$

i.e.,
$$x-y=0$$
$$y-z=0$$
$$z-x=0.$$

From these equations we obtain $x = y = z$.

Thus, ker T consists all vectors of the form

$$\begin{bmatrix} r \\ r \\ r \end{bmatrix} = r\begin{bmatrix} 1 \\ 1 \\ 1 \end{bmatrix}, \text{ where } r \text{ is any real number.}$$

9. *Determine the linear transformation $T : R^3 \to R^2$ which maps the basis vector (1, 0, 0), (0, 1, 0), (0, 0, 1) of R^3 to the vectors (1, 1), (2, 3), (3, 2) respectively. Find ker T and Im T.*

Sol. Let $u = (x, y, z)$ be any vector in R^3.

Then $u = x\,(1, 0, 0) + y\,(0, 1, 0) + z\,(0, 0, 1)$

Since T is a linear mapping,

$$\therefore \quad T\,(u) = x\,T\,(1, 0, 0) + y\,T\,(0, 1, 0) + z\,T\,(0, 0, 1)$$
$$= x\,(1, 1) + y\,(2, 3) + z\,(3, 2) = (x + 2y + 3z, x + 3y + 2z).$$

Thus T is defined by

$$T\,(x, y, z) = (x + 2y + 3z, x + 3y + 2z), x, y, z \in R^3.$$

Let $T\,(x, y, z) \in \text{Ker}\,T.$

Then $T\,(x, y, z) = (0, 0)$

$\therefore \quad x + 2y + 3z = 0$

and $\quad x + 3y + 2z = 0$

$$\therefore \quad \frac{x}{-5} = \frac{y}{1} = \frac{z}{1} = k, \text{ say}$$

$$\therefore \quad (x, y, z) = k\,(-5, 1, 1), \text{ where } k \in R$$

$$\text{Ker T} = L\{\alpha\}$$

where $\alpha = (-5, 1, 1)$

Image of T (Im T) is the linear span of vectors

$T(\alpha_1), T(\alpha_2), T(\alpha_3)$ where $\{\alpha_1, \alpha_2, \alpha_3\}$ is any basis of R^3.

Since $\{(1,0,0), (0,1,0), (0,0,1)\}$ is a basis of R^3,

therefore, $Im\, T = L\{(1,1), (2,3), (3,2)\}$

10. *Define the linear transformation T on R^3 that changes the basis vectors (1, 0, 0), (0, 1, 0), (0, 0 1) to (1, 1, 1) (0, 1, –1), (1, 2, 0) respectively. Find T (1, 1, –1) and T(2, 2, –2).*

Sol. Let $u = (x, y, z)$ be any vector of R^3.

Then $u = x(1,0,0) + y(0,1,0) + z(0,0,1)$.

Since T is a linear transformation therefore,

$$T(u) = x\,T(1,0,0) + y\,T(0,1,0) + z\,(0,0,1)$$
$$= x(1,1,1) + y(0,1,-1) + z(1,2,0)$$
$$= (x+z, x+y+2z, x-y).$$

Hence, the linear transformation is defined by

$$T(x,y,z) = (x+z, x+y+2z, x-y)$$

$\therefore$ $T(1,1,-1) = (1-1, 1+1-2, 1-1) = (0,0,0)$

and $T(2,2,-2) = (2-2, 2+2-4, 2-2) = (0,0,0)$.

EXERCISE 2.6

1. If T $(x, y, z) = (x, y, 0)$ for all $(x, y, z) \in R^3$ is a linear transformation, then Ker T is
   (*i*) (0, 0, 0) (*ii*) x-axis (*iii*) y-axis (*iv*) z-axis
   (W.B.U.T, BCA, 2006)
2. Let $W = \{(x, y, z) \in R^3 : x + y - 2z = 0\}$ show that W is a subspace of R^3, and obtain a basis in W. *(W.B.U.T, 2006)*
3. Find the Kernel of the following linear transformations :
   (*i*) $T : V_4(R) \to V_4(R)$ defined by $T(x_1, x_2, x_3, x_4) = (x_1, 0, x_3, 0)$.
   (*ii*) $T : V_3(R) \to V_3(R)$ defined by $T(x, y, 0) = (x, y, 0)$.
4. Find the basis and dimension of subspace S of R^3, where $S = \{(x, y, z) \in R^3 : x + 2y + z = 0, 2x + y + 3z = 0\}$.
5. If $W_2 = \left\{\begin{pmatrix} a & b \\ c & d \end{pmatrix} : a, b, c, d \text{ are reals with } a + b = 0\right\}$, show that W_2 is a subspace of M_2 and determine dimension (W_2).
6. Let $T : R^2 \to R^3$ be defined by $T(x, y) = (x, x + y, y)$.
   (*i*) Find ker T. (*ii*) Is T one-to-one (*iii*) Is T onto ?

7. Let $T : R^4 \to R^3$ be defined by

$$T\left(\begin{bmatrix} a \\ b \\ c \\ d \end{bmatrix}\right) = \begin{bmatrix} a+b \\ b-d \\ c-d \end{bmatrix}.$$

(*i*) Is T onto? (*ii*) Find the dimension of Ker T.

8. Let $T : M_{22} \to M_{22}$ be a linear transformation defined by

$$T\left(\begin{bmatrix} a & b \\ c & d \end{bmatrix}\right) = \begin{bmatrix} a+b & b+c \\ a+d & b+d \end{bmatrix}.$$

(*i*) Find a basis for Ker T (*ii*) Find a basis for range T.

9. Let $T : M_{22} \to M_{22}$ be a linear transformation defined by

$$T(u) = \begin{bmatrix} 1 & 2 \\ 1 & 1 \end{bmatrix} u - u \begin{bmatrix} 1 & 2 \\ 1 & 1 \end{bmatrix}.$$

(*i*) Find a basis for Ker T (*ii*) Find a basis for range T.

10. Determine which of the following statements are true.

(*a*) Any line passing through the origin is a subspace of V_3 (R).

(*b*) Any plane passing through the origin is a subspace of V_3 (R).

(*c*) If A and B are subspaces of V then

(*i*) A + B is a subspace of V.

(*ii*) A is a subspace of A + B.

(*iii*) B is a subspace of A + B.

(*iv*) Every element of A + B can be uniquely written in the form $x + y$ where $x \in A$ and $y \in B$.

(*v*) (*iv*) is true iff $A \cap B = \{0\}$.

11. Let $T : R^2 \to R^2$ be a linear transformation defined by $T(x, y) = (x, 0)$.

(*i*) Is (0, 2) in Ker T? (*ii*) Is (2, 2) in Ker T?

(*iii*) Is (3, 0) in range T? (*iv*) Is (3, 2) in range T?

(*v*) Find Ker T (*vi*) Find range T.

12. Let $T : R^2 \to R^2$ be a linear transformation defined by

$$T\left(\begin{bmatrix} x \\ y \end{bmatrix}\right) = \begin{bmatrix} 1 & 2 \\ 2 & 4 \end{bmatrix}\begin{bmatrix} x \\ y \end{bmatrix}.$$

(*i*) Is $\begin{bmatrix} 1 \\ 2 \end{bmatrix}$ in Ker T? (*ii*) Is $\begin{bmatrix} 2 \\ -1 \end{bmatrix}$ in Ker T?

(*iii*) Is $\begin{bmatrix} 3 \\ 6 \end{bmatrix}$ in range T? (*iv*) Is $\begin{bmatrix} 2 \\ 3 \end{bmatrix}$ in range T?

(*v*) Find Ker T (*vi*) Find a set of vectors spanning range L.

Answers

1. (*i*)

3. (*i*) $\text{Ker T} = \{(0, a, 0, b) : a, b \in \text{R}\}$

   (*ii*) $\text{Ker T} = \{(0, 0, c) : c \in \text{R}\}$.

4. $\{5, -1, 3\}, 1$

6. (*i*) $\{(0, 0)\}$, (*ii*) yes, (*iii*) No.

7. (*i*) yes (*ii*) 1

8. (*i*) $\ker \text{T} = \left\{\begin{bmatrix} 0 & 0 \\ 0 & 0 \end{bmatrix}\right\}$, so Ker T has no basis

   (*ii*) $\left\{\begin{bmatrix} 1 & 0 \\ 1 & 0 \end{bmatrix}, \begin{bmatrix} 1 & 1 \\ 0 & 1 \end{bmatrix}, \begin{bmatrix} 0 & 1 \\ 0 & 0 \end{bmatrix}, \begin{bmatrix} 0 & 0 \\ 1 & 1 \end{bmatrix}\right\}$

9. (*i*) $\left\{\begin{bmatrix} 1 & 0 \\ 0 & 1 \end{bmatrix}, \begin{bmatrix} 0 & 1 \\ \frac{1}{2} & 0 \end{bmatrix}\right\}$ (*ii*) $\left\{\begin{bmatrix} 0 & -2 \\ 1 & 0 \end{bmatrix}, \begin{bmatrix} -1 & 0 \\ 0 & 1 \end{bmatrix}\right\}$

10. (*a*) T (*b*) T (*c*) (*i*) T (*ii*) T (*iii*) T (*iv*) F (*v*) T

11. (*i*) yes (*ii*) No (*iii*) yes (*iv*) No.

    (*v*) $\{(0, r)\}$, r is any real number (*vi*) $\{(r, 0)\}$, r is any real number.

2.14 REPRESENTATION OF TRANSFORMATIONS BY MATRICES

Let U be an m-dimensional vector space over the field F and let V be an n-dimensional vector space over F. Let $\beta = \{u_1, u_2, ..., u_m\}$ be an ordered basis for U and $\beta' = \{v_1, v_2, ..., v_n\}$ be an ordered basis for V.

Let $\text{T} : \text{U} \to \text{V}$ be a linear transformation. Then T is determined by its action on the vector u_i. Each of m vectors $\text{T}u_i$ is uniquely expressible as a linear combination.

$$\text{T}\,u_i = \sum_{j=1}^{n} a_{ij}\, v_j, \text{ where } a_{ij} \text{ are scalars, } i = 1, 2, 3, \dots m.$$

that is

$$\left.\begin{aligned} \text{T}u_1 &= a_{11}\, v_1 + a_{12}\, v_2 + \dots + a_{1n}\, v_n \\ \text{T}u_2 &= a_{21}\, v_1 + a_{22}\, v_2 + \dots + a_{2n}\, v_n \\ &\quad \dots \qquad \dots \qquad \dots \\ \text{T}u_m &= a_{m1}\, v_1 + a_{m2}\, v_2 + \dots + a_{mn}\, v_n \end{aligned}\right\} \qquad \dots(1)$$

Thus, $\text{T}u_1, \text{T}u_2, ..., \text{T}u_m$ are completely specified by the mn elements a_{ij} of the field F. These a_{ij} elements can be conveniently arranged in the form of m rows and n columns as follows :

$$\begin{pmatrix} a_{11} & a_{12} & \cdots & a_{1n} \\ a_{21} & a_{22} & \cdots & a_{2n} \\ \dots & \dots & \dots & \dots \\ a_{m1} & a_{m2} & \cdots & a_{mn} \end{pmatrix}$$

or, in brief (a_{ij}).

Such an array of mn element arranged in m rows and n columns is known as $m \times n$ matrix over the field F.

The $m \times n$ matrix $A = (a_{ij})$, is called the *matrix of T relative to the pair of ordered bases β and β'.*

Now our task to show explicitly how the matrix A determines the linear transformation T.

If $$u = \lambda_1 u_1 + \lambda_2 u_2 + ... \lambda_m v_m \text{ is a vector in., then}$$

$$Tu = T\left(\sum_{i=1}^{m} \lambda_i u_i\right) = \sum_{i=1}^{m} \lambda_i (Tu_i)$$

$$= \sum_{i=1}^{m} \lambda_i \sum_{j=1}^{n} a_{ij} v_j = \sum_{j=1}^{n} \left(\sum_{i=1}^{m} a_{ij} \lambda_i\right) v_j$$

If X is the coordinate matrix of u in the ordered basis β, then the above computation shows that AX is the coordinate matrix of the vector Tu in the ordered basis β', because the scalar $\sum_{i=1}^{m} a_{ij} \lambda_i$ is the entry in the ith row of the column matrix AX.

Thus, if A is any $m \times n$ matrix over the field F, then

$$T\left(\sum_{i=1}^{m} \lambda_i u_i\right) = \sum_{j=1}^{n} \left(\sum_{i=1}^{m} a_{ij} \lambda_i\right) u_j$$

defines a linear transformation T from U into V, the matrix of which is A, relative to the bases β, β'.

Remark. To every linear transformation T an $m \times n$ matrix, over the field F, is associated with it.

Conversely, any $m \times n$ matrix over F defines a linear transformation $T: U \to V$ given by (1).

Examples :

1. *Let F be a field and T be an operator (mapping) on F^2 defined by $T(x_1, x_2) = (x_1, 0)$. Show that T is a linear mapping on F^2.*

Sol. Let β be the standard ordered basis for F^2,

that is, $$\beta = \{e_1, e_2\}.$$

Now $$T(e_1) = T(1, 0) = (1, 0) = 1e_1 + 0e_2$$

$$T(e_2) = T(0, 1) = (0, 0) = 0e_1 + 0e_2.$$

So the matrix of T in the ordered basis β is

$$[T]_\beta = \begin{bmatrix} 1 & 0 \\ 0 & 0 \end{bmatrix}.$$

2. *Let $T : V_2(R) \to V_2(R)$ be a linear transformation defined by $T(a, b) = (a, a + b)$. Find the matrix of T, considering $\{e_1, e_2\}$ as a basis both for the domain and range.*

Sol.

$$T(e_1) = T(1, 0) = (1, 1 + 0) = (1, 1) = 1e_1 + 1e_2$$
$$T(e_2) = T(0, 1) = (0, 0 + 1) = (0, 1) = 0e_1 + 1e_2$$

so the matrix of T is $\begin{bmatrix} 1 & 1 \\ 0 & 1 \end{bmatrix}$.

3. *Let $T : V_2(R) \to V_2(R)$ be a linear transformation defined by $T(a, b) = (a, a + b)$. Find the matrix of T, taking $\{e_1, e_2\}$ as a basis for the domain and $\{(1, 1), (1, -1)\}$ as a basis for the range.*

Sol. Let $v_1 = (1, 1), v_2 = (1, -1)$.

Then $Te_1 = T(1, 0) = (1, 1 + 0) = (1, 1) = v_1 + 0v_2$

$$Te_2 = T(0, 1) = (0, 0 + 1) = (0, 1) = \frac{1}{2}v_1 - \frac{1}{2}v_2$$

[working procedure : Let

$$(1, 1) = \lambda_1(1, 1) + \lambda_2(1, -1) = (\lambda_1 + \lambda_2, \lambda_1 - \lambda_2)$$

$\therefore$ $\lambda_1 + \lambda_2 = 1$ and $\lambda_1 - \lambda_2 = 1$

solving, we get $\lambda_1 = 1$ and $\lambda_2 = 0$

$\therefore$ $(1, 1) = 1.v_1 + 0.v_2$.

Next, let $(0, 1) = \mu_1 v_1 + \mu_2 v_2 = \mu_1(1, 1) + \mu_2(1, -1) = (\mu_1 + \mu_2, \mu_1 - \mu_2)$

$\therefore$ $\mu_1 + \mu_2 = 0$ and $\mu_1 - \mu_2 = 1$

Solving we obtain, $\mu_1 = \frac{1}{2}$ and $\mu_2 = -\frac{1}{2}$

Hence, $(0, 1) = \frac{1}{2}v_1 - \frac{1}{2}v_2$]

So the matrix of $T = \begin{bmatrix} 1 & 0 \\ \frac{1}{2} & -\frac{1}{2} \end{bmatrix}$.

4. *Obtain the matrix representing the linear transformation $T : V_3(R) \to V_3(R)$ defined by*

$$T(a, b, c) = (3a, a - b, 2a + b + c)$$

with respect to the standard basis for the domain and range.

Sol. The basis is $\{e_1, e_2, e_3\}$.

$\because$ $T(a, b, c) = (3a, a - b, 2a + b + c)$,

$\therefore$ $T(e_1) = T(1, 0, 0) = (3.1, 1 - 0, 2.1 + 0 + 0) = (3, 1, 2) = 3e_1 + 1e_2 + 2e_3$

$$T(e_2) = T(0, 1, 0) = (0, 0 - 1, 0 + 1 + 0) = (0, -1, 1) = 0e_1 - 1e_2 + 1e_3$$
$$T(e_3) = T(0, 0, 1) = (0, 0, 0 + 0 + 1) = (0, 0, 1) = 0e_1 + 0e_2 + 1e_3$$

so the matrix representing T is $\begin{bmatrix} 3 & 1 & 2 \\ 0 & -1 & 1 \\ 0 & 0 & 1 \end{bmatrix}$.

5. *If $T : R^3 \to R^3$ is given by $T(x, y, z) = (x - 2y, y - 2z, z - 2x)$ for $(x, y, z) \in R^3$, obtain a representative matrix for the linear transformation T.* (W.B.U.T, BCA, 2006)

Sol. Given $T(x, y, z) = (x - 2y, y - 2z, z - 2x)$.

Since the basis for domain and range are not given we assume standard basis both for domain and range. The standard basis is $\{e_1, e_2, e_3\}$.

Now,
$$T(e_1) = T(1, 0, 0) = (1-0, 0-0, 0-2) = (1, 0, -2) = 1e_1 + 0e_2 - 2e_3$$
$$T(e_2) = T(0, 1, 0) = (0-2, 1-0, 0-0) = (-2, 1, 0) = -2e_1 + 1e_2 + 0e_3$$
$$T(e_3) = T(0, 0, 1) = (0, 0-2, 1-0) = (0, -2, 1) = 0e_1 - 2e_2 + 1e_3$$

so the matrix representing T is $\begin{bmatrix} 1 & 0 & -2 \\ -2 & 1 & 0 \\ 0 & -2 & 1 \end{bmatrix}$.

6. *Find the linear transformation $T : V_3(R) \to V_3(R)$ determined by the matrix $\begin{bmatrix} 1 & 2 & 1 \\ 0 & 1 & 1 \\ -1 & 3 & 4 \end{bmatrix}$ with respect to standard basis $\{e_1, e_2, e_3\}$.*

Sol. The transformations are
$$T(e_1) = 1e_1 + 2e_2 + 1e_3 = (1, 2, 1)$$
$$T(e_2) = 0e_1 + 1e_2 + 1e_3 = (0, 1, 1)$$
$$T(e_3) = -1e_1 + 3e_2 + 4e_3 = (-1, 3, 4)$$
(using the elements of the rows of the given matrix),

Now,
$$(a, b, c) = a(1, 0, 0) + b(0, 1, 0) + c(0, 0, 1)$$
$$= ae_1 + be_2 + ce_3$$
$\therefore$
$$T(a, b, c) = T(ae_1) + T(be_2) + T(ce_3)$$
$$= aT(e_1) + bT(e_2) + cT(e_3)$$
$$= a(1, 2, 1) + b(0, 1, 1) + c(-1, 3, 4)$$
$$= (a - c, 2a + b + 3c, a + b + 4c)$$

Hence, $T(a, b, c) = (a - c, 2a + b + 3c, a + b + 4c)$,

is the required linear transformation.

7. *Obtain the matrix representing the linear transformation $T : V_2(R) \to V_2(R)$ defined by $T(a, b) = (-b, a)$ with respect to the basis $\{(1, 2), (1, -1)\}$ for both domain and range.*

Sol. Let $\{u_1, u_2\}$ and $\{v_1, v_2\}$ represent the basis for domain and range respectively.

Here $u_1 = (1, 2), u_2 = (1, -1); v_1 = (1, 2), v_2 = (1, -1)$.

Given $T(a, b) = (-b, a)$

$\therefore$
$$Tu_1 = T(1, 2) = (-2, 1)$$
$$Tu_2 = T(1, -1) = (1, 1)$$

Let $(-2, 1) = \lambda_1 v_1 + \lambda_2 v_2 = \lambda_1(1, 2) + \lambda_2(1, -1) = (\lambda_1 + \lambda_2, 2\lambda_1 - \lambda_2)$

$\therefore$ $$\lambda_1 + \lambda_2 = -2 \text{ and } 2\lambda_1 - \lambda_2 = 1,$$

solving, we obtain $$\lambda_1 = -\frac{1}{3} \text{ and } \lambda_2 = -\frac{5}{3}.$$

and $$(1, 1) = \mu_1 v_1 + \mu_2 v_2 = \mu_1 (1, 2) + \mu_2 (1, -1) = (\mu_1 + \mu_2, 2\mu_1 - \mu_2)$$

$\therefore$ $$\mu_1 + \mu_2 = 1 \text{ and } 2\mu_1 - \mu_2 = 1$$

Solving these equations, we get $\mu_1 = \frac{2}{3}$ and $\mu_2 = \frac{1}{3}$

$\therefore$ $$T(u_1) = \lambda_1 v_1 + \lambda_2 v_2 = -\frac{1}{3} v_1 - \frac{5}{3} v_2$$

$$T(u_2) = \mu_1 v_1 + \mu_2 v_2 = \frac{2}{3} v_1 + \frac{1}{3} v_2$$

so the matrix representing T is $\begin{pmatrix} -\frac{1}{3} & -\frac{5}{3} \\ \frac{2}{3} & \frac{1}{3} \end{pmatrix}$.

8. *Obtain the matrix representing the linear transformation $T : V_3 (R) \to V_2 (R)$ given by $T (a, b, c) = (a + b, 2c - a)$ w.r.t.*

(i) Standard basis

(ii) $\{(1, 0, -1), (1, 1, 1), (1, 0, 0)\}$ as a basis for $V_3 (R)$ and $\{(0, 1), (1, 0)\}$ for $V_2 (R)$.

Sol. (*i*) Given $T (a, b, c) = (a + b, 2c - a)$; and Basis : $\{e_1, e_2, e_3\}$

$\therefore$ $$Te_1 = T (1, 0, 0) = (1 + 0, 2.0 - 1) = (1, -1) = 1e_1 - 1e_2$$

$$Te_2 = T(0, 1, 0) = (0 + 1, 2.0 - 0) = (1, 0) = 1e_1 + 0e_2$$

$$Te_3 = T (0, 1, 0) = (0 + 0, 2.1 - 0) = (0, 2) = 0e_1 + 2e_2.$$

Hence, the matrix representing T is $\begin{pmatrix} 1 & -1 \\ 1 & 0 \\ 0 & 2 \end{pmatrix}$.

(*ii*) Given $T (a, b, c) = (a + b, 2c - a)$

Basis for V_3: $\{(1, 0, -1), (1, 1, 1), (1, 0, 0)\}$

and basis for V_2 : $\{(0, 1), (1, 0)\}$

Now, $$T (1, 0, -1) = (1 + 0, 2.(-1) - 1) = (1, -3)$$

$$T (1, 1, 1) = (1 + 1, 2.1 - 1) = (2, 1)$$

$$T (1, 0, 0) = (1 + 0, 2.0 - 1) = (1, -1)$$

Let $$(1, -3) = \lambda_1 (0, 1) + \lambda_2 (1, 0) = (\lambda_2, \lambda_1)$$

$\therefore$ $$\lambda_2 = 1, \lambda_1 = -3$$

$$(2, 1) = a_1 (0, 1) + a_2 (1, 0) = (a_2, a_1)$$

$\therefore$ $$a_2 = 2, a_1 = 1$$

and $\quad (1,-1) = b_1 (0,1) + b_2 (1,0) = (b_2, b_1)$

$\therefore \quad b_2 = 1, b_1 = -1.$

Thus,
$$T(1,0,-1) = -3(0,1) + 1(1,0)$$
$$T(1,1,1) = 1(0,1) + 2(1,0)$$
$$T(1,0,0) = -1(0,1) + 1(1,0)$$

so the matrix representing T is $\begin{pmatrix} -3 & 1 \\ 1 & 2 \\ -1 & 1 \end{pmatrix}$.

9. *Obtain the linear transformation* $T : V_3(R) \to V_3(R)$ *determined by the matrix* $\begin{bmatrix} a & b & c \\ b & c & a \\ c & a & b \end{bmatrix}$ *w.r.t. the standard basis.*

Sol. The standard basis is $\{e_1, e_2, e_3\}$
$$T(e_1) = ae_1 + be_2 + ce_3 = (a, b, c)$$
$$T(e_2) = be_1 + ce_2 + ae_3 = (b, c, a)$$
$$T(e_3) = ce_1 + ae_2 + be_3 = (c, a, b)$$
$$\therefore \quad T(x, y, z) = x(a, b, c) + y(b, c, a) + z(c, a, b)$$
$$= (ax + by + cz, bx + cy + az, cx + ay + bz).$$

2.15 MATRIX ASSOCIATED WITH A LINEAR MAP

Let U and V be vector spaces of dimensions n and m respectively. Let $\beta_1 = \{u_1, u_2, ..., u_n\}$ and $\beta_2 = \{v_1, v_2, ..., v_m\}$ be the ordered bases of U and V, respectively.

Let $T : U \to V$ be a linear map (or transformation) defined by
$$T(u_j) = \alpha_{1j} v_1 + \alpha_{2j} v_2 + ... + \alpha_{mj} v_m, \; j = 1, 2, ..., n$$

so that the coordinate vectors of $T(u_j)$ written as column vector is $\begin{bmatrix} \alpha_{1j} \\ \alpha_{2j} \\ \vdots \\ \alpha_{mj} \end{bmatrix}$.

Write the coordinate vectors of $T(u_1), T(u_2), ..., T(u_j), ... T(u_n)$ successively as column vectors in the form of rectangular array as follows:
$$\begin{bmatrix} \alpha_{11} & \alpha_{12} & \cdots & \alpha_{1j} & \cdots & \alpha_{1n} \\ \alpha_{21} & \alpha_{22} & \cdots & \alpha_{2j} & \cdots & \alpha_{2n} \\ \vdots & \vdots & & \vdots & & \vdots \\ \alpha_{m1} & \alpha_{m2} & \cdots & \alpha_{mj} & \cdots & \alpha_{mn} \end{bmatrix}.$$

The $m \times n$ rectangular array is the matrix of T relative to the ordered bases β_1 and β_2 and is denoted by $[T : \beta_1, \beta_2]$.

Examples :

1. *Let a linear transformation* $T : V_2 \to V_3$ *be defined by* $T(x_1, x_2) = (x_1 + x_2, 2x_1 - x_2, 7x_2)$. *If*
$$\beta_1 = \{e_1, e_2\} \text{ and } \beta_2 = \{f_1, f_2, f_3\}$$
are the standard bases of V_2 *and* V_3 *respectively, find the matrix of T relative to* β_1 *and* β_2.

Sol.
$$T(e_1) = T(1, 0) = (1 + 0, 2 - 0, 0) = (1, 2, 0)$$
$$T(e_2) = T(0, 1) = (0 + 1, 0 - 1, 7) = (1, -1, 7)$$
$\therefore$
$$T(e_1) = f_1 + 2f_2 + 0f_3$$
and
$$T(e_2) = f_1 - f_2 + 7f_3$$

Hence, the coordinate vectors of T (e_1) and T (e_2) are respectively $\begin{bmatrix} 1 \\ 2 \\ 0 \end{bmatrix}$ and $\begin{bmatrix} 1 \\ -1 \\ 7 \end{bmatrix}$. Therefore, the matrix of T relative to β_1 and β_2 is $\begin{bmatrix} 1 & 1 \\ 2 & -1 \\ 0 & 7 \end{bmatrix}$.

2. *Let* $T : R^3 \to R^2$ *be defined by* $T\left(\begin{bmatrix} x \\ y \\ z \end{bmatrix}\right) = \begin{bmatrix} x + y \\ y - z \end{bmatrix}$.

Let $S = \{v_1, v_2, v_3\}$ *and* $W = \{w_1, w_2\}$ *be the bases for* R^3 *and* R^2, *respectively, where*
$$v_1 = \begin{bmatrix} 1 \\ 0 \\ 1 \end{bmatrix}, v_2 = \begin{bmatrix} 0 \\ 1 \\ 1 \end{bmatrix}, v_3 = \begin{bmatrix} 1 \\ 1 \\ 1 \end{bmatrix};$$
$$w_1 = \begin{bmatrix} 1 \\ 2 \end{bmatrix} \quad \text{and } w_2 = \begin{bmatrix} -1 \\ 1 \end{bmatrix}.$$

Find the matrix of T w.r.t. S and W.

Sol. We have
$$T(v_1) = T\begin{bmatrix} 1 \\ 0 \\ 1 \end{bmatrix} = \begin{bmatrix} 1 + 0 \\ 0 - 1 \end{bmatrix} = \begin{bmatrix} 1 \\ -1 \end{bmatrix}$$
$$T(v_2) = T\begin{bmatrix} 0 \\ 1 \\ 1 \end{bmatrix} = \begin{bmatrix} 0 + 1 \\ 1 - 1 \end{bmatrix} = \begin{bmatrix} 1 \\ 0 \end{bmatrix}$$

$$T(v_3) = T\begin{bmatrix}1\\1\\1\end{bmatrix} = \begin{bmatrix}1+1\\1-1\end{bmatrix} = \begin{bmatrix}2\\0\end{bmatrix}$$

To find the coordinate vectors $[T(v_1)]_w$, $[T(v_2)]_w$ and $[T(v_3)]_w$, we write

$$T(v_1) = \begin{bmatrix}1\\-1\end{bmatrix} = a_1 w_1 + a_2 w_2 = a_1\begin{bmatrix}1\\2\end{bmatrix} + a_2\begin{bmatrix}-1\\1\end{bmatrix} = \begin{bmatrix}a_1 - a_2\\2a_1 + a_2\end{bmatrix}$$

$$T(v_2) = \begin{bmatrix}1\\0\end{bmatrix} = b_1 w_1 + b_2 w_2 = b_1\begin{bmatrix}1\\2\end{bmatrix} + b_2\begin{bmatrix}-1\\1\end{bmatrix} = \begin{bmatrix}b_1 - b_2\\2b_1 + b_2\end{bmatrix}$$

$$T(v_3) = \begin{bmatrix}2\\0\end{bmatrix} = c_1 w_1 + c_2 w_2 = c_1\begin{bmatrix}1\\2\end{bmatrix} + c_2\begin{bmatrix}-1\\1\end{bmatrix} = \begin{bmatrix}c_1 - c_2\\2c_1 + c_2\end{bmatrix}$$

Thus, we get

$$\begin{bmatrix}a_1 - a_2\\2a_1 + a_2\end{bmatrix} = \begin{bmatrix}1\\-1\end{bmatrix} \quad \text{...}(i)$$

$$\begin{bmatrix}b_1 - b_2\\2b_1 + b_2\end{bmatrix} = \begin{bmatrix}1\\0\end{bmatrix} \quad \text{...}(ii)$$

and

$$\begin{bmatrix}c_1 - c_2\\2c_1 + c_2\end{bmatrix} = \begin{bmatrix}2\\0\end{bmatrix} \quad \text{...}(iii)$$

from (*i*), we get $a_1 - a_2 = 1$ and $2a_1 + a_2 = -1$

solving, we obtain $a_1 = 0, a_2 = -1$

From (*ii*), we get $b_1 - b_2 = 1, 2b_1 + b_2 = 0$

solving, we have $b_1 = \frac{1}{3}, b_2 = -\frac{2}{3}$

From (*iii*), we obtain $c_1 - c_2 = 2, 2c_1 + c_2 = 0$

Solving, we get $c_1 = \frac{2}{3}, c_2 = -\frac{4}{3}$

Hence, the matrix representing T w.r.t. S and W is

$$\begin{bmatrix}a_1 & b_1 & c_1\\a_2 & b_2 & c_2\end{bmatrix} = \begin{bmatrix}0 & \frac{1}{3} & \frac{2}{3}\\-1 & -\frac{2}{3} & -\frac{4}{3}\end{bmatrix}.$$

3. *Let a linear map* $T : V_3 \to V_2$ *be defined by*

$$T(e_1) = 2f_1 - f_2, T(e_2) = f_1 + 2f_2, T(e_3) = of_1 + of_2,$$

where $\{e_1, e_2, e_3\}$ and $\{f_1, f_2\}$ are standard bases in V_3 and V_2 respectively. What is the matrix of T relative to these bases ?

Sol. The matrix of T is $\begin{bmatrix} 2 & 1 & 0 \\ -1 & 2 & 0 \end{bmatrix}$.

4. *Determine the matrix $(T : \beta_1, \beta_2)$ for the given linear transformation T and the bases β_1 and β_2 :*

$$T : V_2 \to V_2, T(x, y) = (x, -y)$$
$$B_1 = \{e_1, e_2\}, B_2 = \{(1, 1), (1, -1)\}$$

Sol.
$$T(e_1) = T(1, 0) = (1, 0)$$
$$T(e_2) = T(0, 1) = (0, -1)$$

Let
$$(1, 0) = a_1 (1, 1) + a_2 (1, -1) = (a_1 + a_2, a_1 - a_2) \quad \text{.... }(i)$$

and
$$(0, -1) = b_1 (1, 1) + b_2 (1, -1) = (b_1 + b_2, b_1 - b_2) \quad \text{... }(ii)$$

From (*i*),
$$a_1 + a_2 = 1;\ a_1 - a_2 = 0$$

Solving the equations, we get

$$a_1 = \frac{1}{2}, a_2 = \frac{1}{2}.$$

From (*ii*),
$$b_1 + b_2 = 0, b_1 - b_2 = -1$$

Solving these equations, we obtain

$$b_1 = -\frac{1}{2}, b_2 = \frac{1}{2}$$

$$\therefore \quad (T : \beta_1, \beta_2) = \begin{bmatrix} \frac{1}{2} & -\frac{1}{2} \\ \frac{1}{2} & \frac{1}{2} \end{bmatrix}.$$

EXERCISE 2.7

1. Let L : $R^2 \to R^2$ be the linear transformation given by L (x, y) = $(2x, 2y)$. Describe the image under L of the point lying on the circle $x^2 + y^2 = 1$.
2. Let T : $R^2 \to R^2$ be defined by T $(x, y) = (x - 2y, x + 2y)$.
   Let S = {(1, –1), (0, 1)} be a basis for R^2 and let P be natural basis for R. Find the matrix representing T with respect to (*i*) S (*ii*) S and P (*iii*) P and S (*iv*) P.
3. Let a linear mapping T : $R^2 \to R^3$ be defined by
   $$T(x, y) = (x + y, 2x - y, 7y).$$
   If S = $\{e_1, e_2\}$ and P = $\{u_1, u_2, u_3\}$ are the bases of R^2 and R^3, find the matrix of T relative to S and P.

4. Let the matrix of T : $R^2 \to R^2$ with respect to the basis

$$S = \left\{\begin{bmatrix}1\\-1\end{bmatrix}, \begin{bmatrix}0\\1\end{bmatrix}\right\} \text{ be } \begin{bmatrix}1 & 2\\-2 & 3\end{bmatrix}.$$

Find the matrix of T with respect to the natural basis for R^3.

5. The linear transformation T has, relative to the basis $\{e_1, e_2, e_3\}$, the matrix $\begin{pmatrix}15 & -11 & 5\\20 & -15 & 8\\8 & -7 & 6\end{pmatrix}$. Find the matrix relative to the basis

$$f_1 = 2e_1 + 3e_2 + e_3, f_2 = 3e_1 + 4e_2 + e_3, f_3 = e_1 + 2e_2 + 2e_3.$$

6. The linear transformation T has, relative to the basis

$$u_1 = (8,-6,7), u_2 = (-16,7,-13), u_3 = (9,-3,7),$$

the matrix $\begin{pmatrix}1 & -18 & 15\\-1 & -22 & 20\\1 & -25 & 22\end{pmatrix}$. Find the matrix with respect to the basis $v_1 = (1,-2,1)$, $v_2 = (3,-1,2), v_3 = (2,1,2)$.

7. Let the linear transformations T : $R^2 \to R^2$ and L : $R^2 \to R^2$ be defined by

$$T(x, y) = (x + y, 0) \text{ and}$$
$$L(x, y) = (2x, 3x + 4y).$$

Find the linear transformations (*i*) 2L + 3T (ii) 3L – 7T.

8. Let the linear transformation T has matrix $\begin{pmatrix}3 & 5\\4 & 3\end{pmatrix}$ relative to the basis $\{u_1, u_2\}$, where $u_1 = (1, 2)$ and $u_2 = (2, 3)$. The linear transformation L has the matrix $\begin{pmatrix}4 & 6\\6 & 9\end{pmatrix}$ relative to the basis $\{v_1, v_2\}$, where v_1 = (3, 1), v_2 = (4, 2).
Find the matrix of the linear transformation T + L relative to the basis v_1, v_2.

9. Let T : V_3 (R) → V_3 (R) be a linear transformation given by T (*a*, *b*, *c*) = (3*a* + *c*, – 2*a* + *b*, *a* + 2*b* + 4*c*) w.r.t. (*i*) Standard basis (*ii*) The basis {(1, 0, 1), (–1, 2, 1), (2, 1, 1)} for both domain and range.

10. Determine the matrix (T : β_1, β_2) for the given linear transformation T and the bases β_1 and β_2 defined as

$$T : V_3 \to V_2, T(x, y, z) = (x + y, y + z)$$

(*i*) $\beta_1 = \{(1,1,1), (1,0,0), (1,1,0)\}, \beta_2 = \left\{(4,-3), \left(\frac{3}{2}, -1\right)\right\}.$

Answers

1. Image is the circle of radius 2.

2. (i) $\begin{bmatrix} 3 & -2 \\ 2 & 0 \end{bmatrix}$ (ii) $\begin{bmatrix} 3 & -2 \\ -1 & 0 \end{bmatrix}$

(iii) $\begin{bmatrix} 1 & -2 \\ 2 & 0 \end{bmatrix}$ (iv) $\begin{bmatrix} 1 & -2 \\ 1 & 2 \end{bmatrix}$

3. $\begin{bmatrix} 1 & 1 \\ 2 & -1 \\ 0 & 7 \end{bmatrix}$ 5. $\begin{bmatrix} 1 & 0 & 0 \\ 0 & 2 & 0 \\ 0 & 0 & 3 \end{bmatrix}$

6. $\begin{bmatrix} 1 & 2 & 2 \\ 3 & -1 & -2 \\ 2 & -3 & 1 \end{bmatrix}$

7. (i) (2L + 3T) : $(x, y) \to (7x + 3y, 6x + 8y)$
(ii) (3L − 7T) : $(x, y) \to (-x -7y, 9x + 12y)$

8. $\begin{bmatrix} 44 & 44 \\ -\frac{59}{2} & -25 \end{bmatrix}$ 9. (i) $\begin{bmatrix} 3 & 2 & -1 \\ 0 & 1 & 2 \\ 1 & 0 & 4 \end{bmatrix}$ (ii) $\begin{bmatrix} \frac{17}{4} & -\frac{3}{4} & -\frac{1}{2} \\ \frac{35}{4} & \frac{15}{4} & -\frac{7}{2} \\ \frac{17}{2} & -\frac{3}{2} & 0 \end{bmatrix}$

10. (i) $\begin{bmatrix} 2 & 1 & 2 \\ 2 & 0 & 1 \end{bmatrix}$ (ii) $\begin{bmatrix} -\frac{7}{3} & -1 & -\frac{13}{2} \\ \frac{26}{3} & 4 & 23 \end{bmatrix}$.

Chapter 3

Sequences

3.1 SEQUENCE

Definition. *A sequence in a set S is a function, whose domain is the set* **N** *of natural numbers and whose range is a subset of S.*

A sequence whose range is a subset of **R** is called a real sequence. Here we deal with only real sequences, therefore, we shall use the term sequence to denote a real number sequence.

From the above definition, we can say that a sequence assigns to each natural number n, a unique real number a_n. Thus the real number a_n assigned to the natural number n by a sequence a. A sequence is thus the set $\{(n, a_n) : n \in \mathrm{N}\}$

Since the domain of all sequences is the same set, namely N, therefore, a sequence is determined completely if we know a_n for each n. A sequence is usually denoted by $\{a_n\}$ or $<a_n>$, and its range is denoted by $\{a_n : n \in \mathrm{N}\}$.

The numbers $a_1, a_2, \ldots,$ are called the terms (or elements) of the sequence $\{a_n\}$. Thus, a_1 is the first term, a_2 is the second term, ..., a_n is the nth term of the sequence $\{a_n\}$.

3.2 METHODS FOR DESCRIBING A SEQUENCE

A sequence can be described in several different ways.

In one method, we may list in order the first few elements of a sequence, till the rule for writing down different elements becomes obvious.

For example, $\{1, 8, 27, 81, \ldots\}$ is the sequence whose n^{th} term is n^3.

Another method of describing a sequence is to provide a formula for its nth term.

For example, the sequence $\{2, 4, 6, 8, \ldots\}$ can also be written as $\{2, 4, 6, \ldots, 2n, \ldots\}$ or as $\{2n : n \in \mathrm{N}\}$.

Another convenient method of describing a sequence is to specify a_1 and a rule for obtaining a_{n+1} (for each $n \geq 1$) in terms of $a_1, a_2, ..., a_n$.

Such a definition is called a **recursive definition.**

As an example, $a_n = 3a_{n-1}$, $n \geq 2$ and $a_1 = 3$.

This sequence can also be written as $\{a_n\} = \{3^n\}$, $n \in N$.

3.3 FINITE AND INFINITE SEQUENCES

A sequence is said to be **finite** or **infinite** according as the number of elements is finite or infinite.

3.3.1 Illustrative Examples

1. $\{2, 2^2, 2^3, ..., 2^n\}$ is a finite sequence (n number of elements)
2. $\left\{1, \frac{1}{2}, \frac{1}{3}, \frac{1}{4}, ..., \frac{1}{n}, ...\right\}$ is an infinite sequence (here the elements are not finite, *i.e.*, not countable).

Remark. Here we shall consider only infinite sequences and we use the word sequence to represent an infinite sequence.

Examples of Sequences

(*i*) $a_n = 2n - 1$, $n \geq 1$, gives the sequence of numbers $\{1, 3, 5, 7, ...\}$

(*ii*) $a_n = \frac{1}{n}$, represents the sequence $\left\{1, \frac{1}{2}, \frac{1}{3}, \frac{1}{4}, ...\right\}$

(*iii*) $a_n = (-1)^n$, represents the sequence, $\{-1, +1, -1, +1, ...\}$

(*iv*) $a_n = 2$, represents the sequence $\{2, 2, 2, ...\}$

(*v*) $a_n = \frac{1}{2}\left[1-(-1)^n\right]$, represents the sequence $\{1, 0, 1, 0, 1, 0, ...\}$

(*vi*) $F_n = F_{n-1} + F_{n-2}$ $(n \geq 3)$ and $F_1 = 1$, $F_2 = 2$, represents the sequence $\{1, 1, 2, 3, 5, 8, 13, ...\}$, which is well known **Fibonacci sequence.**

3.4 BOUNDS OF A SEQUENCE

3.4.1 Rough upper bound

If there exists a number M such that, for all values of n, $|\ a_n\ | < M$, then the sequence $\{a_n\}$ is said to be *bounded above* (or *on the right*). M is called a *rough upper bound.*

3.4.2 Rough lower bound

If there exists a number m such that, for all values of n, $|\ a_n\ | > m$, then $\{a_n\}$ is said to be *bounded below* (or *on the left*), m is called a *rough lower bound.*

A sequence $\{a_n\}$ which is bounded above and below, is said to be *bounded*, that is, if $m < a_n < M$.

3.4.3 Least upper bound (exact upper bound)

If M is a number such that

(*i*) $| a_n | \le M$, for all n,

(*ii*) $| a_n | + \varepsilon > M$, for some value of n, where ε is an arbitrary positive number, as small as we please, then M is called the *least upper bound.*

3.4.4 Greatest lower bound (exact lower bound)

If m is a number such that

(*i*) $| a_n | \ge m$, for all n,

(*ii*) $| a_n | - \varepsilon < m$, for some value of n, where ε is an arbitrary small positive number, then m is said to be the *greatest lower bound.*

3.4.5 Illustrative Examples

1. The sequence $\left\{1, \frac{1}{2}, \frac{1}{3}, \ldots, \frac{1}{n}, \ldots\right\}$ is bounded ; because $0 < \left|\frac{1}{n}\right| < 1$ for all $n \in N$.

   Its upper bound is 1 and lower bound is 0.
2. The sequence $\{a_n\}$, where $a_n = (-1)^n$, is bounded because $| a_n | \le 1$ for all $n \in N$. The upper bound is 1 and lower bound is -1.
3. The sequence $\{a_n\}$, where $a_n = n^2$, is bounded below but is not bounded above (lower bound is 0).
4. The sequence $\{a_n\}$, where $a_n = -n^2$, is bounded above but is not bounded below (upper bound is 0).
5. The sequence $\{(-1)^n n\}$ is neither bounded above nor bounded below.
6. *Show that the sequence* $\{u_n\} = \left\{\frac{n+1}{n}\right\}$ *is a bounded sequence.*

   Sol.
   $$\{u_n\} = \left\{2, \frac{3}{2}, \frac{4}{3}, \frac{5}{4}, \ldots\right\}.$$

   Here every term of the sequence, except the first term u_1, is less than 2. Therefore, 2 is the upper bound.

   Again $u_n > 1$ for every n. If m is a number $1 + \delta$, so that $m > 1$, then $u_n < m$ for all values of $n > \frac{1}{\delta}$. Thus, the lower bound is equal to 1 and hence, the sequence is bounded.
7. *Prove that the sequence* $\{u_n\}$, *where* $u_n = \frac{n^2 - 3}{n}$, *is bounded below,* $n \in N$.

   Sol. The explicit form of the sequence $\{u_n\}$ is $\left\{-2, \frac{1}{2}, 2, \frac{13}{4}, \frac{22}{5}, \ldots\right\}$.

There is no number in $\{u_n\}$ which is greater than every term of $\{u_n\}$. Therefore, $\{u_n\}$ is not bounded above. But every term of $\{u_n\}$ except u_1 is greater than – 2. Thus, – 2 is the lower bound. Hence, the sequence is bounded below.

8. *Show that the sequence $\{u_n\}$, whose n^{th} term is $\dfrac{2n-7}{3n+2}$ is bounded, $(n \in N)$.*

Sol. The explicit form of the sequence u_n is $\left\{-1, -\dfrac{3}{8}, -\dfrac{1}{11}, \dfrac{1}{14}, \dfrac{3}{17}, \ldots\right\}$

Here $\qquad u_1 = -1 \le u_n$ for all n.

Also, $$u_n = \frac{2n-7}{3n+2} = \frac{(3n+2)-(n+9)}{3n+2} = 1 - \frac{n+9}{3n+2} < 1, \text{ for all } n.$$

$\therefore \qquad -1 \le u_n < 1$, for all n.

Hence, $\{u_n\}$ is a bounded sequence.

3.4.6 Illustrative Examples of Multiple Choice Questions (MCQ)

1. *If $u_n = (-1)^n\, 2$, $n \in N$, then the sequence $\{u_n\}$ is*

(a) bounded below but not bounded above,

(b) bounded above but not bounded below,

(c) not bounded above and not bounded below,

(d) bounded.

Sol. (*d*)

The explicit form of the sequence is {– 2, 2, – 2, 2, ...}.

There is no term of the sequence which is greater than 2 and less than – 2. Therefore 2 and – 2 are the upper and lower bounds of $\{u_n\}$ respectively and both of the bounds are attained by the sequence. Hence, the sequence is bounded.

2. The sequence $\{a_n\}$ is bounded above if there exists a real number M such that

(*a*) $a_n = M$, (*b*) $a_n \ge M$ (*c*) $a_n \le M$ (*d*) none of these.

Sol. (*c*). By definition

3. The bounds of the sequence $\left\{\dfrac{1}{n^2}\right\}$, $n \in N$ are

(*a*) 0, 1; (*b*) 1, 2; (*c*) $1, \dfrac{1}{4}$; (*d*) none of these.

Sol. (*a*).

4. The sequence $\{n^2\}$ is

(*a*) bounded (*b*) bounded above (*c*) bounded below (*d*) none of these.

Sol. (*c*)

Here upper bound tends to ∞ (not a finite number) and lower bound is 1. Hence, the sequence is bounded below.

3.5 MONOTONIC SEQUENCES

Definition. *In a sequence $\{u_n\}$, if $u_{n+1} \geq u_n$, for every value of n, then the sequence $\{u_n\}$ is said to be* ***monotonic increasing*** *or* ***monotone ascending.***

If $u_{n+1} > u_n$ for every value of n, then $\{u_n\}$ is said to be monotone increasing in the stricter sense.

For example, the sequence {1, 2, 2, 3, 3, 4, 4, ...} is monotonic increasing but not in the stricter sense because some elements are repeated.

The sequence {1, 3, 5, 7, 9, ... (2*n* – 1), ...} is monotone increasing in the stricter sense, since $t_1 < t_2 < t_3 < ..., < t_n < ....$

Definition. *In a sequence $\{u_n\}$, if $u_{n+1} \leq u_n$, for every value of n, then the sequence $\{u_n\}$ is said to be* ***monotonic decreasing*** *or* ***monotone descending.***

If $u_{n+1} < u_n$, for every value of n, then the sequence $\{u_n\}$ is said to be **monotonic decreasing** in the **stricter sense.**

For example, the sequence {– 1, – 2, – 2, – 3, – 3, ...} is monotone decreasing but not in the stricter sense.

The sequence $\left\{1, \frac{1}{2}, \frac{2}{3}, \frac{2}{4,}, ..., \frac{1}{n}, ...\right\}$ is monotone decreasing in the stricter sense,

since $t_1 > t_2 > t_3 ..., > t_n > ...$.

3.5.1 Illustrative Examples

1. *Show that the sequence $\{a_n\}$ is bounded monotonic increasing sequence, given*

$$a_n = \frac{1}{1.2} + \frac{1}{2.3} + \frac{1}{3.4} + ... + \frac{1}{n(n+1)}$$

Sol.

$$a_n = \frac{1}{1.2} + \frac{1}{2.3} + \frac{1}{3.4} + ... + \frac{1}{n(n+1)}$$

$$= \left(1 - \frac{1}{2}\right) + \left(\frac{1}{2} - \frac{1}{3}\right) + \left(\frac{1}{3} - \frac{1}{4}\right) + ... + \left(\frac{1}{n} - \frac{1}{n+1}\right)$$

$$= 1 - \frac{1}{n+1} = \frac{n}{n+1}$$

$$\therefore \quad a_{n+1} - a_n = \frac{n+1}{n+2} - \frac{n}{n+1} = \frac{(n+1)^2 - n(n+2)}{(n+2)(n+1)}$$

$$= \frac{1}{(n+1)(n+2)} > 0, \text{ for all } n.$$

$$\therefore \quad a_{n+1} > a_n$$

Hence, $\{a_n\}$ is monotonic increasing.

Again $n < n + 1$ for all positive integer n

$$\therefore \qquad \frac{n}{n+1} < 1 \text{ but } \frac{n}{n+1} > 0,$$

Therefore, $0 < a_n < 1$, for $n > 0$

Hence, $\{a_n\}$ is bounded.

2. *If* $a_n = 2 + \frac{(-1)^n}{n^2}$, *find m, such that* $| a_{n-2} | < \frac{1}{10^4}$, *for* $n > m$.

Sol. Given $$a_n = 2 + \frac{(-1)^n}{n^2}$$

$$\therefore \qquad a_n - 2 = \frac{(-1)^n}{n^2}$$

Hence, $$| a_n - 2 | = \left| \frac{(-1)^n}{n^2} \right| = \frac{1}{n^2}$$

But $$| a_n - 2 | < \frac{1}{10^4}$$

$$\therefore \qquad \frac{1}{n^2} < \frac{1}{10^4}, \text{ or } n^2 > 10^4 \Rightarrow n > 10^2$$

i.e., $$n > 100$$

Taking $m = 100$, we find that $| a_n - 2 | < \frac{1}{10^4}$, for $n > m$.

3.6 CONCEPT OF LIMIT OF A SEQUENCE

The concept of limit of a sequence is a fundamental aspect in Analysis. The concept of limit is necessary to study the ***convergency*** and ***divergency*** of sequences. Let us try to develop it by an example :

Let us consider the sequence $\{a_n\}$ of numbers

$$a_1 = 1, a_2 = 0.1, a_3 = 0.01, a_4 = 0.001, \ldots,$$

where all the numbers are positive, greater than zero and the numbers decrease as their position in the numerical order increases. These numbers form a decreasing sequence. The term a_n can be made 'as small as we please' provided '*n* is sufficiently large'. This can be expressed in mathematical form : for any positive number t there exists a number N so that a_n is a neighbourhood of 0 of order t provided we choose $n > N$, that is, for all $t > 0$, there exists

$$N : n > N \Rightarrow a_n < t$$

Thus, when t is chosen, there exists terminating decimals smaller than t and one of these can be made a a_n by replacing all its digits by zero with the exception of the right-hand extreme digit which is replaced by 1. Then, for $n > N$, all the a_n are still smaller.

This fact can be expressed in mathematical language, **'as n tends to infinity, a_n tends to 0'** or in otherwords, **'a_n has a limit 0'.**

Now, we are in a position to define the limits of a sequence.

Definition. *If, corresponding to any positive number ε that we may choose, no matter how small, there is a positive integer m such that, for every integer n, greater than or equal to m,*

$$| a_n - l | < \varepsilon,$$

that is, $$l - \varepsilon < a_n < l + \varepsilon,$$

where l is a fixed number, then l is called the limit of the sequence $\{a_n\}$.

It can also be expressed as

$$\lim_{n \to \infty} a_n = l \text{ or simply, limit } a_n = l.$$

Definition. *If corresponding to any positive number N, however large, that may choose, there is a positive integer m, such that for every integer n, greater than or equal to m, $a_n > N$, then the sequence $\{a_n\}$ is said to tend to infinity.*

Mathematically, it can be written as $\lim_{n \to \infty} a_n \to \infty$.

Remark 1. The number l is called the limit of the sequence.

Remark 2. If $l = 0$, *i.e.*, $a_n \to 0$ as $n \to \infty$, then the sequence $\{a_n\}$ is called the **null sequence.**

Now, we state without proof some properties of limit of sequences :

3.6.1 Properties of Limit of Sequences

P_1. If $\{a_n\}$ and $\{b_n\}$ are null sequences, then $\{a_n \pm b_n\}$ are also null sequences.

P_2. If $\{a_n\}$ is a null sequence and $\{u_n\}$ is a bounded sequence, then $\{a_n u_n\}$ is a null sequence.

P_3. If $\{a_n\}$ and $\{b_n\}$ are two sequences such that

$$\lim_{n \to \infty} a_n = l \text{ and } \lim_{n \to \infty} b_n = m$$

then (*i*) $\lim_{n \to \infty} \{a_n \pm b_n\} = l \pm m$

(*ii*) $\lim_{n \to \infty} \{a_n b_n\} = lm$

(*iii*) $\lim_{n \to \infty} \left\{\frac{a_n}{b_n}\right\} = \frac{l}{m}$, provided $b_n \neq 0$ for every n, and $m \neq 0$.

3.6.2 Illustrative Examples

1. *Show that the sequence $\{u_n\}$, where $u_n = \sqrt{n+1} - \sqrt{n}$ is a null sequence.*

Sol. $$u_n = \sqrt{n+1} - \sqrt{n} = \frac{\left(\sqrt{n+1} - \sqrt{n}\right)\left(\sqrt{n+1} + \sqrt{n}\right)}{\left(\sqrt{n+1} + \sqrt{n}\right)}$$

$$= \frac{n+1-n}{\sqrt{n+1}+\sqrt{n}} = \frac{1}{\sqrt{n+1}+\sqrt{n}}$$

Now, $\lim_{n\to\infty} u_n = \lim_{n\to\infty} \frac{1}{\sqrt{n+1}+\sqrt{n}} = 0$

Hence, the sequence $\{u_n\}$ is a null sequence.

2. *Show that* $\lim_{n\to\infty} \frac{n^2+3n}{2n^2+n-1} = \frac{1}{2}$.

Sol.

$$\lim_{n\to\infty} \frac{n^2+3n}{2n^2+n-1} = \lim_{n\to\infty} \frac{1+\frac{3}{n}}{2+\frac{1}{n}-\frac{1}{n^2}}$$

$$= \frac{\lim_{n\to\infty}\left(1+\frac{3}{n}\right)}{\lim_{n\to\infty}\left(2+\frac{1}{n}-\frac{1}{n^2}\right)}, \text{ by } P_3\,(iii)$$

$$= \frac{1}{2}.$$

3.7 SOME USEFUL LIMITS

(*i*) $\lim_{n\to\infty} x^n = 0$, if $x < 1$; and $\lim_{x\to\infty} x^n = \infty$ if $x > 1$,

(*ii*) $\lim_{n\to\infty} \frac{x^n}{n!} = 0, \forall x$,

(*iii*) $\lim_{n\to\infty} \frac{\log n}{n} = 0$,

(*iv*) $\lim_{n\to\infty} \left(1+\frac{1}{n}\right)^n = e$,

(*v*) $\lim_{n\to\infty} (n)^{\frac{1}{n}} = 1$,

(*vi*) $\lim_{n\to\infty} (n!)^{\frac{1}{n}} \to \infty$,

(*vii*) $\lim_{n\to\infty} \left[\frac{(n!)}{n}\right]^{\frac{1}{n}} = \frac{1}{e}$,

(*viii*) $\lim_{n\to 0} \left(\frac{a^n-1}{n}\right) = \log a$,

(*ix*) $\lim_{n\to\infty} \left(\frac{a^{\frac{1}{n}}-1}{\frac{1}{n}}\right) = \log a$,

(*x*) $\lim_{n\to\infty} \frac{1}{n^n} = 0$.

3.8 CONVERGENT SEQUENCES

Definition. *A sequence $\{a_n\}$ is said to be convergent if the limit l of the sequence is a finite and definite number, or in otherwords, having given a positive number ε such that*

$$|a_n - l| < \varepsilon \text{ for } n \geq m,$$

then $\{a_n\}$ is said to be convergent to the limit l, (a definite number).

We state a theorem on convergent sequence without proof as below :

Theorem. *A convergent sequence determines its limit uniquely.*

3.8.1 Illustrative Examples

1. *Show that the sequence $\{a_n\}$, where $a_n = \dfrac{2n-1}{2n}$, is convergent and converges to the limit 1.*

Sol. Given $$a_n = \frac{2n-1}{2n}$$

$\therefore$ $$| a_n - 1 | = \left| \frac{2n-1}{2n} - 1 \right| = \left| -\frac{1}{2n} \right| = \frac{1}{2n}$$

and this can be made less than any given positive number ε by choosing

$$\frac{1}{2n} < \varepsilon, \text{ i.e., } n > \frac{1}{2\varepsilon}$$

Thus, the sequence $\{a_n\}$ is convergent and converges to the limit 1.

2. *Show that the sequence $\{u_n\}$, where $u_n = \dfrac{n}{n+1}$, is monotone increasing and convergent.*

Sol. Given $$u_n = \frac{n}{n+1},$$

$\therefore$ $$u_1 = \frac{1}{2}, u_2 = \frac{2}{3}, u_3 = \frac{3}{4}, \ldots$$

Here, $\dfrac{1}{2} < \dfrac{2}{3} < \dfrac{3}{4} < \ldots,$

that is, $u_1 < u_2 < u_3 < \ldots$

Hence, the sequence $\{u_n\}$ is monotone increasing.

Now, $$\lim_{n \to \infty} u_n = \lim_{n \to \infty} \frac{n}{n+1} = \lim_{n \to \infty} \frac{1}{1 + \frac{1}{n}} = 1, \text{ which is finite.}$$

Hence, the given sequence $\{u_n\}$ is convergent.

3. *Show that the sequence $\{a_n\}$, where $a_n = \dfrac{1}{n} \sin \dfrac{n\pi}{2}$, is convergent.*

Sol. Given $$a_n = \frac{1}{n} \sin \frac{n\pi}{2}$$

We know that $$-1 \le \sin \frac{n\pi}{2} \le 1$$

Here $$| a_n - 0 | = \left| \frac{1}{n} \sin \frac{n\pi}{2} - 0 \right| = \left| \frac{1}{n} \sin \frac{n\pi}{2} \right| = \left| \frac{1}{n} \right| \left| \sin \frac{n\pi}{2} \right| < \frac{1}{n} < \varepsilon,$$

if $$n > \frac{1}{\varepsilon}$$

We choose m = positive integral part of $\frac{1}{\varepsilon}$,

if it is itself not an integer.

Therefore, $\{a_n\}$ is convergent and $\lim_{n \to \infty} \frac{1}{n} \sin \frac{n\pi}{2} = 0$.

3.9 NONCONVERGENT (OR DIVERGENT) SEQUENCES

A sequence which is not convergent is called nonconvergent or divergent sequence. The proper definition is as follows :

Definition 1. *The sequence $\{a_n\}$ is said to be* ***divergent*** *if there exists a positive number N, however large, corresponding to which there exists a positive integer m such that $a_n > N$ for all values of $n \geq m$.*

Definition 2. *If corresponding to every arbitrary large positive number M, there exists a positive integer m such that for every $n > m$, $a_n > M$, then the sequence $\{a_n\}$ is said to tend to $+\infty$ or to diverge to $+\infty$ and it is expressed mathematically as*

$$\lim_{n \to \infty} a_n \to +\infty, \text{ or } a_n \to +\infty, \text{ as } n \to \infty$$

Definition 3. *If corresponding to an arbitrary large negative number $-M$ ($M > 0$), there exists positive integer m such that for every $n > m$, $a_n < -M$, then the sequence $\{a_n\}$ is said to tend to $-\infty$, or diverge to $-\infty$ and it is expressed mathematically as*

$$\lim_{n \to \infty} a_n \to -\infty, \text{ or } a_n \to -\infty, \text{ as } n \to \infty.$$

Definition 4. *A sequence $\{a_n\}$, which neither converges nor diverges to $+\infty$ or to $-\infty$, is called an* ***oscillatory sequence.***

If $\{a_n\}$ is an oscillating sequence and there exists a positive number N such that $|u_n| < N$ for every values of n, then $\{a_n\}$ is said to ***oscillate finitely,*** *otherwise, $\{a_n\}$ is said to* ***oscillate infinitely.***

3.9.1 Illustrative Examples

1. *Prove that the divergence $\{u_n\}$, where $u_n = n$, is divergent.*

Sol. Given $$u_n = n,$$

$$\therefore \quad u_1 = 1, u_2 = 2, u_3 = 3, u_4 = 4, \ldots$$

Thus, the sequence increases continuously.

For a given arbitrary large positive number N, we have $u_n > N$, when $n > N$.

Thus, $u_n \to \infty$, as $n \to \infty$.

Hence, the sequence $\{u_n\}$ is divergent and diverges to $+\infty$.

2. *Prove that the sequence $\{a_n\}$, where $a_n = \sqrt{n-1}$ is divergent.*

Sol. Given $$a_n = \sqrt{n-1},$$

$\therefore$ the explicit form of the sequence is $\{0, 1, \sqrt{2}, \sqrt{3}, 2, \ldots\}$ which shows that the terms of the sequences are increasing.

For a given arbitrary large positive number M, we have $a_n > M$, where $n - 1 > M^2$, or $n > M^2 + 1$.

Thus, $a_n \to \infty$, as $n \to \infty$.

Therefore, the sequence $\{a_n\}$ is divergent and diverges to $+\infty$.

3. *Show that the sequence $\{a_n\}$, where $a_n = -n^2$, $(n > 0)$, diverges to $-\infty$.*

Sol. Given $$a_n = -n^2$$

$\therefore$ $$a_1 = -1, a_2 = -4, a_3 = -9, a_4 = -16, \ldots$$

The sequence $\{a_n\}$ decreases negatively.

For a given arbitrarily large negative number $-$M, we have

$$a_n < -M, \text{ where } n > M^2$$

Thus, $a_n \to -\infty$ as $n \to \infty$.

Therefore, the sequence $\{a_n\}$ is divergent and diverges to $-\infty$.

4. *Show that the sequence $\{a_n\}$, where $a_n = \left\{(-1)^n + \dfrac{1}{2^n}\right\}$, is an oscillatory sequence and oscillates finitely.*

Sol. Given $$a_n = \left\{(-1)^n + \frac{1}{2^n}\right\}$$

$\therefore$ $$\lim_{n\to\infty} a_n = \lim_{n\to\infty} \left\{(-1)^n + \frac{1}{2^n}\right\} = 1, \text{ if } n \text{ is even}$$

$$= -1, \text{ if } n \text{ is odd.}$$

Thus, $\lim\limits_{n\to\infty} a_n$ does not tend to a definite limit, hence, $\{a_n\}$ is an oscillatory sequence and it oscillates finitely between -1 and 1.

We are now in a position to state some theorems, which are included in the next article.

3.10 STATEMENT OF THEOREMS (WITHOUT PROOF)

Theorem 1. *A convergent sequence has one and only one limit.*

Theorem 2. *A convergent sequence is bounded.*

Theorem 3. *A monotonic increasing sequence is convergent when it is bounded above.*

Theorem 4. *If a monotonic increasing sequence is not bounded above, it diverges to* $+\infty$.

Theorem 5. *If a monotonic decreasing sequence is bounded below, it is convergent.*

Theorem 6. *If a monotonic decreasing sequence is not bounded below, it diverges to* $-\infty$.

3.11 CAUCHY'S GENERAL PRINCIPLE OF CONVERGENCE OF A SEQUENCE

Statement. *The necessary and sufficient condition for the convergence of any sequence* $\{a_n\}$ *is that corresponding to every arbitrary* ε *there exists a positive integer m such that,*

$$| a_{n+p} - a_n | < \varepsilon,$$

for all values of $n \geq m$ and for all positive integral values of p.

Remark. The sequence which obeys the 'Cauchy Convergence Criterion', named after Augustin Louis Cauchy (1789–1859), is known as **Cauchy Sequence.**

We state some theorems without proof:

Theorem 1. *Every Cauchy sequence is bounded (the converse may not be true).*

Theorem 2. *A sequence converges if and only if it is a Cauchy sequence.*

3.11.1 Illustrative Examples

1. *By using Cauchy's general principle of convergence, show that the sequence* $\{a_n\}$, *where* $a_n = 1 + \frac{1}{2} + \frac{1}{3} + \ldots + \frac{1}{n}$, *is divergent.*

Sol. If possible, let us assume that $\{a_n\}$ is convergent. Then by Cauchy's general principle of convergence, corresponding to $\varepsilon = \frac{1}{2}$, there exists a positive integer m, such that

$$| a_{n+p} - a_n | < \frac{1}{2}, \text{ when } n \geq m \text{ and } p = 0, 1, 2, 3, \ldots$$

$$\therefore \qquad | a_{n+p} - a_n | < \frac{1}{2}, \text{ for all positive integral values of } p.$$

But $$| a_{m+p} - a_m | = \frac{1}{m+1} + \frac{1}{m+2} + \ldots + \frac{1}{m+p} > \frac{p}{m+p} > \frac{1}{2},$$

when $p = m$

This contradicts that

$$| a_{n+p} - a_n | < \frac{1}{2}.$$

Thus, our assumption is not true.

Hence, the sequence $\{a_n\}$ is divergent.

[Here, the sequence $\{a_n\}$ is not a Cauchy sequence]

2. *If $\{a_n\}$ be a sequence of positive numbers such that $a_n = \frac{1}{2}(a_{n-1} + a_{n-2})$, for all $n \geq 2$, then show that the sequence $\{a_n\}$ is a Cauchy sequence. Also find $\lim_{n \to \infty} a_n$.*

Sol. If we consider $a_1 = a_2$, then $a_n = a_1$ for all n, so that $\{a_n\}$ converges to a_1.

We shall consider only the case $a_1 \neq a_2$.

Now,
$$|a_n - a_{n-1}| = \left|\frac{1}{2}\left(a_{n-1} + a_{n-2}\right) - a_{n-1}\right|$$

$$= \frac{1}{2}|a_{n-1} - a_{n-2}|, \text{ whenever } n \geq 2 \qquad \text{... (i)}$$

If $n \geq m$, then

$$|a_n - a_m| = |(a_n - a_{n-1}) + (a_{n-1} - a_{n-2}) + \ldots + (a_{m+1} - a_m)|$$

$$\leq |a_n - a_{n-1}| + |a_{n-1} - a_{n-2}| + \ldots + |a_{m+1} - a_m|$$

$$= \left(\frac{1}{2^{n-2}} + \frac{1}{2^{n-3}} + \ldots + \frac{1}{2^{m-1}}\right)|a_2 - a_1|, \text{ by (i)}$$

$$= \frac{1}{2^{m-1}}\left(1 + \frac{1}{2} + \frac{1}{2^2} + \ldots + \frac{1}{2^{n-m-1}}\right)|a_2 - a_1|$$

$$< \frac{1}{2^{m-2}}|a_2 - a_1| \qquad \text{... (ii)}$$

If $\varepsilon > 0$ be given, it is possible to find a positive integer m such that $\frac{1}{2^{m-2}}|a_2 - a_1| < \varepsilon$

For such a choice of m, we find from (*ii*) that

$$|a_n - a_m| < \varepsilon, \text{ whenever } n \geq m$$

Therefore, $\{a_n\}$ is a Cauchy sequence and by Cauchy's convergence criterion it follows that $\{a_n\}$ is convergent.

Let
$$\lim_{n \to \infty} a_n = l$$

To determine l, we put $n = 3, 4, 5, \ldots, k$, in the given relation

$$a_n = \frac{1}{2}(a_{n-1} + a_{n-2}).$$

Thus, we get

$$\left.\begin{aligned} a_3 &= \frac{1}{2}(a_2 + a_1) \\ a_4 &= \frac{1}{2}(a_3 + a_2) \\ &\vdots \quad \ldots \ldots \quad \vdots \ldots \\ a_k &= \frac{1}{2}(a_{k-1} + a_{k-2}) \end{aligned}\right\} \qquad \ldots (iii)$$

Adding the left and right hand sides of all the relations (*iii*), we get

$$a_k + \frac{1}{2}a_{k-1} = \frac{1}{2}(a_1 + 2a_2)$$

Now, taking $\lim k \to \infty$, we have

$$\frac{3}{2}l = \frac{1}{2}(a_1 + 2a_2)$$

i.e.,

$$l = \frac{1}{3}(a_1 + 2a_2).$$

Now, we state two important limit theorems (without proof) which are known as **Cauchy's limit theorems.**

3.12 CAUCHY'S LIMIT THEOREMS

(*i*) **First limit theorem**

If the sequence $\{a_n\}$ converges to the limit l, then

$$\frac{a_1 + a_2 + \ldots + a_n}{n} \to l \text{ as } n \to \infty.$$

(*ii*) **Second limit theorem**

If $\{a_n\}$ is a sequence of positive terms and if

$$\lim_{n \to \infty} \frac{a_{n+1}}{a_n} \text{ exists}$$

and is equal to l, then

$\lim\limits_{n \to \infty} a_n^{\frac{1}{n}}$ also exists and is equal to l.

3.12.1 Illustrative Examples

1. *Show that* $\lim_{n\to\infty}\left[\frac{1}{\sqrt{n^2+1}}+\frac{1}{\sqrt{n^2+2}}+\ldots+\frac{1}{\sqrt{n^2+n}}\right]=1$

Sol. Let
$$a_n=\frac{n}{\sqrt{n^2+n}}$$

$$\therefore \quad \lim_{n\to\infty} a_n=\lim_{n\to\infty}\frac{n}{\sqrt{n^2+n}}=\lim_{n\to\infty}\frac{1}{\sqrt{1+\frac{1}{n}}}=1$$

Hence, by Cauchy's first theorem on limits, we get $\lim_{n\to\infty}\frac{a_1+a_2+\ldots+a_n}{n}=1$

$\therefore$ by Cauchy's first theorem, $\lim_{n\to\infty}\frac{1}{n}\left[\frac{n}{\sqrt{n^2+1}}+\frac{n}{\sqrt{n^2+2}}+\ldots+\frac{n}{\sqrt{n^2+n}}\right]=1$

Hence, $\lim_{n\to\infty}\left[\frac{1}{\sqrt{n^2+1}}+\frac{1}{\sqrt{n^2+2}}+\ldots+\frac{1}{\sqrt{n^2+n}}\right]=1$

2. *Prove that* $\lim_{n\to\infty}\frac{1}{n}\left[1+2^{\frac{1}{2}}+3^{\frac{1}{3}}+\ldots+n^{\frac{1}{n}}\right]=1$

Sol. Let
$$a_n=n^{\frac{1}{n}}$$

$$\therefore \quad \lim_{n\to\infty} a_n=\lim_{n\to\infty} n^{\frac{1}{n}}=1$$

Hence, by Cauchy's first theorem on limits, we have

$$\lim_{n\to\infty}\frac{a_1+a_2+\ldots+a_n}{n}=1$$

or $\lim_{n\to\infty}\frac{1}{n}\left[1+2^{\frac{1}{2}}+2^{\frac{1}{3}}+\ldots+n^{\frac{1}{n}}\right]=1$

3. *Show that* $\lim_{n\to\infty}\left[\left(\frac{2}{1}\right)^1\left(\frac{3}{2}\right)^2\left(\frac{4}{3}\right)^3\ldots\left(\frac{n+1}{n}\right)^n\right]^{\frac{1}{n}}=e$

Sol. Let
$$a_n=\left(\frac{2}{1}\right)^1\left(\frac{3}{2}\right)^2\left(\frac{4}{3}\right)^3\ldots\left(\frac{n}{n-1}\right)^{n-1}\left(\frac{n+1}{n}\right)^n$$

Replacing n by $n+1$, we get

$$a_{n+1} = \left(\frac{2}{1}\right)^1 \left(\frac{3}{2}\right)^2 \left(\frac{4}{3}\right)^3 \cdots \left(\frac{n+1}{n}\right)^n \left(\frac{n+2}{n+1}\right)^{n+1}$$

$$\therefore \quad \frac{a_{n+1}}{a_n} = \left(\frac{n+2}{n+1}\right)^{n+1} = \left(\frac{n+1+1}{n+1}\right)^{n+1} = \left(1+\frac{1}{n+1}\right)^{n+1}$$

$$\therefore \quad \lim_{n\to\infty} \frac{a_{n+1}}{a_n} = \lim_{n\to\infty}\left(1+\frac{1}{n+1}\right)^{n+1} = e$$

By Cauchy's second theorem on limits, we have

$$\lim_{n\to\infty} (a_n)^{\frac{1}{n}} = \lim_{n\to\infty} \frac{a_{n+1}}{a_n}$$

$$\text{Hence, } \lim_{n\to\infty}\left[\left(\frac{2}{1}\right)^1 \left(\frac{3}{2}\right)^2 \left(\frac{4}{3}\right)^3 \cdots \left(\frac{n+1}{n}\right)^n\right]^{\frac{1}{n}} = e.$$

3.12.2 Worked-Out Examples

1. *Whether the following statements are true or false ?*

   (a) *A convergent sequence has more than one limit.*

   Answer is F. [by theorem 1, 3.19]

   (b) *Every Cauchy sequence is convergent.*

   Answer is T. [by theorem 2, 3.20].

   (c) *If* $a_n = \dfrac{n+1}{n}$, *then the sequence is convergent.*

   Answer is T. $\left[\because a_n = \dfrac{n+1}{n} = 1+\dfrac{1}{n}\right.$

   $\therefore$ as $n \to \infty, a_n \to 1$,

   hence, $\{a_n\}$ is convergent]

   (d) *If* $a_n = \sqrt{n}\left(\sqrt{n+1}-\sqrt{n}\right)$, *then* $\{a_n\}$ *is divergent.*

   Answer is F. (*i.e.*, $\{a_n\}$ is convergent)

$$a_n = \sqrt{n}\left(\frac{1}{\sqrt{n+1}+\sqrt{n}}\right) = \frac{1}{\sqrt{1+\frac{1}{n}}+1};$$

$$\therefore \quad \lim_{n\to\infty} a_n = \lim_{n\to\infty}\left(\frac{1}{\sqrt{1+\frac{1}{n}}+1}\right) = \frac{1}{2},$$

hence, $\{a_n\}$ is convergent.

(e) *The sequence $\{a_n\}$; where $a_n = (-1)^n\, n$, oscillates finitely.*

Answer is F. Oscillates infinitely, *i.e.*, $-\infty$ to $+\infty$.

2. (i) *The sequence $\{x_n\}$, where $x_n = \dfrac{1}{5+6n}$ is*

(a) *increasing and bounded* (b) *increasing and unbounded*
(c) *decreasing and bounded* (d) *none of these.*

Answer. (c)

Justification: $$x_n = \frac{1}{5+6n},\ x_{n+1} = \frac{1}{5+6+6n} = \frac{1}{11+6n}$$

$$\therefore \quad x_{n+1} - x_n = \frac{1}{6n+11} - \frac{1}{6n+5} = \frac{6n+5-6n-11}{(6n+11)(6n+5)}$$

$$= -\frac{6}{(6n+11)(6n+5)}$$

$$< 0, \text{ for } n > 0$$

$$\therefore \quad x_{n+1} - x_n < 0, \text{ for } n > 0$$

i.e., $x_{n+1} < x_n$, *i.e.*, $\{x_n\}$ is a decreasing sequence.

Again, $$\lim_{n\to\infty} x_n = \lim_{n\to\infty} \frac{1}{5+6n} = 0$$

$\therefore$ $[x_n]$ is bounded

(ii) *The sequence $\{a_n\}$, where $a_n = n$ is*

(a) *monotone increasing and divergent*

(b) *monotone increasing and convergent*

(c) *monotone decreasing and convergent*

(d) *none of these*

Answer is (a).

Justification : $\because$ $a_n = n,\ \therefore\ a_{n+1} - a_n = (n+1) - n = 1$

i.e., $a_{n+1} > 1 + a_n \Rightarrow a_{n+1} > a_n$

Hence, $\{a_n\}$ is monotone increasing.

Also $\lim_{n\to\infty} a_n = \lim_{n\to\infty} n \to \infty$. Hence, $\{a_n\}$ is divergent.

(iii) *The sequence* $\left\{\frac{1}{2}, \frac{2}{3}, \frac{3}{4}, \frac{4}{5}, \ldots\right\}$ *converges to*

(a) $\frac{1}{2}$ (b) $\frac{4}{5}$ (c) 1 (d) 0.

Answer. (c)

Justification : Here $a_n = \frac{n}{n+1} = \frac{1}{1+\frac{1}{n}}$

$$\therefore \quad \lim_{n\to\infty} a_n = \lim_{n\to\infty} \frac{1}{1+\frac{1}{n}} = 1$$

Hence, the sequences converges to 1.

(iv) *If the sequence* $\{a_n\}_n$ *converges, then the sequence* $\{|a_n|\}_n$ *is*

(a) *divergent* (b) *convergent* (c) *nothing can be said.*

Answer. (b)

Justification : Since $\{a_n\}_n$ is convergent, therefore, it is bounded (since every convergent sequence is bounded).

Since, $\{a_n\}_n$ is bounded, there exists a definite number, l, say, such that $|a_n| \le l$ for all n ;

$\therefore$ $\{|a_n|\}_n$ is also bounded.

Hence, $\{|a_n|\}_n$ is convergent.

3. **Subjective Problems**

(i) *Discuss the convergence of the sequence* $\{u_n\}$, *where* $u_n = \frac{(-1)^n}{n}$.

Sol. Here, $u_n = \frac{(-1)^n}{n}$

$$\therefore \quad u_{2n} = \frac{(-1)^{2n}}{2n} = \frac{1}{2n}$$

And, $\lim_{n\to\infty} u_{2n} = \lim_{n\to\infty} \frac{1}{2n} = 0$

Again, $u_{2n+1} = \frac{(-1)^{2n+1}}{2n+1} = \frac{-1}{2n+1}$

$$\therefore \qquad \lim_{n\to\infty} u_{2n+1} = \lim_{n\to\infty} \frac{-1}{2n+1} = 0.$$

Thus, $\lim_{n\to\infty} u_n = 0.$

Hence, $\{u_n\}$ converges to zero.

(ii) *Discuss the convergence of the sequence* $\left\{a-b, a+\frac{b}{2}, a-\frac{b}{3}, a+\frac{b}{4}, \ldots, a+\frac{(-1)^n b}{n}, \ldots\right\}$.

Also find its limit.

Sol. Here $\qquad u_n = a + \frac{(-1)^n b}{n},$

n may be even or odd

$$\therefore \qquad u_{2n} = a + \frac{(-1)^{2n} b}{2n} = a + \frac{b}{2n}$$

$$\therefore \qquad \lim_{n\to\infty} u_{2n} = \lim_{n\to\infty}\left(a + \frac{b}{2n}\right) = a$$

Again $\qquad u_{2n+1} = a + \frac{(-1)^{2n+1} b}{2n+1} = a - \frac{b}{2n+1}$

$$\therefore \qquad \lim_{n\to\infty} u_{2n+1} = \lim_{n\to\infty}\left(a - \frac{b}{2n+1}\right) = a$$

Therefore, for all (integral) values of n, $\lim_{n\to\infty} u_n = a$

Hence the sequence converges to a.

(iii) *Show that the sequence* $\{a_n\}$, *where* $a_n = \sqrt{n^2+4n} - n$ *is convergent.*

Sol. Here $\qquad a_n = \sqrt{n^2+4n} - n = \frac{n^2+4n-n^2}{\sqrt{n^2+4n}+n} = \frac{4n}{\sqrt{n^2+4n}+n}$

$$= \frac{4}{\sqrt{\left(1+\frac{4}{n}\right)}+1}$$

$$\therefore \qquad \lim_{n\to\infty} a_n = \lim_{n\to\infty} \frac{4}{\sqrt{1+\frac{4}{n}}+1} = \frac{4}{2} = 2.$$

Hence, $\{a_n\}$ converges to 2.

(iv) *Prove that the sequence $\{a_n\}$ defined by*

$$a_n = \frac{1}{n+1} + \frac{1}{n+2} + \ldots + \frac{1}{n+n} \text{ converges.}$$

Sol. Given $$a_n = \frac{1}{n+1} + \frac{1}{n+2} + \ldots + \frac{1}{n+n}$$

$$\therefore \quad a_{n+1} - a_n = \left(\frac{1}{n+2} + \frac{1}{n+3} + \ldots + \frac{1}{2n+2}\right) - \left(\frac{1}{n+1} + \frac{1}{n+2} + \ldots + \frac{1}{2n}\right)$$

$$= \frac{1}{2n+1} + \frac{1}{2n+2} - \frac{1}{n+1}$$

$$= \frac{1}{2n+1} - \frac{1}{2n+2} > 0 \text{ for all } n$$

i.e., $a_{n+1} > a_n$, for all n.

Therefore, the sequence $\{a_n\}$ is monotonically increasing.

Also, $$|a_n| = \frac{1}{n+1} + \frac{1}{n+2} + \ldots + \frac{1}{n+n} < \frac{1}{n} + \frac{1}{n} + \ldots + \frac{1}{n} = \frac{n}{n} = 1$$

i.e., $|a_n| < 1$ for all n.

Therefore, $\{a_n\}$ is bounded.

Since, $\{a_n\}$ is bounded, monotonically increasing sequence, therefore, $\{a_n\}$ converges.

(v) *Show that the sequence* $\left\{\sqrt{2}, \sqrt{2\sqrt{2}}, \sqrt{2\sqrt{2\sqrt{2}}}, \ldots\right\}$ *converges to the limit 2.*

Sol. Here $$u_1 = \sqrt{2},\ u_2 = \sqrt{2\sqrt{2}} = \sqrt{2u_1},\ u_3 = \sqrt{2\sqrt{2\sqrt{2}}} = \sqrt{2u_2}, \ldots$$

$$\therefore \quad u_{n+1} = \sqrt{2u_n} \text{ or } u_{n+1}^2 = 2u_n$$

$$\therefore \quad u_n^2 = 2u_{n-1}$$

Now, $$u_{n+1}^2 - u_n^2 = 2(u_n - u_{n-1})$$

or $$(u_{n+1} + u_n)(u_{n+1} - u_n) = 2(u_n - u_{n-1})$$

$$\therefore \quad u_{n+1} > u_n, \text{ if } u_n > u_{n-1}$$

We have, $u_2 = \sqrt{2\sqrt{2}}, u_1 = \sqrt{2}$.

This shows that $u_2 > u_1$

Consequently, $u_3 > u_2 > u_4 > u_3$ and so on.

Hence, $\{u_n\}$ is monotonic increasing.

Again, $u_{n+1}^2 = 2u_n$ and $u_{n+1} > u_n$

$\therefore$ $u_n^2 < 2u_n$

or, $u_n(u_n - 2) < 0$

$\because$ $u_n > 0$, we have $u_n < 2$.

$\therefore$ $\{u_n\}$ is bounded and therefore, it is convergent.

Next, let $u_n \to l$ as $n \to \infty$

Then $u_{n+1} \to l$ as $n \to \infty$

Now, $\lim_{n\to\infty} u_{n+1}^2 = \lim_{n\to\infty} 2u_n$

or $l^2 = 2l$, i.e., $l(l-2) = 0$

$\because$ $u_n > 0, l \neq 0, \therefore l = 2.$

Hence, the sequence converges to the limit 2.

(vi) *Prove that the sequence $\{a_n\}$, whose n^{th} term is $\frac{2n-7}{3n+2}$, is*

(a) *monotone increasing*

(b) *bounded and*

(c) *tends to the limit $\frac{2}{3}$.*

Sol. (a) Here, $a_n = \frac{2n-7}{3n+2}$

$\therefore$ $$a_{n+1} = \frac{2(n+1)-7}{3(n+1)+2} = \frac{2n-5}{3n+5}$$

$\therefore$ $$a_{n+1} - a_n = \frac{2n-5}{3n+5} - \frac{2n-7}{3n+2} = \frac{(2n-5)(3n+2)-(2n-7)(3n+5)}{(3n+2)(3n+5)}$$

$$= \frac{25}{(3n+2)(3n+5)} > 0, \text{ for all } n$$

$\therefore$ $$a_{n+1} > a_n, \text{ for all } n.$$

Hence, the sequence $\{a_n\}$ is monotone increasing.

(*b*) Substituting $n = 1, 2, 3, 4, \ldots$, in a_n, we have

$$a_1 = -1, a_2 = -\frac{3}{8}, a_3 = -\frac{1}{11}, a_4 = \frac{1}{14}, \ldots$$

This shows that $-1 \le u_n$, for all n.

Also, $$a_n = \frac{2n-7}{3n+2} = \frac{(3n+2)-(n+9)}{(3n+2)} = 1 - \frac{n+9}{3n+2} < 1, \text{ for all } n.$$

Thus, $$-1 \le u_n < 1, \text{ for all } n.$$

Hence $\{a_n\}$ is a bounded sequence.

(*c*) Since the sequence is monotone increasing and bounded above, therefore it must converge.

Now, $$\lim_{n\to\infty} a_n = \lim_{n\to\infty} \frac{2n-7}{3n+2} = \lim_{n\to\infty} \frac{2-\frac{7}{n}}{3+\frac{2}{n}} = \frac{2}{3}.$$

Therefore, the sequence $\{a_n\}$ converges to the limit $\frac{2}{3}$.

EXERCISE 3.1

1. State whether the following statements are **true** or **false** by writing T or F :
   (*i*) The sequence $\{a_n\}$ whose n^{th} term is $2n-1$, $n \ge 1$, is a finite sequence.
   (*ii*) The sequence $\{a_n\}$ whose n^{th} term, $a_n = 2$, is a finite sequence.
   (*iii*) The sequence $\left\{\frac{1}{2}\left(1-(-1)^n\right)\right\}$ is an oscillatory sequence.
   (*iv*) The sequence $\left\{1, \frac{1}{2}, \frac{1}{3}, \ldots\right\}$ is bounded.
   (*v*) The sequence $\left\{\frac{n^2+3}{n+2}\right\}$ is divergent.
   (*vi*) If $a_n = \left(\sqrt{n+1} - \sqrt{n}\right)$, then $\{a_n\}$ is convergent.

(*vii*) Let $\{a_n\}$ be a sequence such that $\{a_n^2\}$ converges, then $\{a_n\}$ also converges to zero.

(*viii*) The sequence $\left\{\dfrac{3n}{n+1}\right\}$ is a monotonic decreasing sequence.

(*ix*) The sequence $\left\{\dfrac{3n+1}{n+2}\right\}$ is strictly increasing.

(*x*) A monotonic increasing sequence is convergent when it is bounded above.

2. Fill up the blanks :

(*i*) A sequence is said to oscillates if it neither ________ nor ________.

(*ii*) The sequence $\{a_n\}$, where $a_n = (-1)^n$, is bounded because $|\ a_n\ |$ ________ for all $n \in N$.

(*iii*) The sequence $\left\{-1, -\dfrac{1}{2}, -\dfrac{1}{3}, -\dfrac{1}{4}, \ldots\right\}$ is monotonically ________.

(*iv*) Every bounded monotonic sequence ________.

(*v*) The sequence $\{x_n\}$, where $x_n = \sqrt{n}$ diverges to ________.

(*vi*) If $\lim\limits_{n \to \infty} x_n = 0$, the sequence $\{x_n\}$ is a ________ sequence.

(*vii*) The nth term of the sequence $\left\{1, -\dfrac{1}{2}, \dfrac{1}{3}, -\dfrac{1}{4}, \ldots\right\}$ is ________.

(*viii*) The monotone increasing sequence $\{u_n\}$, where $u_n = \dfrac{n}{n+1}$, is ________.

(*ix*) The nth term of the sequence $\left\{\dfrac{1}{3}, -\dfrac{2}{3^2}, \dfrac{3}{3^3}, -\dfrac{4}{3^4}, \ldots\right\}$ is ________.

(*x*) $\lim\limits_{n \to \infty} \dfrac{5n}{n + 5\sqrt{n}}$ is ________.

3. Prove that if $a_1 = 1$ and $a_{n+1} = \dfrac{5 + 4a_n}{4 + a_n}$, the sequence $\{a_n\}$ is monotonic increasing if $a_n^2 < 5$ for all values of n.

Also, show that $a_n \to \sqrt{5}$.

4. Discuss the convergence of the sequence $\{u_n\}$, where $u_n = \dfrac{(-1)^n}{n}$.

5. Show that the sequence $\sqrt{3}, \sqrt{3\sqrt{3}}, \sqrt{3\sqrt{3\sqrt{3}}}, \ldots$ converges to the limit 3.

6. Show that the sequence $\sqrt{2}, \sqrt{2+\sqrt{2}}, \sqrt{2+\sqrt{2+\sqrt{2}}}, \ldots$ is monotonic increasing, bounded and converges to the limit 2.

7. If $a_n = \frac{1}{\sqrt{n}} + \frac{1}{\sqrt{n+1}} + \ldots + \frac{1}{\sqrt{2n}}$, show that $\{a_n\}$ is divergent.

8. Show that the sequence $\{a_n\}$, where $a_n = 2 + (-1)^n \frac{1}{n}$, is convergent and converges to limit 2.

9. Show that the sequence $\{a_n\}$ whose n^{th} term is given by (i) $a_n = (-1)\frac{2n-1}{n}$, (ii) $a_n = \frac{2}{n} + (-1)^n 3$, oscillates finitely.

10. Examine the convergence of the sequences $\{a_n\}$ whose n^{th} terms are given below :

(i) $a_n = 3n^2$; (ii) $a_n = 1 + \frac{1}{n}$; (iii) $a_n = [n + (-1)^n]^{-1}$ (iv) $a_n = \log \frac{1}{n}$.

11. Show that the sequence $\{(n + 1)^{3n} (1 + n^3)^{-n}\}$ converges to 1.

12. If $a_n = 3 + \left(-\frac{1}{2}\right)^n$, then prove that $|\ a_n - 3\ | < 0.001$ if $n > 9$.

13. Let $a_1 = 1$, $a_{n+1} = \sqrt{2 + a_n}$, whenever $n \geq 1$. Show that the sequence $\{a_n\}$ is monotonically increasing and bounded. Find also $\lim\limits_{n \to \infty} a_n$.

14. Prove that the sequence $\{a_n\}$ defined by the relation $a_1 = 1$ and $a_n = 1 + \frac{1}{1!} + \frac{1}{2!} + \frac{1}{3!} + \ldots + \frac{1}{(n-1)!}$, $n \geq 2$, converges.

15. Prove that $\lim\limits_{n \to \infty} \left(\frac{n^n}{n!}\right)^{\frac{1}{n}}$ exists and equals e.

16. Evaluate the following limits :

(i) $\lim\limits_{n \to \infty} \frac{(2n+1)^2}{2n^2}$; (ii) $\lim\limits_{n \to \infty} \frac{\sqrt[n]{n!}}{n}$

17. Show that $\lim\limits_{n \to \infty} n^{\frac{1}{n}} = 1$ and hence show that

$$\lim_{n \to \infty} \frac{1}{n}\left(1 + 2^{\frac{1}{2}} + 3^{\frac{1}{3}} + \ldots + n^{\frac{1}{n}}\right) = 1$$

18. Show that $\lim\limits_{n \to \infty} \left[\frac{(2n)!}{n!\,n!}\right]^{\frac{1}{n}} = 4$, using Cauchy's second theorem of limit.

19. Show that $\lim\limits_{n \to \infty} \left(\frac{(n+1)(n+2)\ldots(2n)}{(n).n!}\right)^{\frac{1}{n}} = 4$.

20. Show by Cauchy's Criteria, that each of the sequences $\{a_n\}$ and $\{b_n\}$, where

$$a_n = 1-\frac{1}{2}+\frac{1}{3}-\frac{1}{4}+\ldots+(-1)^{n+\frac{1}{n}} \text{ and}$$

$$b_n = 1+\frac{1}{1!}+\frac{1}{2!}+\ldots+\frac{1}{n!} \text{ converges.}$$

21. Choose the correct answers for the following :

(*i*) The sequence $\{|\ a_n\ |\}$ converges to zero iff the sequence $\{a_n\}$ converges to

(*a*) 0 (*b*) 1 (*c*) $\frac{1}{2}$ (*d*) Any value > 0.

(*ii*) If $\lim\limits_{n\to\infty} a_n \to a$, $\lim\limits_{n\to\infty} b_n \to b$ and $a_n < b_n$ for all n, then

(*a*) $a = b$ (*b*) $a \le b$ (*c*) $a \ge b$ (*d*) $a \ne b$

(*iii*) If $\lim\limits_{n\to\infty} a_n \to l$, $\lim\limits_{n\to\infty} b_n \to l$ and $a_n \le c_n \le b_n$ for all n, then

(*a*) $c_n = l$ (*b*) $c_n \to 0$ (*c*) $c \to l$ (*d*) None of these.

Answers

1. (*i*) F (*ii*) F (*iii*) T (*iv*) T (*v*) T
(*vi*) T (*vii*) T (*viii*) F (*ix*) T (*x*) T

2. (*i*) Converges, diverges (*ii*) 1 (*iii*) increasing (*iv*) converges
(*v*) $+\infty$ (*vi*) null (*vii*) $(-1)^{n-1}\frac{1}{n}$ (*viii*) Convergent (*ix*) $(-1)^{n-1}\cdot\frac{n}{3^n}$
(*x*) 5

4. Converges to 0.

10. (*i*) divergent (*ii*) convergent (*iii*) convergent (*iv*) divergent

13. 2

16. (*i*) 2 (*ii*) $\frac{1}{e}$

21. (*i*) *a* (*ii*) *b* (*iii*) c.

Chapter **4**

Infinite Series

4.1 INTRODUCTION

We are already familiar with few series, viz., Arithmetic series, Geometric series, Exponential series, Logarithmic series, etc. The terms of these series follow some definite rule. For example, in an arithmetic series, each term after the first term is either increased or decreased by a fixed number; in a geometric series, each term after the first is formed by multiplying the preceding term by a fixed number. A series can be formed in other ways also. For example, the series 1 + 4 + 9 + 16 + ... is formed by squaring the natural numbers.

Now, we define formal definition of an infinite series.

Definition. *If $u_1, u_2, u_3, ..., u_n, ...$ be the terms of an infinite sequence of real numbers, then $u_1 + u_2 + ... + u_n + ...$ is called an infinite series (number of terms are infinite).*

An infinite series is denoted by $\sum_{n=1}^{\infty} u_n$ or simply $\sum u_n$. The sum of first n terms of an infinite series is denoted by S_n.

4.2 SEQUENCE OF PARTIAL SUMS OF AN INFINITE SERIES

Let $u_1 + u_2 + ... + u_n + ...$ be an infinite series. If S_n denotes the sum of the first n terms of this series, such that

$$S_n = u_1 + u_2 + ... + u_n = \sum_{k=1}^{n} u_k,$$

then $\{S_n\}$ is called the sequence of partial sums of the given series.

For example, let us consider the series $1 + 3 + 3^2 + 3^3 + ... + 3^{n-1} + ...$

The sum of the first n terms of the series is

$$S_n = \frac{3^n - 1}{3 - 1} = \frac{1}{2}(3^n - 1)$$

Therefore, the sequence of partial sums of this series is $\{S_n\}$, where $S_n = \frac{1}{2}(3^n - 1)$.

Remark. If the sequence of partial sums S_n is known, then we know x_n and hence the series (because $x_n = S_n - S_{n-1}$). Thus, a sequence $\{S_n\}$ of partial sums is associated to each given series $\sum_{n=1}^{\infty} x_n$ and vice-versa.

Hence, we can assign the same behaviour to the series, $\sum_{n=1}^{\infty} x_n$ as is exhibited by its $\{S_n\}$ of partial sums as $n \to \infty$.

This leads to the following definition of convergence of an infinite series.

4.3 CONVERGENCE OF AN INFINITE SERIES

Definition. *An infinite series* $\sum u_n$ *is said to be convergent, divergent or oscillatory according as the sequence* $\{S_n\}$ *of its partial sum converges, diverges or oscillates.*

If the sequence $\{S_n\}$ of partial sums of a series $\sum u_n$ converges to S, then S is said to be the sum of the series $\sum u_n$. We state two theorems (without proof) which are direct consequences of the above definition.

Theorem 4.1. *The alternation or omission of a finite number of terms of a series has no affect on its convergence, divergence or oscillation.*

Theorem 4.2. *Multiplication of all the terms of a series by a fixed non-zero number has no affect on the convergence, divergence or oscillation of the series.*

Remark. The partial sum S_n of a series $\sum u_n$ is a function of n and as n increases indefinitely the following possibilities arise :

(*i*) If $S_n \to S$, as $n \to \infty$, where S is finite, then the series $\sum u_n$ is convergent.

(*ii*) If $S_n \to \infty$ as $n \to \infty$, then $\sum u_n$ is divergent and diverges to $+\infty$.

(*iii*) If $S_n \to -\infty$ as $n \to \infty$, then $\sum u_n$ is divergent and diverges to $-\infty$.

(*iv*) If S_n does not tend to a unique limit as $n \to \infty$, then $\sum u_n$ is oscillatory or non-convergent.

4.3.1 Illustrative Examples

1. *Discuss the convergence of the series*

$$\sum u_n = 1+\frac{1}{2}+\frac{1}{4}+\frac{1}{8}+\frac{1}{16}+\ldots+\frac{1}{2^{n-1}}+\ldots$$

Sol. Let $\{S_n\}$ be the sequence of partial sums of the given series $\sum u_n$, then

$$S_n = 1+\frac{1}{2}+\frac{1}{4}+\ldots+\frac{1}{2^{n-1}} = \frac{1-\frac{1}{2^n}}{1-\frac{1}{2}} = 2-\frac{1}{2^{n-1}}.$$

$$\therefore \qquad \lim_{n\to\infty} S_n = \lim_{n\to\infty}\left(2-\frac{1}{2^{n-1}}\right) = 2-\lim_{n\to\infty}\frac{1}{2^{n-1}} = 2-0 = 2,$$

which is a finite quantity.

Hence $\{S_n\}$ converges to 2 and consequently the given series $\sum u_n$ is convergent.

2. *Show that the series* $\sum u_n = 1+\frac{2}{3}+\frac{2^2}{3^2}+\ldots+\frac{2^{n-1}}{3^{n-1}}+\ldots$ *is convergent.*

Sol. Here,

$$S_n = 1+\frac{2}{3}+\frac{2^2}{3^2}+\ldots+\frac{2^{n-1}}{3^{n-1}}$$

$$= \frac{1-\left(\frac{2}{3}\right)^n}{1-\frac{2}{3}} = 3-3\left(\frac{2}{3}\right)^n$$

$$\therefore \qquad \lim_{n\to\infty} S_n = 3-\lim_{n\to\infty} 3.\left(\frac{2}{3}\right)^n = 3-0 = 3.$$

Since, the sequence $\{S_n\}$ of partial sums converges to 3, therefore, the given series $\sum u_n$ is convergent.

3. *Show that the series*

$$\sum u_n = 1+2+3+\ldots\infty \textit{ is divergent.}$$

Sol. Here,

$$S_n = 1+2+3+\ldots+n = \frac{n(n+1)}{2}$$

Now,

$$\lim_{n\to\infty} S_n = \lim_{n\to\infty}\frac{1}{2}n(n+1) \to +\infty$$

Since $S_n \to +\infty$, as $n\to\infty$, hence the series $\sum u_n$ is divergent and diverges to $+\infty$.

4. *Test the convergence of the series* $\sum u_n = \sum_{n=1}^{\infty}(-1)^{n-1}.2$

Sol. Here, $S_n = 2 - 2 + 2 - 2 + \ldots$ to n terms.

Therefore, $S_n = 0$ or 2 according as n is even or odd.

Hence, $S_n \to 0$ or 2 as $n \to \infty$

Moreover, $|S_n| \le 2$

Therefore, $\{S_n\}$ oscillates finitely.

Hence, $\sum u_n$ is oscillatory and oscillates finitely.

5. *Discuss the convergence of the series*

$$\sum u_n = 5 - 4 - 1 + 5 - 4 - 1 + 5 - 4 - 1 + \ldots \infty$$

Sol. Here $S_n = 5 - 4 - 1 + 5 - 4 - 1 + 5 - 4 - 1 + \ldots$ to n terms

$= 0, 5, 1$ according as $n = 3m, 3m + 1, 3m + 2$

where m is a positive integer.

This shows that S_n does not tend to a unique limit. Hence, the series is oscillatory, not convergent.

6. *Test the convergence of the series*

$$\sum u_n = \frac{1}{1.3} + \frac{1}{3.5} + \frac{1}{5.7} + \ldots + \frac{1}{n.(n+2)} + \ldots$$

Sol. Here

$$S_n = \frac{1}{1.3} + \frac{1}{3.5} + \frac{1}{5.7} + \ldots + \frac{1}{n.(n+2)}$$

$$= \frac{1}{2}\left(1 - \frac{1}{3}\right) + \frac{1}{2}\left(\frac{1}{3} - \frac{1}{5}\right) + \frac{1}{2}\left(\frac{1}{5} - \frac{1}{7}\right) + \ldots + \frac{1}{2}\left(\frac{1}{n} - \frac{1}{n+2}\right)$$

$$= \frac{1}{2}\left(1 - \frac{1}{n+2}\right)$$

$$\therefore \quad \lim_{n \to \infty} S_n = \frac{1}{2}\left(1 - \lim_{n \to \infty}\frac{1}{n+2}\right) = \frac{1}{2}(1 - 0) = \frac{1}{2}$$

Thus, $S_n \to \frac{1}{2}$ as $n \to \infty$, Hence $\{S_n\}$ is convergent. Therefore $\sum u_n$ is also convergent.

We now state and prove a theorem on convergence of a series.

Theorem 4.3. A necessary condition for a series. $\sum u_n$ to converge is that $u_n \to 0$ as $n \to \infty$.

Proof. Let $S_n = u_1 + u_2 + \ldots + u_n$

Then, $$S_{n-1} = u_1 + u_2 + ... + u_{n-1}$$

$\therefore$ $$S_n - S_{n-1} = u_n \qquad ...(1)$$

Since $\sum u_n$ is convergent, so let $S_n \to S$ as $n \to \infty$, then S_{n-1} will also tend to S as $n \to \infty$.

Taking limits of both sides of (1), we get

$$\lim_{n\to\infty} u_n = \lim_{n\to\infty} (S_n - S_{n-1}) = \lim_{n\to\infty} S_n - \lim_{n\to\infty} S_{n-1} = S - S = 0$$

Hence, the condition is necessary for convergence of $\sum u_n$.

4.3.2 Illustrative Examples

1. *Show that the series* $\sum u_n = \frac{1}{2} + \frac{2}{3} + \frac{3}{4} + ...$

is not convergent.

Sol. Here $$u_n = \frac{n}{n+1}$$

$\therefore$ $$\lim_{n\to\infty} u_n = \lim_{n\to\infty} \frac{n}{n+1} = \lim_{n\to\infty} \frac{1}{1+\frac{1}{n}} = 1 \neq 0$$

Hence, the series $\sum u_n$ is not convergent.

2. *Show that the series*

$1 + \frac{1}{1!} + \frac{1}{2!} + \frac{1}{3!} + ... + \frac{1}{(n-1)!} + ...$ *converges.*

Sol. Here $$S_n = 1 + 1 + \frac{1}{2} + \frac{1}{2.3} + \frac{1}{2.3.4} + ... + \frac{1}{2.3.4....(n-1)}$$

$$< 1 + 1 + \frac{1}{2} + \frac{1}{2.3} + \frac{1}{3.4} + ... + \frac{1}{(n-2)(n-1)}, \qquad (\because n > 4)$$

$$< 1 + 1 + \frac{1}{2} + \left(\frac{1}{2} - \frac{1}{3}\right) + \left(\frac{1}{3} - \frac{1}{4}\right) + ... + \left(\frac{1}{n-2} - \frac{1}{n-1}\right)$$

$$< 1 + 1 + \frac{1}{2} + \left(\frac{1}{2} - \frac{1}{n+1}\right) = 3 - \frac{1}{n+1} < 3$$

This shows that S_n steadily increases as n increases but is always less than 3.

Thus, S_n tends to a limit when $n \to \infty$ and this limit is 3.

Hence, the series converges.

3. *Show that the series log 2 + $\log\frac{3}{2}+\log\frac{4}{3}+\log\frac{5}{4}+\dots$ diverges.*

Sol. Here,

$$S_n = \log 2+\log\frac{3}{2}+\log\frac{4}{3}+\log\frac{5}{4}+\dots+\log\left(\frac{n+1}{n}\right)$$

$$= \log 2+\log 3-\log 2+\log 4-\log 3+\log 5-\log 4+\dots+\log(n+1)-\log n$$

$$= \log(1+n)$$

Now, $\lim_{n\to\infty} S_n = \lim_{n\to\infty}\log(1+n)\to\infty$

Since $S_n\to\infty$, as $n\to\infty$

Hence $\{S_n\}$ is divergent and thus the given series is divergent.

4. *Test the convergence of the series* $\sum\left(\frac{2^n-1}{2^n+1}\right)$.

Sol. Here,
$$u_n = \frac{2^n-1}{2^n+1}$$

$\therefore$
$$\lim_{n\to\infty} u_n = \lim_{n\to\infty}\frac{2^n-1}{2^n+1} = \lim_{n\to\infty}\frac{2^n\left(1-\frac{1}{2^n}\right)}{2^n\left(1+\frac{1}{2^n}\right)}$$

$$= \lim_{n\to\infty}\frac{\left(1-\frac{1}{2^n}\right)}{\left(1+\frac{1}{2^n}\right)} = \frac{1}{1} = 1\neq 0.$$

Hence, the series is not convergent, that is the series is divergent.

5. *Test for the convergence of the series* $(1-x)+(x-x^2)+(x^2-x^3)+\dots$

Sol. Here,
$$S_n = (1-x)+(x-x^2)+(x^2-x^3)+\dots+(x^{n-1}-x^n)$$

$$= (1+x+x^2+\dots+x^{n-1})-x(1+x+x^2+\dots+x^{n-1})$$

$$= (1-x)(1+x+x^2+\dots+x^{n-1})$$

$$= (1-x).\frac{(1-x^n)}{1-x} = (1-x^n), \text{ if } x\neq 1$$

$\therefore$
$$\lim_{n\to\infty} S_n = \lim_{n\to\infty}\left(1-x^n\right) = 1, \text{ if } |x|<1$$

$$= -\infty, \text{ if } x>1$$

$\therefore$ $\{S_n\}$ convergent for $x = 1$ and divergent for $x > 1$

Hence, the given series is convergent for $x = 1$ and diverges to $-\infty$ for $x > 1$.

6. *Test the convergence of the series*

$$\frac{1}{1+x}+\frac{x}{1+x^2}+\frac{x^2}{1+x^3}+\ldots, (x>0)$$

Sol. Let $\sum u_n$ be the given series.

Then $$u_n = \frac{x^{n-1}}{1+x^n}$$

Now, the following cases to be considered:

Case I. Let $x > 1$

Then, $$u_n = \frac{\frac{1}{x}}{\frac{1}{x^n}+1}$$

$\therefore$ $$\lim_{n\to\infty} u_n = \lim_{n\to\infty} \frac{\frac{1}{x}}{\frac{1}{x^n}+1} = \frac{1}{x} \neq 0$$

Hence, the given series $\sum u_n$ is divergent, when $x > 1$.

Case II. Let $x < 1$.

Then, $$\lim_{n\to\infty} \frac{x^{n-1}}{1+x^n} = \frac{0}{1+0} = 0$$

Hence the given series $\sum u_n$ is convergent, when $x < 1$.

Case III. Let $x = 1$

Then, $$u_n = \frac{1}{2}+\frac{1}{2}+\frac{1}{2}+\ldots,$$

which is divergent.

$\therefore$ The given series is convergent when $x < 1$ and divergent when $x \geq 1$.

7. *Show that the series* $\sqrt{\frac{1}{4}}+\sqrt{\frac{2}{6}}+\sqrt{\frac{3}{8}}+\ldots+\sqrt{\frac{n}{2(n+1)}}+\ldots$ *does not converge.*

Sol. Here, $$u_n = \sqrt{\frac{n}{2(n+1)}} = \frac{1}{\sqrt{2}}\sqrt{\frac{1}{1+\frac{1}{n}}}$$

$$\therefore \quad \lim_{n\to\infty} u_n = \frac{1}{\sqrt{2}} \lim_{n\to\infty} \sqrt{\frac{1}{1+\frac{1}{n}}} = \frac{1}{\sqrt{2}}.1 = \frac{1}{\sqrt{2}} \neq 0$$

Since u_n does not tend to zero as $n \to \infty$, therefore, the given series does not converge.

Remark. The condition stated for convergence in Theorem 2.3 is not sufficient. We shall prove it by an example in the next article.

In the foregoing examples, we considered series for which the sequences of partial sums could be easily computed. In many cases, it is either impossible or difficult to compute the sequence of partial sums of a given series and yet we wish to know whether the series converges or not. For dealing with such series, we shall develop certain tests which would enable us to determine their behaviour. The fundamental principle involved in determining the behaviour is *Cauchy's general principle of convergence,* which we state and prove below.

4.4 CAUCHY'S GENERAL PRINCIPLE OF CONVERGENCE FOR SERIES

Statement *A necessary and sufficient condition for a series* $\sum u_n$ *to converge is that to each* $\varepsilon > 0$ *there exists a positive integer m such that*

$$|u_{m+1} + u_{m+2} + \ldots + u_n| < \varepsilon, \text{ whenever } n > m.$$

Proof. The series $\sum u_n$ converges if and only if the sequence $\{S_n\}$ of its partial sums converges. By Cauchy's general principle of convergence for sequence, a necessary and sufficient condition for the convergence of $\{S_n\}$ is that given any $\varepsilon > 0$, there exists a positive integer m such that

$$|S_n - S_m| < \varepsilon \text{ whenever } n > m,$$

that is, $|u_{m+1} + u_{m+2} + \ldots + u_n| < \varepsilon$ whenever $n > m$,

which is Cauchy's general principle of convergence for series.

As an illustrative example, let

$$\sum u_n = 1 + \frac{1}{2} + \frac{1}{3} + \ldots + \frac{1}{n} + \ldots$$

Here, $\quad u_n = \frac{1}{n}$, which tends to zero as $n \to \infty$

$\therefore \quad \sum u_n$ satisfies the necessary condition for convergence. So, let us suppose that the series $\sum u_n$ converges.

Taking $\varepsilon = \frac{1}{4}$, by Cauchy's general principle of convergence, there must exist a positive integer m such that,

$$\left|\frac{1}{m+1}+\frac{1}{m+2}+\ldots+\frac{1}{n}\right|<\frac{1}{4} \text{ for all } n>m$$

But by taking $n = 2m$, we have

$$\frac{1}{m+1}+\frac{1}{m+2}+\ldots+\frac{1}{n}=\frac{1}{m+1}+\frac{1}{m+2}+\ldots+\frac{1}{2m}>m.\frac{1}{2m}=\frac{1}{2}$$

$$\left[\because \frac{1}{m+1}, \frac{1}{m+2}, \ldots \text{ each} > \frac{1}{2m}\right]$$

Thus, we have a contradiction.

Hence, the given series $\sum u_n$ does not converge.

Remark. Two convergent series, can be added (or subtracted) term by term to give a convergent series. But it is not true for two non-convergent series, because the sum of two non-convergent series may form a convergent series.

Often we come across the series with non-negative terms. We shall now consider this types of series, called series of positive terms.

4.5 SERIES OF POSITIVE TERMS

Let us suppose that $\sum u_n$ is a series of positive terms.

If $$S_n = u_1 + u_2 + \ldots + u_n,$$

is the partial sum of the sequence $\{S_n\}$, then

$$S_n - S_{n-1} = u_n.$$

Since, $u_n > 0$ for all n, therefore, $S_n - S_{n-1} > 0$, for all n, that is, $S_n > S_{n-1}$ for all n.

Thus, the sequence $\{S_n\}$ of partial sums of a series of positive terms is a monotonic increasing sequence. Since a monotonic increasing sequence can either converge or diverge to ∞, but cannot oscillate, therefore, for a series of positive terms, we have only two possibilities, that is, either the series converges or it diverges to $+\infty$.

We now state a theorem (without proof) for series of positive terms as below :

Theorem 4.4. *A positive term series converges, if, and only if, the sequence of its partial sums is bounded above.*

4.6 GEOMETRIC SERIES

The power series $\sum u_n = 1 + x + x^2 + x^3 + \ldots$ is known as *geometric series* with common ratio x.

This series converges if $0 < x < 1$, diverges if $x \geq 1$ and oscillatory if $x \leq -1$.

4.6.1 Illustrative Examples

1. *Show that the series* $1 + \frac{1}{2} + \frac{1}{2^2} + \frac{1}{2^3} + \ldots$ *is convergent.*

Sol. Here,

$$S_n = 1 + \frac{1}{2} + \frac{1}{2^2} + \frac{1}{2^3} + \ldots + \frac{1}{2^n}.$$

$$= \frac{1 - \frac{1}{2^n}}{1 - \frac{1}{2}} = 2\left(1 - \frac{1}{2^n}\right)$$

$$\therefore \quad \lim_{n \to \infty} S_n = \lim_{n \to \infty} 2\left(1 - \frac{1}{2^n}\right) = 2(1 - 0) = 2$$

$\because$ $S_n \to 2$ as $n \to \infty$, therefore, $\{S_n\}$ is convergent.

Hence, the given series converges.

[or here $x = \frac{1}{2} < 1$, so the series converges]

2. *Show that the series* $1 + 2 + 2^2 + 2^3 + \ldots$ *diverges.*

Sol. Here,

$$S_n = 1 + 2 + 2^2 + \ldots + 2^n = \frac{(2^n - 1)}{2 - 1} = 2^n - 1$$

$$\therefore \quad \lim_{n \to \infty} S_n = \lim_{n \to \infty} (2^n - 1) \to +\infty$$

Therefore, the series diverges to $+\infty$.

[or, here $x = 2 > 1$, so the series diverges].

3. *Test the convergence of the series*

$$1 - \frac{1}{2} + \frac{1}{3^2} - \frac{1}{4} + \frac{1}{5^2} - \frac{1}{6} + \frac{1}{7^2} - \frac{1}{8} + \ldots \infty$$

Sol. Let S_n and S'_n denote the sums of the first n positive terms and the first n negative terms respectively.

Then,

$$S_n = 1 + \frac{1}{3^2} + \frac{1}{5^2} + \ldots + \frac{1}{(2n-1)^2}$$

$$\therefore \quad S_n < 1 + \frac{1}{2^2} + \frac{1}{3^2} + \frac{1}{4^2} + \frac{1}{5^2} + \ldots + \frac{1}{n^2}$$

That is, the sum S_n is less than the sum of a convergent series.

Hence, S_n tends to a finite limit.

Again $$S'_n = -\left(\frac{1}{2}+\frac{1}{4}+\frac{1}{6}+\ldots \text{ to } n \text{ terms}\right)$$

$\therefore \quad \lim_{n\to\infty} S'_n \to -\infty$

If P_{2n} denote the sum of the given series upto $2n$ terms, then

$$P_{2n} = S_n + S'_n$$

Then, as $n \to \infty$, $\quad P_{2n} \to -\infty$.

Hence, the given series is divergent and diverges to $-\infty$.

4.7 SOME PROPERTIES OF INFINITE SERIES

The following are some useful properties of infinite series :

(i) If $\sum u_n$ is convergent and converges to the limit l, then $\sum ku_n$, $(k > 0)$, is also convergent and converges to the limit kl.

(ii) If $\sum u_n$ is divergent, then $\sum ku_n$ diverges to $+\infty$ or $-\infty$, according as k is positive or negative.

(iii) If $\sum u_n$ and $\sum v_n$ are two convergent series and converges to S and l respectively, then $\sum (k_1 u_n + k_2 v_n)$ is convergent and converges to $k_1 S + k_2 l$, where $k_1 > 0$, $k_2 > 0$.

As an example, we shall show that the series

$$\sum u_n = \frac{1}{2}+\frac{1}{3}+\frac{1}{2^2}+\frac{1}{3^2}+\frac{1}{2^3}+\frac{1}{3^3}+\ldots \text{ is convergent.}$$

Here, $\sum u_n$ is the sum of two geometric series, viz.,

$$\sum v_n = \frac{1}{2}+\frac{1}{2^2}+\frac{1}{2^3}+\ldots \text{ and}$$

$$\sum w_n = \frac{1}{3}+\frac{1}{3^2}+\frac{1}{3^3}+\ldots$$

Now, $$v_n = \frac{1}{2}+\frac{1}{2^2}+\frac{1}{2^3}+\ldots+\frac{1}{2^n} = \frac{\frac{1}{2}\left(1-\frac{1}{2^n}\right)}{1-\frac{1}{2}} = 1-\frac{1}{2^n}$$

$$\therefore \quad \lim_{n\to\infty} v_n = \lim_{n\to\infty}\left(1-\frac{1}{2^n}\right) = 1$$

$\therefore \quad \sum v_n$ converges to 1.

And $$w_n = \frac{1}{3}+\frac{1}{3^2}+\frac{1}{3^3}+\ldots+\frac{1}{3^n} = \frac{\frac{1}{3}\left(1-\frac{1}{3^n}\right)}{1-\frac{1}{3}} = \frac{1}{2}\left(1-\frac{1}{3^n}\right)$$

$\therefore$ $$\lim_{n\to\infty} w_n = \lim_{n\to\infty} \frac{1}{2}\left(1-\frac{1}{3^n}\right) = \frac{1}{2}$$

$\therefore$ $\sum w_n$ converges to $\frac{1}{2}$.

Therefore, $\sum v_n + \sum w_n$ is convergent and converges to $1+\frac{1}{2}=\frac{3}{2}$.

Hence, the series $\sum u_n$ is convergent and converges to $\frac{3}{2}$. [by property (*iii*)]

EXERCISE 4.1

1. If the following statements are true, write 'T', if falls, write 'F'.
   (*i*) If the partial sums of the sequence of an infinite series converges then the series diverges.
   (*ii*) The alteration or omission of a finite number of terms of an infinite series does not affect the convergence or divergence of the series.
   (*iii*) Multiplication of all terms of a series by a non-zero constant, has affect on its convergence or divergence.
   (*iv*) The series $1 + 2 + 3 + \ldots \infty$ is oscillatory.
   (*v*) '$u_n \to 0$ as $n \to \infty$, is sufficient condition for convergence of the series $\sum u_n$.
   (*vi*) The series $1 + \frac{1}{2}+\frac{1}{3}+\ldots+\frac{1}{n}+\ldots$ is convergent.
   (*vii*) A series of positive terms converges if, the sequence of its partial sums is bounded.
   (*viii*) The term by term addition of convergent series is not convergent.
   (*ix*) A positive term series converges, if and only if, the sequence of its partial sums is bounded above.
   (*x*) The series $\sum \frac{1}{n}$ converges.
2. Show that the series $\sum_{1}^{\infty} \frac{1}{n(n+1)}$ is convergent and find its limit.
3. Test the convergence of the following series :
   (*i*) $\frac{1}{1.3}+\frac{1}{3.5}+\frac{1}{5.7}+\ldots$ (*ii*) $\sum \frac{1}{n^2}$

(iii) $\sum \frac{1}{n!}$ (iv) $\sum \frac{1}{(2n-1)}$

(v) $\sum \frac{2n+1}{3n+5}$ (vi) $\sum \frac{2^n-1}{2^n}$.

4. For the series $\sum \frac{1}{n(n+1)(n+2)}$, find the partial sums S_n and then show that the series converges to $\frac{1}{4}$.

5. Test the convergence of the following series

(i) $\frac{1}{2!}+\frac{2}{3!}+\frac{3}{4!}+\frac{4}{5!}+\ldots$ (ii) $\frac{1}{1.5}+\frac{1}{5.9}+\frac{1}{9.13}+\frac{1}{13.17}+\ldots$.

6. Show that the series $1-\frac{1}{2}+\frac{1}{3}-\frac{1}{4}+\ldots$ is convergent.

7. Show that the $\frac{3}{5}+\frac{4}{5^2}+\frac{3}{5^3}+\frac{4}{5^4}+\ldots\infty$ is convergent.

Answers

1. (i) F (ii) T (iii) F (iv) F
   (v) F (vi) F (vii) T (viii) F
   (ix) T (x) F.
2. 1.
3. (i) Convergent (ii) Convergent (iii) Convergent (iv) Not convergent
   (v) Not convergent (vi) Divergent
5. (i) Convergent (ii) Convergent.

In the above lines we studied some properties of infinite series. Now, we need to test for convergency and divergency of a series. There are different tests to study the property, convergence and divergence of a series. We shall discuss some of them in the next articles.

4.8 COMPARISON TESTS

Statement. If $\sum u_n$ and $\sum v_n$ are two series of non-negative terms and

A. Let $\sum v_n$ is convergent, then $\sum u_n$ is also convergent if,

(i) $u_n \le v_n$ for all $n \ge m$, m being some fixed finite number,

(ii) $\frac{u_n}{v_n} < k$ for all $n \ge m$, and k being a fixed positive number,

(iii) $\lim_{n\to\infty} \frac{u_n}{v_n} = l$, a finite positive non-zero number.

B. Let $\sum v_n$ is divergent, then $\sum u_n$ is also divergent if,

(*i*) $u_n \geq v_n$ for all $n \geq m$, m being some fixed finite number,

(*ii*) $\dfrac{u_n}{v_n} > N$ for all $n \geq m$, N being some fixed positive number,

(*iii*) $\lim\limits_{n \to \infty} \dfrac{u_n}{v_n} = k$, a finite positive non-zero number.

4.9 AN AUXILIARY SERIES [HYPER HARMONIC SERIES]

Statement *The auxiliary series* $\sum \dfrac{1}{n^p} = \dfrac{1}{1^p} + \dfrac{1}{2^p} + \dfrac{1}{3^p} + \ldots + \dfrac{1}{n^p} + \ldots$ *is convergent if* $p > 1$, *and divergent if* $p \leq 1$.

Proof. Case I. Let $p > 1$, then

$$\frac{1}{1^p} = \frac{1}{1^p},$$

$$\frac{1}{2^p} + \frac{1}{3^p} < \frac{2}{2^p} \qquad \left[\because 3^p > 2^p, \therefore \frac{1}{3^p} < \frac{1}{2^p}\right]$$

$$\frac{1}{4^p} + \frac{1}{5^p} + \frac{1}{6^p} + \frac{1}{7^p} < \frac{4}{4^p}$$

$$\ldots \qquad \ldots \qquad \ldots$$

$$\frac{1}{\left(2^k\right)^p} + \ldots + \frac{1}{\left(2^{k+1} - 1\right)^p} < \frac{2k}{\left(2^k\right)^p}$$

Adding the corresponding sides of the above relations, we obtain

$$\frac{1}{1^p} + \frac{1}{2^p} + \ldots + \frac{1}{\left(2^{k+1} - 1\right)^p} < \frac{1}{1^p} + \frac{2}{2^p} + \frac{4}{4^p} + \ldots + \frac{2^k}{\left(2^k\right)^p}$$

$$= \frac{1 - \left(\frac{1}{2}\right)^{(p-1)(k+1)}}{1 - \left(\frac{1}{2}\right)^{p-1}} < \frac{1}{1 - \left(\frac{1}{2}\right)^{p-1}}$$

Thus, $\qquad S_2^{k+1} - 1 < A$ for all k,

where, $\qquad S_m = \dfrac{1}{1^p} + \dfrac{1}{2^p} + \ldots + \dfrac{1}{m^p}$, $A = \dfrac{1}{1 - 2^{1-p}}$

Since, all terms of the given series are positive, therefore, $S_n > S_m$, whenever $n > m$.

Also, if n be any positive integer, whatever, it is always possible to find a positive integer k, such that $n < 2^{k+1} - 1$. With such a choice of k,

$$S_n < S_2^{(k+1)} - 1 < A$$

Thus, $$S_n < A \text{ for all } n.$$

Since the sequence S_n of partial sums of the given series is bounded, therefore the given series is convergent.

Case II. Let $p = 1$.

Then, $$1 + \frac{1}{2} = 1 + \frac{1}{2}$$

$$\frac{1}{3} + \frac{1}{4} > \frac{2}{4} \qquad \left[\because 3 < 4, \therefore \frac{1}{3} > \frac{1}{4}\right]$$

$$\frac{1}{5} + \frac{1}{6} + \frac{1}{7} + \frac{1}{8} > \frac{4}{8} \qquad \left[\because \frac{1}{5} > \frac{1}{8}, \frac{1}{6} > \frac{1}{8}, \frac{1}{7} > \frac{1}{8}\right]$$

$$\ldots \qquad \ldots \qquad \ldots$$

$$\frac{1}{2^k + 1} + \ldots + \frac{1}{2^{k+1}} < \frac{2^k}{2^{k+1}}$$

Adding the corresponding sides of the above relations, we get

$$1 + \frac{1}{2} + \frac{1}{3} + \ldots + \frac{1}{2^{k+1}} > 1 + \left(\frac{1}{2} + \frac{1}{2} + \ldots (k+1) \text{ terms}\right) > 1 + \frac{1}{2}(k+1)$$

or, $$S_{2^{k+1}} < \frac{3}{2} + \frac{1}{2}k, \text{ for all } k.$$

Whatever number K we may have, it is possible to find a positive integer k such that $\frac{3}{2} + \frac{1}{2}k > K$.

With such a choice of k, $S_{2^{n+1}} > K$,

So that $S_n > K$, whenever $n > 2^{k+1}$

This shows that the sequence $\{S_n\}$ of partial sums of the given series is not bounded and hence the series.

$$1 + \frac{1}{2} + \frac{1}{3} + \ldots + \frac{1}{n} + \ldots$$

is divergent.

Case III. Let $p < 1$.

Since $p < 1$, therefore, $\frac{1}{n^p} > \frac{1}{n}$ for all n.

Since the series $\sum \frac{1}{n}$ is divergent, therefore from the above inequality by applying the comparison test it follows that the series $\sum \frac{1}{n^p}$ is divergent when $p < 1$.

Remark. If $\lim_{n \to \infty} \frac{u_n}{v_n}$ is neither zero or infinite, the two series $\sum u_n$ and $\sum v_n$ are both convergent or both divergent. In order to determine the nature of a series $\sum u_n$, we compare u_n, the nth term of $\sum u_n$, with the nth term v_n, of a suitable auxiliary series $\sum v_n$, whose convergence or divergence is already known to us. The series discussed in 4.9 is useful for this purpose.

We choose $v_n = \frac{1}{n^{p-q}}$, where p and q are respectively the highest indices of n in the denominator and numerator of u_n when it is in the form of a fraction. In case, when u_n is expanded in ascending powers of $\frac{1}{n}$, we choose v_n retaining only the lowest power of $\frac{1}{n}$.

4.10 ALGORITHM FOR COMPARISON TEST

Step 1 : Find the nth term u_n, of the given series $\sum u_n$

Step 2 : Consider an auxiliary series $\sum v_n$ from $\sum u_n$ whose n^{th} term is v_n

Step 3 : Divide u_n by v_n

Step 4 : Find the limit of $\frac{u_n}{v_n}$ as $n \to \infty$

Step 5 : If the limit $\frac{u_n}{v_n}$ is finite and non-zero, compare $\sum v_n$ with the auxiliary series $\sum \frac{1}{n^p}$ which is convergent if $p > 1$ and divergent if $p \leq 1$

Step 6 : By comparison test $\sum u_n$ has the same property (convergent or divergent) as that of $\sum v_n$.

4.10.1 Illustrative Examples

1. *Test for convergence or divergence of the series* $\sum \frac{\sqrt{n}}{n^3+1}$.

Sol. Here,

$$u_n = \frac{n^{\frac{1}{2}}}{n^3+1} = \frac{n^{\frac{1}{2}}}{n^3\left(1+\frac{1}{n^3}\right)}$$

$$= \frac{1}{n^{3-\frac{1}{2}}\left(1+\frac{1}{n^3}\right)} = \frac{1}{n^{\frac{5}{2}}\left(1+\frac{1}{n^3}\right)}$$

Let us take $$v_n = \frac{1}{n^{\frac{5}{2}}}$$

$$\therefore \quad \lim_{n\to\infty}\frac{u_n}{v_n} = \lim_{n\to\infty}\frac{1}{n^{\frac{5}{2}}\left(1+\frac{1}{n^3}\right)}\times n^{\frac{5}{2}} = \lim_{n\to\infty}\frac{1}{\left(1+\frac{1}{n^3}\right)} = 1$$

which is a finite number.

$\therefore$ The property of $\sum u_n$ depends on the property of $\sum v_n$.

But, $$\sum v_n = \sum \frac{1}{n^{\frac{5}{2}}} \text{ is convergent, since } p = \frac{5}{2} > 1.$$

Hence, $\sum u_n$ is also convergent.

2. *Test for convergence of the series $\sum u_n$, whose n^{th} term is*

$$u_n = \sqrt{n^4+1} - \sqrt{n^4-1}$$

Sol. $\because$ $$u_n = \sqrt{n^4+1} - \sqrt{n^4-1}$$

$$\therefore \quad u_n = \frac{\left(\sqrt{n^4+1} - \sqrt{n^4-1}\right)\left(\sqrt{n^4+1} + \sqrt{n^4-1}\right)}{\left(\sqrt{n^4+1} + \sqrt{n^4-1}\right)}$$

$$= \frac{n^4+1-n^4+1}{\sqrt{n^4+1}+\sqrt{n^4-1}} = \frac{2}{\sqrt{n^4+1}+\sqrt{n^4-1}}$$

$$= \frac{2}{n^2\left[\sqrt{1+\frac{1}{n^4}}+\sqrt{1-\frac{1}{n^4}}\right]}$$

Let us take $$\sum v_n = \sum \frac{1}{n^2}$$

whose n^{th} term $v_n = \dfrac{1}{n^2}$

$$\therefore \quad \frac{u_n}{v_n} = \frac{2}{n^2\left[\sqrt{1+\frac{1}{n^4}}+\sqrt{1-\frac{1}{n^4}}\right]}\,n^2 = \frac{2}{\sqrt{1+\frac{1}{n^4}}+\sqrt{1-\frac{1}{n^4}}}$$

$$\therefore \quad \lim_{n\to\infty}\frac{u_n}{v_n} = \lim_{n\to\infty}\frac{2}{\sqrt{1+\frac{1}{n^4}}+\sqrt{1-\frac{1}{n^4}}}$$

$$= \frac{2}{\sqrt{1}+\sqrt{1}} = \frac{2}{2} = 1$$

which is a definite number and not equal to zero.

$\therefore \quad \sum u_n$ and $\sum v_n$ have the same property.

But $\sum v_n = \sum \frac{1}{n^2}$ which is convergent, since $p = 2 > 1$.

Therefore, $\sum u_n$ is also convergent.

3. *Test for convergence of the series* $\sum\limits_{n=1}^{\infty}\frac{2n^2+1}{3n^3+5n^2+6}$

Sol. Let
$$\sum u_n = \sum_{n=1}^{\infty}\frac{2n^2+1}{3n^3+5n^2+6}$$

Here,
$$u_n = \frac{2n^2+1}{3n^3+5n^2+6} = \frac{n^2\left(2+\frac{1}{n^2}\right)}{n^3\left(3+\frac{5}{n}+\frac{6}{n^3}\right)} = \frac{\left(2+\frac{1}{n^2}\right)}{n\left(3+\frac{5}{n}+\frac{6}{n^3}\right)}$$

Let
$$\sum v_n = \sum \frac{1}{n}$$

$$\therefore \quad v_n = \frac{1}{n}$$

Now,
$$\frac{u_n}{v_n} = \frac{2+\frac{1}{n^2}}{3+\frac{5}{n}+\frac{6}{n^3}}$$

$$\therefore \quad \lim_{n\to\infty}\frac{u_n}{v_n} = \lim_{n\to\infty}\frac{2+\frac{1}{n^2}}{3+\frac{5}{n}+\frac{6}{n^3}} = \frac{2}{3}$$

which is finite and non-zero.

Hence, by comparison test, $\sum u_n$ and $\sum v_n$ converge or diverge together.

But, $\sum v_n = \sum \frac{1}{n}$ diverges, since $p = 1$ (comparing with $\sum \frac{1}{n^p}$)

$\therefore \quad \sum u_n$ is also divergent.

Hence, $\sum \frac{2n^2+1}{3n^3+5n^2+6}$ is divergent.

4. *Test the convergency of the series*

$$\frac{2}{1}+\frac{7}{15}+\frac{12}{53}+\frac{17}{127}+\ldots+\frac{5n-3}{2n^3-1}+\ldots$$

Sol. Let $$\sum u_n = \frac{2}{1}+\frac{7}{15}+\frac{12}{53}+\frac{17}{127}+\ldots$$

Here, $$u_n = \frac{5n-1}{2n^3-1} = \frac{n\left(5-\frac{1}{n}\right)}{n^3\left(2-\frac{1}{n^3}\right)} = \frac{1}{n^2}\left(\frac{5-\frac{1}{n}}{2-\frac{1}{n^3}}\right)$$

Let $$\sum v_n = \sum \frac{1}{n^2}, \text{ then } v_n = \frac{1}{n^2}$$

$$\therefore \quad \lim_{n\to\infty} \frac{u_n}{v_n} = \lim_{n\to\infty}\left(\frac{5-\frac{1}{n}}{2-\frac{1}{n^3}}\right) = \frac{5}{2}$$

which is non-zero finite number.

Therefore, by comparison test $\sum u_n$ and $\sum v_n$ converge or diverge together.

But $\sum v_n = \sum \frac{1}{n^2}$ is convergent, since $p = 2 > 1$. (Comparing with the auxiliary series $\sum \frac{1}{n^p}$)

$\therefore \quad \sum u_n$ is also convergent.

Hence, the given series is convergent.

5. *Test for convergence of the series*

$$\frac{\sqrt{2}}{2+\sqrt{2}}+\frac{\sqrt{3}}{3+\sqrt{3}}+\ldots+\frac{\sqrt{n+1}}{(n+1)+\sqrt{n+1}}+\ldots$$

Sol. Let $$\sum u_n = \frac{\sqrt{2}}{2+\sqrt{2}}+\frac{\sqrt{3}}{3+\sqrt{3}}+\ldots+\frac{\sqrt{n+1}}{(n+1)+\sqrt{n+1}}+\ldots$$

Here, $$u_n = \frac{\sqrt{n+1}}{(n+1)+\sqrt{n+1}} = \frac{\sqrt{n}\left(\sqrt{1+\frac{1}{n}}\right)}{n\left[\left(1+\frac{1}{n}\right)+\sqrt{\frac{1}{n}+\frac{1}{n^2}}\right]}$$

$$= \frac{\sqrt{1+\frac{1}{n}}}{\sqrt{n}\left[\left(1+\frac{1}{n}\right)+\sqrt{\frac{1}{n}+\frac{1}{n^2}}\right]}$$

Let $$\sum v_n = \sum \frac{1}{\sqrt{n}}$$

Then, $$v_n = \frac{1}{n^{\frac{1}{2}}}$$

$\therefore$ $$\lim_{n\to\infty} \frac{u_n}{v_n} = \lim_{n\to\infty} \frac{\sqrt{1+\frac{1}{n}}}{\left(1+\frac{1}{n}\right)+\sqrt{\frac{1}{n}+\frac{1}{n^2}}} = \frac{1}{1} = 1,$$

which is a non-zero finite number.

Therefore, by comparison test $\sum u_n$ and $\sum v_n$ converge or diverge together.

But $\sum v_n = \sum \frac{1}{n^{\frac{1}{2}}}$ is divergent, since $p = \frac{1}{2} < 1$ (Comparing with the auxiliary series $\sum \frac{1}{n^p}$).

$\therefore$ $\sum u_n$ is also divergent.

Thus, the given series diverges.

6. *Test the convergence of the series*

$$\frac{1.2}{3^2.4^2} + \frac{3.4}{5^2.6^2} + \frac{5.6}{7^2.8^2} + \ldots \infty$$

Sol. Let the given series be $\sum u_n$.

Then, $$u_n = \frac{2n(2n-1)}{(2n+1)^2(2n+2)^2}$$

$$= \frac{2n^2\left(2-\frac{1}{n}\right)}{4n^4\left(2+\frac{1}{n}\right)^2\left(1+\frac{1}{n}\right)^2} = \frac{\left(2-\frac{1}{n}\right)}{2n^2\left(2+\frac{1}{n}\right)^2\left(1+\frac{1}{n}\right)^2}$$

Let $$\sum v_n = \sum \frac{1}{n^2}, \text{ then } v_n = \frac{1}{n^2}$$

$$\therefore \quad \lim_{n\to\infty} \frac{u_n}{v_n} = \lim_{n\to\infty} \frac{\left(2-\frac{1}{n}\right)}{2\left(2+\frac{1}{n}\right)^2\left(1+\frac{1}{n}\right)^2}$$

$$= \frac{2}{2.4.1} = \frac{1}{4},$$

which is a non-zero definite number.

Therefore, by comparison test $\sum u$ and $\sum v_n$ either both convergent or both divergent.

But, $$\sum v_n = \sum \frac{1}{n^2} \text{ is convergent, since } p = 2 > 1$$

[Comparing with the auxiliary series $\sum \frac{1}{n^p}$]

$\therefore \quad \sum u_n$ is also convergent, *i.e.*, the given series is convergent.

7. *Test the convergence of the series*

$$1 + \frac{1}{3^2} + \frac{1}{5^2} + \frac{1}{7^2} + \ldots \infty$$

Sol. Let $$\sum u_n = 1 + \frac{1}{3^2} + \frac{1}{5^2} + \frac{1}{7^2} + \ldots$$

Here, $$u_n = \frac{1}{(2n-1)^2} = \frac{1}{n^2\left(2-\frac{1}{n}\right)^2}$$

Let $$\sum v_n = \sum \frac{1}{n^2}, \text{ then } v_n = \frac{1}{n^2}.$$

$$\therefore \quad \lim_{n\to\infty} \frac{u_n}{v_n} = \lim_{n\to\infty} \frac{1}{\left(2-\frac{1}{n}\right)^2} = \frac{1}{4},$$

which is non-zero definite number.

$\therefore$ By comparison test $\sum u_n$ and $\sum v_n$ converge or diverge together.

But the series $\sum v_n = \sum \frac{1}{n^2}$ is convergent. (Here $p = 2 > 1$)

$\therefore \quad \sum u_n$ is also convergent, *i.e.*, the given series is convergent.

8. *Test the convergence of the series* $\sum u_n$ *whose* n^{th} *term is sin* $\left(\frac{1}{n}\right)$.

Sol. Here,
$$u_n = \sin\left(\frac{1}{n}\right) = \frac{1}{n} - \frac{1}{3!n^3} + \frac{1}{5!n^5} \dots \infty$$
$$\left[\because \sin\theta = \theta - \frac{\theta^3}{3!} + \frac{\theta^5}{5!} - \frac{\theta^7}{7!} + \dots\right]$$

$\therefore$
$$u_n = \frac{1}{n}\left[1 - \frac{1}{3!n^2} + \frac{1}{5!n^4} - \dots\right]$$

Let
$$\sum v_n = \sum \frac{1}{n}, \quad \therefore v_n = \frac{1}{n}$$

$\therefore$
$$\lim_{n\to\infty} \frac{u_n}{v_n} = \lim_{n\to\infty}\left[1 - \frac{1}{3!n^2} + \frac{1}{5!n^4} - \dots\right]$$
$$= 1 \text{ which is non-zero finite number.}$$

$\therefore$ By comparison test, $\sum u_n$ and $\sum v_n$ converge or diverge together.

But $\sum v_n = \sum \frac{1}{n}$ is divergent, since $p = 1$ [comparing with the auxiliary series $\sum \frac{1}{n^p}$]

$\therefore$ $\sum u_n$ is also divergent.

9. *Test the convergence of the series*
$$\sum u_n = 1 + \frac{1}{2^2} + \frac{2^2}{3^3} + \frac{3^3}{4^4} + \frac{4^4}{5^5} + \dots$$

Sol. Here,
$$u_n = \frac{(n-1)^{n-1}}{n^n} = \frac{n^{n-1}\left(1 - \frac{1}{n}\right)^{n-1}}{n^n} = \frac{1}{n}\left(1 - \frac{1}{n}\right)^{n-1}$$

Let
$$\sum v_n = \sum \frac{1}{n}, \quad \therefore v_n = \frac{1}{n}$$

$\therefore$
$$\lim_{n\to\infty} \frac{u_n}{v_n} = \lim_{n\to\infty}\left(1 - \frac{1}{n}\right)^{n-1} = \lim_{n\to\infty}\left(1 - \frac{1}{n}\right)^n \cdot \frac{1}{\left(1 - \frac{1}{n}\right)}$$
$$= \lim_{n\to\infty} \frac{\left[\left\{1 + \left(-\frac{1}{n}\right)\right\}^{-n}\right]^{-1}}{1 - \frac{1}{n}} = \frac{e^{-1}}{1} = \frac{1}{e}$$

which is non-zero and a finite number $(\because 2 < e < 3)$

$\therefore$ By comparison test, $\sum u_n$ and $\sum v_n$ both have like property that is, both convergent or both divergent.

$\therefore$ $\sum u_n$ is convergent.

10. *Examine the convergence of the series*

$\frac{2^a}{1^b}+\frac{3^a}{2^b}+\frac{4^a}{3^b}+\ldots$, *where a and b are positive numbers.*

Sol. Let
$$\sum u_n = \frac{2^a}{1^b}+\frac{3^a}{2^b}+\frac{4^a}{3^b}+\ldots$$

Here,
$$u_n = \frac{(n+1)^a}{n^b}$$

$$= \frac{n^a\left(1+\frac{1}{n}\right)^a}{n^b} = \frac{1}{n^{b-a}}\left(1+\frac{1}{n}\right)^a.$$

Let
$$\sum v_n = \sum \frac{1}{n^{b-a}}, \text{ then } v_n = \frac{1}{n^{b-a}}$$

$\therefore$
$$\lim_{n\to\infty}\frac{u_n}{v_n} = \lim_{n\to\infty}\left(1+\frac{1}{n}\right)^a = (1)^a = 1,$$

which is a non-zero, finite number.

Therefore, by comparison test $\sum u_n$ and $\sum v_n$ both converge or both diverge.

But
$$\sum v_n = \frac{1}{1^{b-a}}+\frac{1}{2^{b-a}}+\frac{1}{3^{b-a}}+\ldots+\frac{1}{n^{b-a}}+\ldots$$

will be convergent or divergent according as $b-a>1$ or $b-a\le 1$, that is $b>1+a$ or $b\le 1+a$.

Therefore, $\sum u$ will converge or diverge according as $b>1+a$ or $b\le 1+a$.

11. *Test for convergence the series*

$$\frac{1}{\log 2}+\frac{1}{\log 3}+\frac{1}{\log 4}+\ldots+\frac{1}{\log n}+\ldots$$

Sol. Since $\log n < n$ for all $n>1$, therefore,

$$\frac{1}{\log n} > \frac{1}{n} \text{ for all } n>2.$$

Let us, compare the given series with the divergent series.

$$\frac{1}{2}+\frac{1}{3}+\frac{1}{4}+\ldots+\frac{1}{n}+\ldots \qquad \ldots(1)$$

Since each term of the given series exceeds the corresponding term of the divergent series (1), therefore, by the comparison test, the given series diverges.

12. *Test for convergence the series*

$$\sum_{n=1}^{\infty} n^{\log x}$$

Sol.
$$\sum_{n=1}^{\infty} n^{\log x} = 1^{\log x} + 2^{\log x} + 3^{\log x} + \ldots$$

$$= \frac{1}{1^{-\log x}} + \frac{1}{2^{-\log x}} + \frac{1}{3^{-\log x}} + \ldots$$

$$= \frac{1}{1^p} + \frac{1}{2^p} + \frac{1}{3^p} + \ldots, \text{ putting } p = -\log x$$

The series is convergent if $p > 1$, *i.e.*, if $-\log x > 1$

i.e., if $\log x^{-1} > 1$, *i.e.*, if $x^{-1} > e$

i.e., if $\frac{1}{x} > e$ or $x < \frac{1}{e}$

The series is divergent, if $p \le 1$,

i.e., if $-\log x \le 1$

i.e., if $\frac{1}{x} \le e$

i.e., if $x \ge \frac{1}{e}$, where $2 < e < 3$.

13. *Discuss the convergence of the series.*

$$\frac{1}{1.2.3} + \frac{1}{2.3.4} + \frac{1}{3.4.5} + \ldots$$

Sol. Let $\sum u_n$ be the given series.

$$\therefore \quad u_n = \frac{1}{n(n+1)(n+2)} = \frac{1}{n^3\left(1+\frac{1}{n}\right)\left(1+\frac{2}{n}\right)}$$

Let
$$\sum v_n = \sum \frac{1}{n^3}, \text{ then } v_n = \frac{1}{n^3}.$$

$$\therefore \quad \lim_{n\to\infty} \frac{u_n}{v_n} = \lim_{n\to\infty} \frac{1}{\left(1+\frac{1}{n}\right)\left(1+\frac{2}{n}\right)} = \frac{1}{1.1} = 1,$$

which is a non-zero definite number.

Therefore, by comparison test $\sum u_n$ and $\sum v_n$ have the same property, that is, both divergent or both convergent.

But,
$$\sum v_n = \sum \frac{1}{n^3} \text{ is convergent, since } p = 3 > 1$$

(comparing with $\sum \frac{1}{n^p}$). Therefore, $\sum u_n$ is convergent, that is, the given series is convergent.

14. *Discuss the convergence of the series*

$$\frac{2}{1^p} + \frac{3}{2^p} + \frac{4}{3^p} + \frac{5}{4^p} + \ldots,$$

where p is a real number.

Sol. Let $$\sum u_n = \frac{2}{1^p} + \frac{3}{2^p} + \frac{4}{3^p} + \ldots$$

Here, $$u_n = \frac{n+1}{n^p} = \frac{1+\frac{1}{n}}{n^{p-1}}$$

Let $$\sum v_n = \sum \frac{1}{n^{p-1}}, \therefore v_n = \frac{1}{n^{p-1}}$$

$\therefore$ $$\lim_{n\to\infty} \frac{u_n}{v_n} = \lim_{n\to\infty} \left(1 + \frac{1}{n}\right) = 1,$$

which is a non-zero definite number.

Therefore, by comparison test $\sum u_n$ and $\sum v_n$ either both convergent or both divergent.

But $$v_n = \frac{1}{n^{p-1}}$$

will be convergent or divergent according as $p - 1 > 1$ or $p - 1 \le 1$

that if $$p > 2 \text{ or } p \le 2.$$

Hence, the given series $\sum u_n$ will be convergent if $p > 2$ and divergent if $p \le 2$.

EXERCISE 4.2

1. Test the convergence of the following series :

(*i*) $\sum\left(\sqrt[3]{n^3+1} - n\right)$ (*ii*) $\sum \frac{\sqrt{n+1}-\sqrt{n}}{n^p}$ (*iii*) $\sum \frac{2n^3+5}{4n^5+1}$ (*iv*) $\sum \frac{\sqrt{n^2-1}}{n^3+1}$

(*v*) $\sum\left(\sqrt[3]{n+1} - \sqrt[3]{n}\right)$ (*vi*) $\sum \frac{1}{\sqrt{n}} \sin\frac{1}{n}$ (*vii*) $\sum \cot^{-1} n^2$ (*viii*) $\sum\left(\frac{1}{n} - \log\frac{n+1}{n}\right)$

(*ix*) $\sum \frac{1}{n^p (n+1)^p}$ (*x*) $\sum\left(\frac{n^2+2}{15+2n^3}\right)^{\frac{1}{3}}$ (*xi*) $\sum \frac{n+1}{n^6}$ (*xii*) $\sum \frac{\log n}{\sqrt{n+1}}$

(xiii) $\sum_{n=2}^{\infty} \frac{1}{n \log n}$.

2. Test for convergence and divergence of the following series :

(i) $\frac{1}{\sqrt{1.2}} + \frac{1}{\sqrt{2.3}} + \frac{1}{\sqrt{3.4}} + \ldots \infty$ (ii) $\sqrt{\frac{1}{2^3}} + \sqrt{\frac{2}{3^3}} + \sqrt{\frac{3}{4^3}} + \ldots \infty$

(iii) $1 + \frac{1}{\sqrt{2}} + \frac{1}{\sqrt{3}} + \frac{1}{\sqrt{4}} + \ldots \infty$ (iv) $\frac{1}{2} + \frac{1}{3} + \frac{1}{2^2} + \frac{1}{3^2} + \frac{1}{2^3} + \frac{1}{3^3} + \ldots \infty$

(v) $\frac{1}{1+1^2} + \frac{1}{1+2^2} + \frac{1}{1+3^2} + \ldots \infty$ (vi) $1 + \frac{1}{1.2} + \frac{1}{2.2^2} + \frac{1}{3.2^3} + \ldots \infty$

(vii) $\frac{2}{1^p} + \frac{3}{2^p} + \frac{4}{3^p} + \ldots \infty$ (viii) $1 + \frac{1}{2.1^2+1} + \frac{1}{2.2^2+1} + \frac{1}{2.3^2+1} + \ldots \infty$

(ix) $\frac{1}{1+\sqrt{2}} + \frac{2}{1+2\sqrt{3}} + \frac{3}{1+3\sqrt{4}} + \ldots \infty$ (x) $1 + \frac{1}{4^{\frac{2}{3}}} + \frac{1}{9^{\frac{2}{3}}} + \frac{1}{16^{\frac{2}{3}}} + \ldots \infty$

(xi) $\frac{1}{(x-d)^2} + \frac{1}{(x-2d)^2} + \frac{1}{(x-3d)^2} + \ldots \infty$

(xii) $\frac{1}{1.3} + \frac{2}{3.5} + \frac{3}{5.7} + \frac{4}{7.9} + \ldots \infty$ (xiii) $\frac{1}{1^{s-1}} + \frac{1}{2^{s-1}} + \frac{1}{3^{s-1}} + \ldots \infty$.

3. Show that the following series are divergent.

(i) $\frac{1}{a.1^2+b} + \frac{2}{a.2^2+b} + \frac{3}{a.3^2+b} + \ldots \infty$

(ii) $\frac{1}{1+2^{-1}} + \frac{2}{1+2^{-2}} + \frac{3}{1+2^{-3}} + \ldots \infty$.

4. Test for convergence of the series

$$\frac{1}{1+x} + \frac{x}{1+x^2} + \frac{x^2}{1+x^3} + \ldots \infty, \text{ when } x > 0$$

5. Show that the series $\frac{\sqrt{2}-1}{3^3-1} + \frac{\sqrt{3}-1}{4^3-1} + \frac{\sqrt{4}-1}{5^3-1} + \ldots$, is convergent.

Answers

1. (i) convergent (ii) convergent when $p > \frac{1}{2}$, divergent when $p \le \frac{1}{2}$

(iii) convergent (iv) convergent (v) divergent (vi) convergent

(vii) convergent (viii) convergent

(ix) convergent when $p > 2$, divergent when $p \le 2$

(x) divergent (xi) convergent. (xii) divergent (xiii) divergent.

2. (i) divergent (ii) divergent. (iii) divergent (iv) convergent

(v) convergent (vi) convergent

(vii) convergent for $p > 2$ and divergent for $p \leq 2$

(viii) convergent (ix) divergent (x) convergent (xi) convergent

(xii) divergent (xiii) convergent if $s > 2$ and divergent if $s \leq 2$

4. Convergent when $x < 1$, divergent when $x \geq 1$.

In the last article we studied the 'comparison test' for convergence and divergence of infinite series. If, it is not possible to apply comparison test we use another important test known as *D' Alembert's ratio test*, named after the famous French mathematician **'Jean Le Rond D'Alembert** (1717-1785)'.

The statement of D'Alembert's ratio test, without proof, is as follows:

4.11 D'ALEMBERT'S RATIO TEST

Statement. If $\sum u_n$ be a series of positive terms such that

(*a*) $\lim\limits_{n \to \infty} \dfrac{u_{n+1}}{u_n} = l$, a definite number,

then (*i*) $\sum u_n$ is convergent if $l < 1$

(*ii*) $\sum u_n$ is divergent if $l > 1$

(*iii*) If $l = 1$, the test fails, that is, $\sum u_n$ may converge or diverge (no definite conclusion can be drawn).

(*b*) If $\dfrac{u_{n+1}}{u_n} \to 0$, as $n \to \infty$, then $\sum u_n$ converges.

Remark. If $l = 1$, then comparison test may be applied for definite conclusion for convergence or divergence.

4.11.1 Illustrative Examples

1. *Test for convergence of the Series* $\dfrac{2!}{3} + \dfrac{3!}{3^2} + \dfrac{4!}{3^3} + \ldots + \dfrac{(n+1)!}{3^n} + \ldots$.

Sol. Let
$$\sum u_n = \sum \frac{(n+1)!}{3^n},$$

then
$$u_n = \frac{(n+1)!}{3^n}, \text{ and } u_{n+1} = \frac{(n+2)!}{3^{n+1}}$$

$$\therefore \quad \frac{u_{n+1}}{u_n} = \frac{(n+2)!}{3^{n+1}} \times \frac{3^n}{(n+1)!} = \frac{(n+2)}{3}$$

Now, $$\lim_{n\to\infty}\frac{u_{n+1}}{u_n}=\lim_{n\to\infty}\left(\frac{n+2}{3}\right)\to\infty\ (>1)$$

$\therefore$ By D' Alembert's ratio test the given series $\sum u_n$ is divergent.

2. *Test for the convergence of the series* $\sum \frac{n^3+a}{2^n+a}$

Sol. Here, $$u_n=\frac{n^3+a}{2^n+a},\ \therefore u_{n+1}=\frac{(n+1)^3+a}{2^{n+1}+a}$$

and $$\frac{u_{n+1}}{u_n}=\frac{(n+1)^3+a}{2^{n+1}+a}\times\frac{2^n+a}{n^3+a}=\frac{n^3\left\{\left(1+\frac{1}{n}\right)^3+\frac{a}{n^3}\right\}}{2^n\left(2+\frac{a}{2^n}\right)}\times\frac{2^n\left(1+\frac{a}{2^n}\right)}{n^3\left(1+\frac{a}{n^3}\right)}$$

$$=\frac{\left(1+\frac{1}{n}\right)^3+\frac{a}{n^3}}{2+\frac{a}{2^n}}\times\frac{1+\frac{a}{2^n}}{1+\frac{a}{n^3}}$$

$$\therefore\quad \lim_{n\to\infty}\frac{u_{n+1}}{u_n}=\lim_{n\to\infty}\left\{\frac{\left(1+\frac{1}{n}\right)^3+\frac{a}{n^3}}{2+\frac{a}{2^n}}\right\}\times\lim_{n\to\infty}\left\{\frac{1+\frac{a}{2^n}}{1+\frac{a}{n^3}}\right\}$$

$$=\frac{1}{2}\times\frac{1}{1}=\frac{1}{2}.$$

$$\because\quad \lim_{n\to\infty}\frac{u_{n+1}}{u_n}=\frac{1}{2}<1,$$

hence by D'Alembert's ratio test the series $\sum u_n$ is convergent.

3. *Test for convergence the series*

$$\frac{2}{1^2}x+\frac{3^2}{2^3}x^2+\frac{4^3}{3^4}x^3+\ldots+\frac{(n+1)^n}{n^{n+1}}x^n+\ldots\infty$$

Sol. Let the given series be $\sum u_n=\sum\frac{(n+1)^n}{n^{n+1}}x^n$

Then, $$u_n = \frac{(n+1)^n}{n^{n+1}} x^n \text{ and } u_{n+1} = \frac{(n+2)^{n+1}}{(n+1)^{n+2}} x^{n+1}$$

$$\therefore \quad \frac{u_{n+1}}{u_n} = \frac{(n+2)^{n+1}}{(n+1)^{n+2}} x^{n+1} \cdot \frac{n^{n+1}}{(n+1)^n \cdot x^n}$$

$$= \frac{(n+2)^{n+1}}{(n+1)^{n+2}} \cdot \frac{n^{n+1}}{(n+1)^n} \cdot x = \frac{(n+2)^{n+1}}{(n+1)^{n+1}} \times \frac{n^{n+1}}{(n+1)^{n+1}} \cdot x$$

$$= \left\{\frac{n+1+1}{n+1}\right\}^{n+1} \frac{1}{\left(\frac{n+1}{n}\right)^{n+1}} \cdot x = \left(1+\frac{1}{n+1}\right)^{n+1} \cdot \frac{1}{\left(1+\frac{1}{n}\right)^n} \cdot \frac{1}{\left(1+\frac{1}{n}\right)} \cdot x$$

Now, $$\lim_{n\to\infty} \frac{u_{n+1}}{u_n} = \lim_{n\to\infty}\left(1+\frac{1}{n+1}\right)^{n+1} \cdot \frac{1}{\lim_{n\to\infty}\left(1+\frac{1}{n}\right)^n} \cdot \frac{1}{\lim_{n\to\infty}\left(1+\frac{1}{n}\right)} \cdot x$$

$$= e \cdot \frac{1}{e} \cdot \frac{1}{1} \cdot x \qquad \left[\because \lim_{n\to\infty}\left(1+\frac{1}{n}\right)^n = e\right]$$

$$= x$$

If, $x < 1$, then $\lim_{n\to\infty} \frac{u_{n+1}}{u_n} < 1$, hence by D'Alembert's ratio test the series $\sum u_n$ converges

If, $x > 1$, the series $\sum u_n$ diverges

and if, $x = 1$, the test fails.

For $x = 1$, we apply the ratio test.

When $x = 1$, $\sum u_n = \sum \frac{(n+1)^n}{n^{n+1}}$

In this case, $$u_n = \frac{(n+1)^n}{n^{n+1}} = \frac{n^n\left(1+\frac{1}{n}\right)^n}{n^n \cdot n} = \frac{\left(1+\frac{1}{n}\right)^n}{n}$$

Let $$\sum v_n = \sum \frac{1}{n}$$

$$\therefore \quad \lim_{n\to\infty} \frac{u_n}{v_n} = \lim_{n\to\infty}\left(1+\frac{1}{n}\right)^n = e, \text{ a non-zero finite quantity.}$$

$\therefore$ By ratio test $\sum u_n$ and $\sum v_n$ are of the same property.

But $\sum v_n = \sum \frac{1}{n}$ is divergent (comparing with $\sum \frac{1}{n^p}, p = 1$)

$\therefore$ $\sum u_n$ is also divergent.

Hence, $\sum u_n$ converges when $x < 1$ and diverges when $x \geq 1$.

4. *Show that the infinite series* $\sum_{n=1}^{\infty} \frac{n^2 - 1}{n^2 + 1} x^n$ *is convergent if* $x < 1$ *and divergent if* $x > 1$.

(W.B.U.T. BCA, 2006)

Sol. Here, $\sum u_n = \sum_{n=1}^{\infty} \frac{n^2 - 1}{n^2 + 1} x^n$

$$\therefore \quad u_n = \frac{n^2 - 1}{n^2 + 1} x^n \text{ and } u_{n+1} = \frac{(n+1)^2 - 1}{(n+1)^2 + 1} x^{n+1}$$

$$\therefore \quad \frac{u_{n+1}}{u_n} = \frac{(n+1)^2 - 1}{(n+1)^2 + 1} x^{n+1} \cdot \frac{(n^2 + 1)}{(n^2 - 1) x^n}$$

$$= \frac{n^2 \left[\left(1 + \frac{1}{n} \right)^2 - \frac{1}{n^2} \right]}{n^2 \left[\left(1 + \frac{1}{n} \right)^2 + \frac{1}{n^2} \right]} \cdot \frac{n^2 \left(1 + \frac{1}{n^2} \right)}{n^2 \left(1 - \frac{1}{n^2} \right)} \cdot x = \frac{\left[\left(1 + \frac{1}{n} \right)^2 - \frac{1}{n^2} \right]}{\left(1 + \frac{1}{n} \right)^2 + \frac{1}{n^2}} \cdot \frac{1 + \frac{1}{n^2}}{1 - \frac{1}{n^2}} \cdot x$$

$$\therefore \quad \lim_{n \to \infty} \frac{u_{n+1}}{u_n} = \frac{\lim_{n \to \infty} \left[\left(1 + \frac{1}{n} \right)^2 - \frac{1}{n^2} \right]}{\lim_{n \to \infty} \left[\left(1 + \frac{1}{n} \right)^2 + \frac{1}{n^2} \right]} \times \frac{\lim_{n \to \infty} \left(1 + \frac{1}{n^2} \right)}{\lim_{n \to \infty} \left(1 - \frac{1}{n^2} \right)} \cdot x$$

$$= \frac{1}{1} \cdot \frac{1}{1} \cdot x = x$$

Now, $\lim_{n \to \infty} \frac{u_{n+1}}{u_n} < 1$,

if $x < 1$ and then the series $\sum u_n$ converges by D'Alembert's ratio test.

And $\lim_{n\to\infty} \frac{u_{n+1}}{u_n} > 1$ if $x > 1$

and then the series $\sum u_n$ diverges by D'Alembert's ratio test.

Hence, the given series $\sum u_n = \sum \frac{n^2-1}{n^2+1} x^n$ converges if $x < 1$ and diverges if $x > 1$.

5. *Test for convergence the hypergeometric series*

$$\frac{\alpha}{\beta} + \frac{1+\alpha}{1+\beta} + \frac{(1+\alpha)(2+\alpha)}{(1+\beta)(2+\beta)} + \ldots$$ (W.B.U.T., BCA, 2006)

Sol. Let the given series be $\sum u_n$,

then $$u_n = \frac{(\alpha+1)(\alpha+2)\ldots(\alpha+n-1)}{(\beta+1)(\beta+2)\ldots(\beta+n-1)}$$

and $$u_{n+1} = \frac{(\alpha+1)(\alpha+2)\ldots(\alpha+n-1)(\alpha+n)}{(\beta+1)(\beta+2)\ldots(\beta+n-1)(\beta+n)}$$

(considering from the second term).

$$\therefore \quad \frac{u_{n+1}}{u_n} = \frac{\alpha+n}{\beta+n} = \frac{n\left(1+\frac{\alpha}{n}\right)}{n\left(1+\frac{\beta}{n}\right)} = \frac{1+\frac{\alpha}{n}}{1+\frac{\beta}{n}}.$$

$$\therefore \quad \lim_{n\to\infty} \frac{u_{n+1}}{u_n} = \lim_{n\to\infty} \frac{1+\frac{\alpha}{n}}{1+\frac{\beta}{n}} = \frac{1}{1} = 1.$$

$$\because \quad \lim_{n\to\infty} \frac{u_{n+1}}{u_n} = 1,$$

so no conclusion can be drawn on the behaviour of the series. [To conform the convergency or divergency we have to apply some other ratio test, which will be considered in the next article].

6. *Test for convergence the series*

$$\sum \frac{(1+x)(2+x)(3+x)\ldots(n+x)}{(2+x)(4+x)(6+x)\ldots(2n+x)}.$$

Sol. Let $$\sum u_n = \sum \frac{(1+x)(2+x)(3+x)\ldots(n+x)}{(2+x)(4+x)(6+x)\ldots(2n+x)}.$$

Then, $$u_n = \frac{(1+x)(2+x)(3+x)\ldots(n+x)}{(2+x)(4+x)(6+x)\ldots(2n+x)}$$

and $$u_{n+1} = \frac{(1+x)(2+x)(3+x)\ldots(n+x)(n+1+x)}{(2+x)(4+x)(6+x)\ldots(2n+x)(2n+2+x)}$$

$$\therefore \quad \frac{u_{n+1}}{u_n} = \frac{n+1+x}{2n+2+x} = \frac{n\left(1+\frac{1}{n}+\frac{x}{n}\right)}{n\left(2+\frac{2}{n}+\frac{x}{n}\right)} = \frac{1+\frac{1}{n}+\frac{x}{n}}{2+\frac{2}{n}+\frac{x}{n}}$$

$$\therefore \quad \lim_{n\to\infty} \frac{u_{n+1}}{u_n} = \lim_{n\to\infty} \left(\frac{1+\frac{1}{n}+\frac{x}{n}}{2+\frac{2}{n}+\frac{x}{n}}\right) = \frac{1}{2} < 1.$$

Hence, by D'Alembert's ratio test the given series $\sum u_n$ is convergent.

7. *Show that* $1+\frac{1}{2}+\frac{1}{3}+\frac{1}{4}+\ldots$, *is a divergent series.* (W.B.U.T., BCA, 2006)

Sol. For solution, see proof of case II. of 4.9 (An auxiliary series).

8. *Test the convergence of the series whose* n^{th} *term is* $\frac{x^n}{2n^2+3}$.

Sol. Let $\sum u_n$ be the series whose n^{th} term is

$$u_n = \frac{x^n}{2n^2+3}$$

$$\therefore \quad u_{n+1} = \frac{x^{n+1}}{2(n+1)^2+3}$$

$$\therefore \quad \frac{u_{n+1}}{u_n} = \frac{x^{n+1}}{2(n+1)^2+3} \times \frac{2n^2+3}{x^n} = \frac{2n^2+3}{2(n+1)^2+3}.x$$

$$= \frac{2+\frac{3}{n^2}}{2\left(1+\frac{1}{n}\right)^2+\frac{3}{n^2}}.x$$

$$\therefore \qquad \lim_{n\to\infty}\frac{u_{n+1}}{u_n}=\lim_{n\to\infty}\left\{\frac{2+\dfrac{3}{n^2}}{2\left(1+\dfrac{1}{n}\right)^2+\dfrac{3}{n^2}}\right\}.x=\frac{2}{2}.x=x$$

Hence, by D'Alembert's ratio test, $\sum u_n$ is convergent when $x < 1$, $\sum u_n$ is divergent when $x > 1$,

when $x = 1$, the test fails

In this case, we apply comparison test,

when $x = 1$, $\qquad \sum u_n = \dfrac{1}{2n^2+3}$

Let $\qquad \sum v_n = \dfrac{1}{n^2}.$

Then, $\qquad \dfrac{u_n}{v_n}=\dfrac{1}{2n^2+3}.n^2=\dfrac{1}{2+\dfrac{3}{n^2}}$

$\therefore \qquad \lim\limits_{n\to\infty}\dfrac{u_n}{v_n}=\lim\limits_{n\to\infty}\dfrac{1}{2+\dfrac{3}{n^2}}=\dfrac{1}{2}$, which is a non-zero finite number.

$\therefore$ By comparison test $\sum u_n$ and $\sum v_n$ converge or diverge together.

But, $\qquad \sum v_n = \sum \dfrac{1}{n^2}$ converges, since $p = 2 > 1$.

Thus, the given series is convergent when $x \le 1$ and divergent when $x > 1$.

9. *Test for convergence, the series* , $\sum u_n = \sum \dfrac{x^n}{\sqrt{n^2+1}}$, $x > 0$.

Sol. Here, $\qquad u_n=\dfrac{x^n}{\sqrt{n^2+1}}$, $\therefore u_{n+1}=\dfrac{x^{n+1}}{\sqrt{(n+1)^2+1}}$.

and $\qquad \dfrac{u_{n+1}}{u_n}=\dfrac{x^{n+1}}{\sqrt{(n+1)^2+1}}\times\dfrac{\sqrt{n^2+1}}{x^n}=\dfrac{\sqrt{1+\dfrac{1}{n^2}}}{\sqrt{\left(1+\dfrac{1}{n}\right)^2+\dfrac{1}{n^2}}}.x$

$$\therefore \qquad \lim_{n\to\infty}\frac{u_{n+1}}{u_n}=\lim_{n\to\infty}\left\{\frac{\sqrt{1+\frac{1}{n^2}}}{\sqrt{\left(1+\frac{1}{n}\right)^2+\frac{1}{n^2}}}\right\}.x$$

$$=\frac{1}{1}.x=x.$$

$\therefore$ By D'Alembert's ratio test the series $\sum u_n$ is convergent when $0<x<1$, divergent when $x>1$.

The test fails when $x=1$.

When $x=1$, $\qquad \sum u_n=\sum\frac{1}{\sqrt{n^2+1}}$

Let $\qquad \sum v_n=\frac{1}{n}$

$$\therefore \qquad \frac{u_n}{v_n}=\frac{1}{\sqrt{n^2+1}}\times n=\frac{1}{\sqrt{1+\frac{1}{n^2}}}$$

Hence, $\qquad \lim_{n\to\infty}\frac{u_n}{v_n}=\lim_{n\to\infty}\frac{1}{\sqrt{1+\frac{1}{n^2}}}=\frac{1}{1}=1.$

So, $\sum u_n$ and $\sum v_n$ have the same property.

But $\qquad \sum v_n=\sum\frac{1}{n}$ is divergent, since $p=1$.

Hence, the series $\sum u_n$ is divergent when $x=1$.

$\therefore$ The series $\sum u_n$ is convergent when $x<1$ and divergent when $x\geq 1$.

10. *Test the series* $x+\frac{1}{2}.\frac{x^3}{3}+\frac{1.3.x^5}{2.4.5}+\ldots$, *for convergence,* $(x\neq 1)$.

Sol. Let $\qquad \sum u_n=x+\frac{1}{2}.\frac{x^3}{3}+\frac{1}{2}.\frac{3}{4}.\frac{x^5}{5}+\ldots$

Here, $\qquad u_n=\frac{1.3.5\ldots(2n-3)\,x^{2n-1}}{2.4.6\ldots(2n-2)(2n-1)},$

$$\therefore \qquad u_{n+1}=\frac{1.3.5\ldots(2n-3)(2n-1)}{2.4.6\ldots(2n-2)\,2n}.\frac{x^{2n+1}}{2n+1}.$$

$$\therefore \quad \frac{u_{n+1}}{u_n} = \frac{(2n-1)^2}{2n\,(2n+1)}\,.\,x^2 = \frac{\left(2-\frac{1}{n}\right)^2}{2\left(2+\frac{1}{n}\right)}\,.\,x^2$$

$$\therefore \quad \lim_{n\to\infty}\frac{u_{n+1}}{u_n} = \lim_{n\to\infty}\frac{\left(2-\frac{1}{n}\right)^2}{2\left(2+\frac{1}{n}\right)}\,.\,x^2 = \frac{4}{4}\,.\,x^2 = x^2$$

$\therefore$ By D' Alembert's ratio test, the series $\sum u_n$ is convergent when $x^2 < 1$, *i.e.*, $|x| < 1$ and divergent when $x^2 > 1$ *i.e.*, $|x| > 1$.

11. *Test for convergence the series,*

$$1+\frac{2^p}{2!}+\frac{3^p}{3!}+\frac{4^p}{4!}+\ldots\ (p>0)$$

Sol. Let
$$\sum u_n = 1+\frac{2^p}{2!}+\frac{3^p}{3!}+\ldots$$

Here,
$$u_n = \frac{n^p}{n!},\ u_{n+1} = \frac{(n+1)^p}{(n+1)!}$$

$$\therefore \quad \frac{u_{n+1}}{u_n} = \frac{(n+1)^p}{(n+1)!}\times\frac{n!}{n^p} = \left(1+\frac{1}{n}\right)^p.\frac{1}{(n+1)}$$

$$\therefore \quad \lim_{n\to\infty}\frac{u_{n+1}}{u_n} = \lim_{n\to\infty}\left(1+\frac{1}{n}\right)^p.\frac{1}{\lim_{n\to\infty}(n+1)} = 1\,.\,0 = 0 < 1.$$

$\therefore$ By D'Alembert's ratio test the series $\sum u_n$ is convergent.

12. *Test the convergence of the series*

$$\frac{x}{1.2}+\frac{x^2}{2.3}+\frac{x^3}{3.4}+\frac{x^4}{4.5}+\ldots\ (x>0)$$

Sol. Let
$$\sum u_n = \frac{x}{1.2}+\frac{x^2}{2.3}+\frac{x^3}{3.4}+\ldots$$

Here,
$$u_n = \frac{x^n}{n(n+1)} \text{ and } u_{n+1} = \frac{x^{n+1}}{(n+1)(n+2)}$$

$$\therefore \quad \frac{u_{n+1}}{u_n} = \frac{x^{n+1}}{(n+1)(n+2)} \times \frac{n(n+1)}{x^n} = \frac{n}{n+2} \cdot x$$

$$= \frac{1}{1+\frac{2}{n}} \cdot x$$

$$\therefore \quad \lim_{x \to \infty} \frac{u_{n+1}}{u_n} = \lim_{n \to \infty} \frac{1}{1+\frac{2}{n}} \cdot x = 1 \cdot x = x$$

Hence, by D'Alembert's ratio test $\sum u_n$ is convergent when $x < 1$, and divergent when $x > 1$

The test fails when $x = 1$.

When $x = 1$, $\quad \sum u_n = \frac{1}{n(n+1)}$

Let $\quad \sum v_n = \sum \frac{1}{n^2}$.

Then, $\quad \lim_{n \to \infty} \frac{u_n}{v_n} = \lim_{n \to \infty} \frac{1}{n(n+1)} \cdot n^2 = \lim_{n \to \infty} \frac{n^2}{n^2\left(1+\frac{1}{n}\right)} = \lim_{n \to \infty} \frac{1}{1+\frac{1}{n}} = 1,$

which is a non-zero finite number.

$\therefore \quad \sum u_n$ and $\sum v_n$ have the same property (by comparison test). But $\sum v_n = \sum \frac{1}{n^2}$ is convergent, (Here $p = 2 > 1$)

$\therefore \quad \sum u_n$ is also convergent.

Thus, $\sum u_n$ is convergent when $x \leq 1$

and divergent when $n > 1$.

EXERCISE 4.3

1. Discuss the convergence of each of the following series :

(i) $\sum \frac{n!}{n^n}$ (ii) $\sum \frac{n!\,2^n}{n^n}$ (iii) $\sum \frac{5^n}{n^2+5}$ (iv) $\sum \frac{2^{n-1}}{3^{n+1}}$

(v) $\sum \frac{n^2}{3^n}$ (vi) $\sum \frac{x^n}{3^n \cdot n^2}, x > 0$ (vii) $\sum \frac{x^n}{(2n)!}$

(viii) $\sum \frac{n^2 (n+1)^2}{n!}$ (ix) $\sum \frac{1}{1+nx^n}$.

2. Test the convergence of each of the following series :

(i) $1+\frac{2^{-p}}{2!}+\frac{3^{-p}}{3!}+\ldots+\frac{n^{-p}}{n!}+\ldots$ (ii) $\frac{1}{4}+\frac{1.3}{4.7}+\frac{1.3.5}{4.7.10}+\frac{1.3.5.7}{4.7.10.13}+\ldots$

(iii) $\frac{1}{3}+\frac{1.2}{3.5}+\frac{1.2.3}{3.5.7}+\frac{1.2.3.4}{3.5.7.9}+\ldots$ (iv) $\frac{1}{5}+\frac{2!}{5^2}+\frac{3!}{5^3}+\frac{4!}{5^4}+\ldots$

(v) $1+\frac{2!}{2^2}+\frac{3!}{3^3}+\frac{4!}{4^4}+\ldots$ (vi) $\frac{1}{1+2}+\frac{2}{1+2^2}+\frac{3}{1+2^3}+\ldots$

(vii) $\left(\frac{1}{3}\right)^2+\left(\frac{1.2}{3.5}\right)^2+\left(\frac{1.2.3}{3.5.7}\right)^2+\ldots$ (viii) $\frac{2}{1.4.5}+\frac{3}{2.5.6}+\frac{4}{3.6.7}+\ldots.$

3. Test for convergence of each of the following series :

(i) $x+\frac{x^3}{3!}+\frac{x^5}{5!}+\frac{x^7}{7!}+\ldots\infty\ (x>0)$ (ii) $1+\frac{x^2}{2}+\frac{x^4}{4}+\frac{x^6}{6}+\ldots, (x>0)$

(iii) $1+\frac{x}{2}+\frac{x^2}{5}+\frac{x^3}{10}+\ldots+\frac{x^n}{n^2+1}+\ldots$ (iv) $\frac{x}{2\sqrt{3}}+\frac{x^2}{3\sqrt{4}}+\frac{x^3}{4\sqrt{5}}+\ldots$

(v) $1+\frac{2^2}{3^2}.x+\frac{2^2.4^2}{3^2.5^2}.x^2+\frac{2^2.4^2.6^2}{3^2.5^2.7^2}.x^3+\ldots\ (x\neq 1)$

(vi) $\frac{1}{2}+\frac{x}{3\sqrt{2}}+\frac{x^2}{4\sqrt{3}}+\frac{x^3}{5\sqrt{4}}+\ldots(x>0)$.

4. Show that the series $1+\frac{\alpha+1}{\beta+1}+\frac{(\alpha+1)(2\alpha+1)}{(\beta+1)(2\beta+1)}+\frac{(\alpha+1)(2\alpha+1)(3\alpha+1)}{(\beta+1)(2\beta+1)(3\beta+1)}+\ldots$ converges, if $\beta > \alpha > 0$ and diverges if $\alpha \geq \beta > 0$.

5. Impose the conditions on 'a' for the convergence and divergence of the series

$$\frac{1}{x+1}+\frac{a}{x+a}+\frac{a^2}{x+a^2}+\frac{a^3}{x+a^3}+\ldots.$$

Answers

1. (i) convergent (ii) convergent (iii) divergent (iv) convergent
(v) convergent (vi) convergent for $x \leq 3$, divergent for $x > 3$
(vii) convergent (viii) convergent
(ix) convergent, when $x > 1$, divergent, when $x \leq 1$

2. (i) convergent (ii) convergent (iii) convergent (iv) divergent
(v) convergent (vi) convergent (vii) convergent (viii) convergent

3. (i) convergent (ii) convergent for $|x| < 1$, divergent for $|x| \geq 1$

(*iii*) convergent when $x \le 1$, divergent when $x > 1$
(*iv*) convergent when $x \le 1$, divergent when $x > 1$
(*v*) convergent when $x < 1$, divergent when $x > 1$
(*vi*) convergent when $x \le 1$, divergent when $x > 1$.

5. convergent if $a < 1$, divergent if $a \ge 1$.

We have seen that in D'Alembert's ratio test if

$$\lim_{n\to\infty} \frac{u_{n+1}}{u_n} = 1,$$

we could not draw any definite conclusion for convergency or divergency. In that case, we had to take help of 'Comparison test' for definite conclusion. Some times, comparison test may not also be applied. Then in case failure of D'Alembert's ratio test some other ratio tests, viz., *Raabe's test, Logarithmic test etc.*, may be applied for definite conclusion.

First, we state (without proof) Raabe's test, which is due to the name of famous Zurich mathematician **Joseph L. Raabe** (1801 – 1859).

4.12 RAABE'S TEST (THEOREM)

Statement. Let $\sum u_n$ be a series of positive terms such that

$$\lim_{n\to\infty} \left\{ n\left(\frac{u_n}{u_{n+1}} - 1 \right) \right\} = l,$$

then (*i*) $\sum u_n$ converges, if $l > 1$

(*ii*) $\sum u_n$ diverges, if $l < 1$

(*iii*) the test fails, if $l = 1$, *i.e.*, $\sum u_n$ may either converge or diverge.

4.12.1 Illustrative Examples

1. *Test for convergence the series*

$$\frac{1}{2}x + \frac{1.3}{2.4}x^2 + \frac{1.3.5}{2.4.6}x^3 + \ldots + \frac{1.3\ldots.(2n-1)}{2.4\ldots.(2n)}x^n + \ldots, (x > 0).$$

Sol. Since $x > 0$, therefore the series is of positive terms. Let us denote the series by $\sum u_n$.

Then, $$u_n = \frac{1.3.5\ldots.(2n-1)}{2.4.6\ldots.(2n)}x^n,$$

and $$u_{n+1} = \frac{1.3.5\ldots.(2n+1)}{2.4.6\ldots.(2n+2)}x^{n+1},$$

$$\therefore \quad \frac{u_n}{u_{n+1}} = \frac{2n+2}{2n+1} \cdot \frac{1}{x} = \frac{2+\frac{2}{n}}{2+\frac{1}{n}} \cdot \frac{1}{x}$$

$$\therefore \quad \lim_{n\to\infty} \frac{u_n}{u_{n+1}} = \lim_{n\to\infty} \frac{2+\frac{2}{n}}{2+\frac{1}{n}} \cdot \frac{1}{x} = \frac{2}{2} \cdot \frac{1}{x} = \frac{1}{x}$$

$\therefore$ By D'Alembert's ratio test $\sum u_n$ converges if $\frac{1}{x} > 1$, *i.e.*, $x < 1$ and $\sum u_n$ diverges if $\frac{1}{x} < 1$, *i.e.*, $x > 1$.

When $x = 1$, the ratio test fails.

Now, when $x = 1$, $\quad \sum u_n = \frac{1.3.5....(2n-1)}{2.4.6....2n}$

In this case, comparison test will not be applicable.
So, we need to apply Raabe's test,

For $x = 1$, $\quad \frac{u_n}{u_{n+1}} = \frac{2n+2}{2n+1}$

$$\therefore \quad \lim_{n\to\infty} \left\{ n\left(\frac{u_n}{u_{n+1}} - 1 \right) \right\} = \lim_{n\to\infty} \left\{ n\left(\frac{2n+2}{2n+1} - 1 \right) \right\}$$

$$= \lim_{n\to\infty} \left\{ n\left(\frac{2n+2-2n-1}{2n+1} \right) \right\}$$

$$= \lim_{n\to\infty} \left\{ \frac{n}{2n+1} \right\} = \lim_{n\to\infty} \left\{ \frac{1}{2+\frac{1}{n}} \right\} = \frac{1}{2} < 1.$$

Hence, by Raabe's test the series $\sum u_n$ is diverges,

i.e., the given series is divergent when $x \geq 1$ and convergent when $x < 1$.

2. *Test for convergence the series*

$$1 + \frac{1}{2} \cdot \frac{x^2}{4} + \frac{1}{2} \cdot \frac{3}{4} \cdot \frac{5}{6} \cdot \frac{x^4}{8} + \frac{1}{2} \cdot \frac{3}{4} \cdot \frac{5}{6} \cdot \frac{7}{8} \cdot \frac{9}{10} \cdot \frac{x^6}{12} + \ldots$$

Sol. Let us denote the given series by $\sum u_n$.

Then,
$$u_n = \frac{1.3.5....(4n-7)}{2.4.6....(4n-6)} \cdot \frac{x^{2n-2}}{4n-4}$$

and
$$u_{n+1} = \frac{1.3.5....(4n-7)(4n-5)(4n-3)}{2.4.6....(4n-6)(4n-4)(4n-2)} \cdot \frac{x^{2n}}{4n}$$

$$\therefore \quad \frac{u_n}{u_{n+1}} = \frac{(4n-4)(4n-2)}{(4n-5)(4n-3)} \cdot \frac{4n}{4n-4} \cdot \frac{1}{x^2} = \frac{16n^2-8n}{16n^2-32n+15} \cdot \frac{1}{x^2}$$

Now,
$$\lim_{n\to\infty} \frac{u_n}{u_{n+1}} = \lim_{n\to\infty} \frac{16n^2-8n}{16n^2-32n+15} \cdot \frac{1}{x^2}$$

$$= \lim_{n\to\infty} \frac{8\left(2-\frac{1}{n}\right)}{16\left(1-\frac{2}{n}+\frac{15}{16n^2}\right)} \cdot \frac{1}{x^2} = \frac{16}{16} \cdot \frac{1}{x^2} = \frac{1}{x^2}.$$

Therefore, by D'Alembert's ratio test, the series diverges if $\frac{1}{x^2} < 1$, *i.e.*, if $x^2 > 1$, *i.e.*, if $|x| > 1$.

and converges if $\frac{1}{x^2} > 1$ *i.e.*, if $x^2 < 1$, *i.e.*, if $|x| < 1$.

If $|x| = 1$, *i.e.*, $x^2 = 1$, then the test fails.

In this case, we apply Raabe's test.

We have
$$\lim_{n\to\infty} \left\{ n\left(\frac{u_n}{u_{n+1}} - 1\right) \right\} = \lim_{n\to\infty} \left\{ n\left(\frac{16n^2-8n}{16n^2-32n+15} - 1\right) \right\}$$

$$= \lim_{n\to\infty} \left\{ n\left(\frac{24n-15}{16n^2-32n+15}\right) \right\}$$

$$= \lim_{n\to\infty} \left\{ \frac{3\left(8-\frac{5}{n}\right)}{16\left(1-\frac{2}{n}+\frac{15}{16n^2}\right)} \right\} = \frac{3}{2} > 1$$

Hence, by Raabe's test the series, $\sum u_n$ is convergent, that is, the given series is convergent when $|x| = 1$.

Thus, the given series diverges if $|x| > 1$ and converges if $|x| \le 1$.

3. *Test the convergence of the series whose* n^{th} *term* u_n *is given by*

$$u_n = \frac{2^2 . 4^2 . 6^2 .... (2n-2)^2}{3.4.5....(2n-2)(2n-1)} x^{2n}.$$

Sol. Let $\sum u_n$ be the series.

Since,
$$u_n = \frac{2^2 . 4^2 . 6^2 .... (2n-2)^2}{3.4.5....(2n-2)(2n-1)} x^{2n}$$

$\therefore$
$$u_{n+1} = \frac{2^2 . 4^2 . 6^2 .... (2n-2)^2 . (2n)^2}{3.4.5....(2n-2)(2n-1)(2n+1)} . x^{2n+2}$$

$\therefore$
$$\frac{u_n}{u_{n+1}} = \frac{2n+1}{2n} . \frac{1}{x^2}$$

$\therefore$
$$\lim_{n\to\infty} \frac{u_n}{u_{n+1}} = \lim_{n\to\infty} \frac{2n+1}{2n} . \frac{1}{x^2} = \lim_{n\to\infty} \frac{2+\frac{1}{n}}{2} . \frac{1}{x^2}$$

$$= \frac{2}{2} . \frac{1}{x^2} = \frac{1}{x^2}$$

Now, by D'Alembert's ratio test the series $\sum u_n$ converges if $\frac{1}{x^2} > 1$, *i.e.*, if $x^2 < 1$, *i.e.*, if $|x| < 1$ and diverges if $\frac{1}{x^2} < 1$, *i.e.*, if $x^2 > 1$ *i.e.*, if $|x| > 1$.

If $x^2 = 1$, *i.e.*, $|x| = 1$, the test fails.

Now, applying Raabe's test, we have

$$\lim_{n\to\infty} \left\{ n\left(\frac{u_n}{u_{n+1}} - 1 \right) \right\} = \lim_{n\to\infty} \left\{ n . \left(\frac{2n+1}{2n} - 1 \right) \right\}$$

$$= \lim_{n\to\infty} \left\{ n\left(\frac{2n+1-2n}{2n} \right) \right\}$$

$$= \lim_{n\to\infty} \left(\frac{1}{2} \right) = \frac{1}{2} < 1$$

Hence, by Raabe's test, the given series is divergent when $x^2 = 1$ or $|x| = 1$.

Therefore, the given series converges, when $x^2 < 1$ *i.e.*, $|x| < 1$, and diverges, when $x^2 \geq 1$ *i.e.*, $|x| \geq 1$.

4. *Test the convergence, for positive real values of x, of the series*

$$x^2 (\log_e 2)^p + x^3 (\log_e 3)^p + x^4 (\log_e 4)^p + \ldots$$

Sol. Let the series be $\sum u_n$,

Then, $\quad u_n = x^{n+1} \{\log_e (n+1)\}^p$

and $\quad u_{n+1} = x^{n+2} \{\log_e (n+2)\}^p$.

$$\therefore \quad \frac{u_n}{u_{n+1}} = \left\{\frac{\log_e (n+1)}{\log_e (n+2)}\right\}^p \cdot \frac{1}{x} = \left\{\frac{\log_e (n+2-1)}{\log_e (n+2)}\right\}^p \cdot \frac{1}{x}$$

$$= \left[\frac{\log_e \left\{(n+2)\left(1-\frac{1}{n+2}\right)\right\}}{\log (n+2)}\right]^p \cdot \frac{1}{x}$$

$$= \left[\frac{\log_e (n+2) + \log_e \left(1-\frac{1}{n+2}\right)}{\log_e (n+2)}\right]^p \cdot \frac{1}{x}$$

$$= \left[1 - \frac{1}{(n+2)\log_e (n+2)} - \frac{1}{2(n+2)^2 \log_e (n+2)} - \ldots\right]^p \cdot \frac{1}{x}$$

$$\left[\because \log (1-x) = -x - \frac{x^2}{2} - \frac{x^3}{3} - \ldots\right]$$

$$= \left[1 - \left\{\frac{1}{(n+2)\log_e (n+2)} + \frac{1}{2(n+2)^2 \log_e (n+2)} + \ldots\right\}\right]^p \cdot \frac{1}{x}$$

$$= \left[1 - p\left\{\frac{1}{(n+2)\log_e (n+2)} + \frac{1}{2(n+2)^2 \log_e (n+2)} + \ldots\right\}\right] \cdot \frac{1}{x}$$

$$\left[\because (1-x)^n = 1 - {}^nC_1 x + {}^nC_2 x^2 - \ldots + {}^nC_n x^n\right]$$

$$\therefore \lim_{n\to\infty} \frac{u_n}{u_{n+1}} = \lim_{n\to\infty} \left[1 - p\left\{\frac{1}{(n+2)\log_e (n+2)} + \frac{1}{2(n+2)^2 \log_e (n+2)} + \ldots\right\}\right] \cdot \frac{1}{x}$$

$$= \frac{1}{x}$$

Therefore, by D'Alembert's ratio test, the given series is convergent or divergent according as $x < 1$ or $x > 1$.

When, $x = 1$, $\lim_{n \to \infty} \frac{u_n}{u_{n+1}} = 1.$

$\therefore$ When $x = 1$, the ratio test fails.

Now,
$$\frac{u_n}{u_{n+1}} - 1 = -\frac{p}{(n+2)\log_e(n+2)} - \frac{p}{2(n+2)^2 \log_e(n+2)} - \ldots$$

$$\therefore \quad n\left(\frac{u_n}{u_{n+1}} - 1\right) = -\frac{p}{\left(1+\frac{2}{n}\right)\log_e(n+2)} - \frac{\frac{p}{n}}{2\left(1+\frac{2}{n}\right)^2 \log_e(n+2)} - \ldots$$

$$\therefore \quad \lim_{n \to \infty}\left[n\left(\frac{u_n}{u_{n+1}} - 1\right)\right] = 0 < 1$$

Hence, by Raabe's test, the given series is divergent when $x = 1$.

Therefore, the given series is convergent when $x < 1$ and divergent when $x \geq 1$.

5. *Test for convergence the hypergeometric series*

$$\frac{\alpha}{\beta} + \frac{1+\alpha}{1+\beta} + \frac{(1+\alpha)(2+\alpha)}{(1+\beta)(2+\beta)} + \ldots \qquad \text{(W.B.U.T., BCA, 2006)}$$

Sol. Here,
$$u_n = \frac{(1+\alpha)(2+\alpha)\ldots(n-1+\alpha)}{(1+\beta)(2+\beta)\ldots(n-1+\beta)} \qquad \text{[considering from second term]}$$

$$\therefore \quad u_{n+1} = \frac{(1+\alpha)(2+\alpha)\ldots(n-1+\alpha)(n+\alpha)}{(1+\beta)(2+\beta)\ldots(n-1+\beta)(n+\beta)}$$

$$\therefore \quad \frac{u_n}{u_{n+1}} = \frac{n+\beta}{n+\alpha}$$

$$\therefore \quad \lim_{n \to \infty} \frac{u_n}{u_{n+1}} = \lim_{n \to \infty}\left(\frac{1+\frac{\beta}{n}}{1+\frac{\alpha}{n}}\right) = 1.$$

$\therefore$ D' Alembert's ratio test fails.

Now, we use Raabe's test.

$$\lim_{n \to \infty}\left[n\left(\frac{u_n}{u_{n+1}} - 1\right)\right] = \lim_{n \to \infty}\left[n\left(\frac{n+\beta}{n+\alpha} - 1\right)\right]$$

$$= \lim_{n\to\infty}\left[n\left(\frac{n+\beta-n-\alpha}{n+\alpha}\right)\right]$$

$$= \lim_{n\to\infty}\left[\frac{n(\beta-\alpha)}{n+\alpha}\right] = \lim_{n\to\infty}\left[(\beta-\alpha)\times\frac{1}{1+\frac{\alpha}{n}}\right]$$

$$= \beta-\alpha.$$

$\therefore$ By Raabe's test, the series is convergent. when $\beta-\alpha>1$ and divergent when $\beta-\alpha<1$.

Remark. Though we have omitted the first term, $\left(\frac{\alpha}{\beta}\right)$, the property of the series remains the same, if we add the first term $\frac{\alpha}{\beta}$, [by theorem 4.1]

EXERCISE 4.4

1. Test for convergence the series

$$\frac{x}{1}+\frac{1}{2}\cdot\frac{x^2}{3}+\frac{1.3}{2.4}\cdot\frac{x^3}{5}+\frac{1.3.5}{2.4.6}\cdot\frac{x^4}{7}+\ldots, x>0.$$

2. Examine, the convergence of the series

$$\sum u_n = \sum_{n=1}^{\infty}\frac{(n+1)(n+2)}{(n+3)(n+4)}.$$

3. Impose the conditions on 'a' for which the series

$$1+a+\frac{a(a+1)}{1.2}+\frac{a(a+1)(a+2)}{1.2.3}+\ldots\infty,$$

converges or diverges.

4. Test the convergence of the series

$$1+\frac{3}{7}x+\frac{3.6}{7.10}x^2+\frac{3.6.9}{7.10.13}x^3+\ldots,\ x>0.$$

5. Test the convergence of the series

$$\sum\frac{4.7.10\ldots(3n+1)}{1.2.3\ldots n}x^n.$$

6. Test the convergence of the series

$$\sum\frac{(n!)^2}{(2n)!}x^{2n},\ (x>0).$$

7. Test for convergence of the series

$$1+\frac{(1!)^2}{2!}x^2+\frac{(2!)^2}{4!}x^4+\frac{(3!)^2}{6!}x^6+\ldots\ (x>0).$$

Answers

1. convergent when $x \le 1$, divergent when $x > 1$ 2. divergent
3. convergent if $a \le 0$, divergent if $a > 0$
4. convergent for $x \le 1$ and divergent for $x > 1$
5. convergent for $x < \frac{1}{3}$, divergent for $x \ge \frac{1}{3}$
6. convergent for $x^2 < 4$, divergent for $x^2 \ge 4$
7. convergent for $x^2 < 4$, divergent for $x^2 \ge 4$.

Remark. Raabe's test is applied when D'Alembert's ratio test fails, particularly, when the exponential number e does not occur in the ratio test $\lim_{n\to\infty}\frac{u_n}{u_{n+1}}$. If $\lim_{n\to\infty}\frac{u_n}{u_{n+1}}$ involves e, we apply *Logarithmic test* after D'Alembert's ratio test and not the Raabe's test. Some times D'Alembert's ratio test and Raabe's test both fail, in this case another test, called *D'Morgan and Bertrand's test* to be applied. We shall not discuss the test as it is beyond the syllabus.

We now state (without proof) the logarithmic test as follows :

4.13. LOGARITHMIC TEST

Statement. Let $\sum u_n$ be a series of positive terms such that $\lim_{n\to\infty}\left(n\log\frac{u_n}{u_{n+1}}\right)=1$, then the series $\sum u_n$ converges whenever $l > 1$ and diverges whenever $l < 1$. The test fails when $l = 1$.

4.13.1 Illustrative Examples

1. *Examine the convergence of the series*

$$x+\frac{2^2x^2}{2!}+\frac{3^3x^3}{3!}+\frac{4^4x^4}{4!}+\ldots,\ x>0$$

Sol. Let the series be denoted by $\sum u_n$.

Then,
$$u_n=\frac{n^n x^n}{n!} \text{ and } u_{n+1}=\frac{(n+1)^{n+1}x^{n+1}}{(n+1)!}$$

$$=\frac{(n+1)^n.(n+1)x^n.x}{(n+1).n!}=\frac{(n+1)^n x^n.x}{n!}$$

$$\therefore \qquad \frac{u_{n+1}}{u_n} = \frac{(n+1)^n}{n^n}.x = \left(1+\frac{1}{n}\right)^n.x$$

$$\therefore \qquad \lim_{n\to\infty}\frac{u_{n+1}}{u_n} = \lim_{n\to\infty}\left(1+\frac{1}{n}\right)^n.x = ex$$

Hence, by D'Alembert's ratio test, $\sum u_n$ is convergent when $ex < 1$ *i.e.*, when $x < \frac{1}{e}$ and divergent when $ex > 1$, *i.e.*, when $x > \frac{1}{e}$. But when $x = \frac{1}{e}$, the test fails.

We now apply logarithmic test.

When $x = \frac{1}{e}$, $\qquad \frac{u_n}{u_{n+1}} = \frac{e}{\left(1+\frac{1}{n}\right)^n}$

$$\therefore \qquad \log\left(\frac{u_n}{u_{n+1}}\right) = \log e - n\log\left(1+\frac{1}{n}\right)$$

$$= \left[1 - n.\frac{1}{n} + n.\frac{1}{2n^2} - n.\frac{1}{3n^3} + \ldots\right]$$

$$\therefore \qquad n\log\left(\frac{u_n}{u_{n+1}}\right) = n\left[1-1+\frac{1}{2n}-\frac{1}{3n^2}+\ldots\right]$$

$$= \frac{1}{2} - \frac{1}{3n} + \ldots$$

$$\therefore \qquad \lim_{n\to\infty}\left(n\log\frac{u_n}{u_{n+1}}\right) = \frac{1}{2} < 1.$$

Hence, by logarithmic ratio test, the given series is divergent, when $x = \frac{1}{e}$.

2. *Test the convergence of the series*

$$\frac{a+x}{1!} + \frac{(a+2x)^2}{2!} + \frac{(a+3x)^3}{3!} + \ldots, x > 0$$

Sol. Here the nth term of the given series is

$$u_n = \frac{(a+nx)^n}{n!}, \therefore u_{n+1} = \frac{\{a+(n+1)x\}^{n+1}}{(n+1)!}$$

$$\therefore \quad \frac{u_{n+1}}{u_n} = \frac{x^{n+1}(n+1)^{n+1}\left\{1+\frac{a}{(n+1)x}\right\}^{n+1}}{n!(n+1)} \cdot \frac{n!}{n^n x^n \left(1+\frac{a}{nx}\right)^n}$$

$$= \left(1+\frac{1}{n}\right)^n \left[\left\{1+\frac{1}{\frac{(n+1)x}{a}}\right\}^{\frac{(n+1)x}{a}}\right]^{\frac{a}{x}} \cdot \frac{1}{\left[\left\{1+\frac{1}{\frac{nx}{a}}\right\}^{\frac{nx}{a}}\right]^{\frac{a}{x}}} \cdot x$$

$$\therefore \quad \lim_{n\to\infty} \frac{u_{n+1}}{u_n} = e \,.\, e^{\frac{a}{x}} \cdot \frac{1}{e^{\frac{a}{x}}} x = ex \qquad \left[\text{using } \lim_{x\to\infty}\left(1+\frac{1}{x}\right)^x = e\right].$$

Hence, by D'Alembert's ratio test, the given series is convergent or divergent according as $ex < 1$ or $ex > 1$, that is, $x < \frac{1}{e}$ or $x > \frac{1}{e}$.

When $x = \frac{1}{e}$, the test fails.

We now apply logarithmic test.

$$\frac{u_n}{u_{n+1}} = \left(1+\frac{ae}{n}\right)^n \cdot \frac{1}{\left(1+\frac{1}{n}\right)^n} \cdot \frac{1}{\left(1+\frac{ae}{n+1}\right)^{n+1}} \cdot e$$

$$\therefore \quad \log \frac{u}{u_{n+1}} = 1 + n\log\left(1+\frac{ae}{n}\right) - n\log\left(1+\frac{1}{n}\right) - (n+1)\log\left(1+\frac{ae}{n+1}\right)$$

$(\because \log e = 1)$

$$= 1 + n\left[\frac{ae}{n} - \frac{a^2e^2}{2n^2} + \frac{a^3e^3}{3n^3} - \ldots\right] - n\left[\frac{1}{n} - \frac{1}{2n^2} + \frac{1}{3n^3} - \ldots\right]$$

$$- (n+1)\left[\frac{ae}{n+1} - \frac{a^2e^2}{2(n+1)^2} + \frac{1}{3}\frac{a^3e^3}{3(n+1)^3} - \ldots\right]$$

$$= \left(-\frac{a^2e^2}{2n} + \frac{a^3e^3}{3n^2} - \ldots\right) + \left(\frac{1}{2n} - \frac{1}{3n^2} + \ldots\right) + \left(\frac{a^2e^2}{2(n+1)} - \frac{a^3e^3}{2(n+1)^2} + \ldots\right)$$

$$\therefore \quad \lim_{n\to\infty}\left(n\log\frac{u_n}{u_{n+1}}\right) = \lim_{n\to\infty}\left[-\frac{a^2e^2}{2}+\frac{a^3e^3}{3n}-\ldots\right]$$

$$+\lim_{n\to\infty}\left[\frac{1}{2}-\frac{1}{3n}+\ldots\right]+\lim_{n\to\infty}\left[\frac{a^2e^2}{2n\left(1+\frac{1}{n}\right)}-\frac{a^3e^3}{3n\left(1+\frac{1}{n}\right)^2}-\ldots\right]$$

$$= -\frac{a^2e^2}{2}+\frac{1}{2}+\frac{a^2e^2}{2} = \frac{1}{2} < 1.$$

Hence, by logarithmic ratio test, the given series is divergent when $x = \frac{1}{e}$.

$\therefore$ The given series is convergent when $x < \frac{1}{e}$ and divergent when $x \geq \frac{1}{e}$.

3. *Discuss the convergence of the series*

$$\frac{x}{1}+\frac{1}{2}\cdot\frac{x^3}{3}+\frac{1}{2}\cdot\frac{3}{4}\cdot\frac{x^5}{5}+\frac{1}{2}\cdot\frac{3}{4}\cdot\frac{5}{6}\cdot\frac{x^7}{7}+\ldots,(x>0)$$

Sol. Let $\sum u_n$ be the given series.

Here, $$u_n = \frac{1.3.5\ldots(2n-1)}{2.4.6\ldots 2n}\cdot\frac{x^{2n+1}}{2n+1} \text{ (neglecting the first term)}$$

$$\therefore \quad u_{n+1} = \frac{1.3.5\ldots(2n-1)(2n+1)}{2.4.6\ldots 2n\,(2n+2)}\cdot\frac{x^{2n+3}}{2n+3}.$$

$$\therefore \quad \frac{u_n}{u_{n+1}} = \frac{2n+2}{2n+1}\cdot\frac{2n+3}{2n+1}\cdot\frac{1}{x^2} = \frac{\left(1+\frac{1}{n}\right)\cdot\left(1+\frac{3}{2n}\right)}{\left(1+\frac{1}{2n}\right)\cdot\left(1+\frac{1}{2n}\right)}\cdot\frac{1}{x^2}$$

$$\therefore \quad \lim_{n\to\infty}\frac{u_n}{u_{n+1}} = \lim_{n\to\infty}\frac{\left(1+\frac{1}{n}\right)\left(1+\frac{3}{2n}\right)}{\left(1+\frac{1}{2n}\right)\left(1+\frac{1}{2n}\right)}\cdot\frac{1}{x^2} = \frac{1}{x^2}$$

Therefore, by D'Alembert's ratio test, the given series $\sum u_n$ is convergent if $\frac{1}{x^2} > 1$ *i.e.*, $x^2 < 1$ and divergent, if $\frac{1}{x^2} < 1$ *i.e.*, if $x^2 > 1$.

If $x^2 = 1$, the test fails.

Now, we apply logarithmic test

$$\log \frac{u_n}{u_{n+1}} = \log \left\{ \frac{\left(1+\frac{1}{n}\right)\left(1+\frac{3}{2n}\right)}{\left(1+\frac{1}{2n}\right)^2} \right\}$$

$$= \log\left(1+\frac{1}{n}\right) + \log\left(1+\frac{3}{2n}\right) - 2\log\left(1+\frac{1}{2n}\right)$$

$$= \left(\frac{1}{n} - \frac{1}{2n^2} + \frac{1}{3n^3} - \ldots\right) + \left(\frac{3}{2n} - \frac{1}{2}\cdot\frac{9}{4n^2} + \ldots\right) - 2\left(\frac{1}{2n} - \frac{1}{2.4}\cdot\frac{1}{n^2} + \ldots\right)$$

$$\therefore \lim_{n\to\infty} n \log \frac{u_n}{u_{n+1}} = \lim_{n\to\infty} n\left[\frac{1}{n}\left(1+\frac{3}{2}-1\right) + \frac{1}{n^2}\left(-\frac{1}{2}-\frac{9}{8}+\frac{1}{4}+\ldots\right)\right]$$

$$= \lim_{n\to\infty}\left[\frac{3}{2} + \frac{1}{n}\left(-\frac{1}{2}-\frac{9}{8}+\frac{1}{4}+\ldots\right)\right]$$

$$= \frac{3}{2} > 1.$$

Hence, the series is convergent.

Thus, the given series converges if $x^2 \le 1$ and diverges if $x^2 > 1$.

EXERCISE 4.5

1. Test the convergence of the series $\sum\left(\frac{nx}{n+1}\right)^n$, $x > 0$.

2. Test the convergence of the series

$$\frac{1}{(\log 2)^2} + \frac{1}{(\log 3)^p} + \ldots + \frac{1}{(\log n)^p} + \ldots$$

3. Discuss the convergence of the series

$$1 + \frac{x}{2} + \frac{2!}{3^2}x^2 + \frac{3!}{4^2}x^3 + \frac{4!}{5^4}x^4 + \ldots,\ (x > 0)$$

4. Discuss the convergence of the series

$$\frac{x}{1.2} + \frac{x^2}{3.4} + \frac{x^3}{5.6} + \frac{x^4}{7.8} + \ldots,\ (x > 0)$$

5. Discuss the convergence of the series

$$\frac{x}{1.2} + \frac{x^2}{2.3} + \frac{x^3}{3.4} + \frac{x^4}{4.5} + \ldots,\ (x > 0).$$

Answers

1. Convergent when $x < 1$, divergent when $x \geq 1$.
2. Divergent for all values of p.
3. Convergent if $x < e$ and divergent if $x \geq e$.
4. Convergent for $x \leq 1$ and divergent for $x > 1$.
5. Convergent for $x \leq 1$, divergent for $x > 1$.

4.14 CAUCHY'S ROOT TEST (nTH ROOT TEST)

Statement. If Σu_n is a series of positive terms and $\lim\limits_{n \to \infty} u^{\frac{1}{n}} = l$, then

(i) Σu_n is convergent if $l < 1$.

(ii) Σu_n is divergent if $l > 1$.

If $l = 1$, the test fails, *i.e.*, no definite conclusion can be drawn about the convergence or divergence of the series, *i.e.*, the series may converge or it may diverge.

4.14.1 Illustrative Examples

1. *Test for convergence of the series* $\Sigma \dfrac{n^{n^2}}{(n+1)^{n^2}}$

Sol. Here

$$u_n = \frac{n^{n^2}}{(n+1)^{n^2}} = \left(\frac{n}{n+1}\right)^{n^2} = \frac{1}{\left(1+\frac{1}{n}\right)^{n^2}}$$

$$\therefore \quad u_n^{\frac{1}{n}} = \frac{1}{\left(1+\frac{1}{n}\right)^{n}} = \left[\left(1+\frac{1}{n}\right)^{n}\right]^{-1}$$

$$\therefore \quad \lim_{n \to \infty} u_n^{\frac{1}{n}} = \lim_{n \to \infty} \left[\left(1+\frac{1}{n}\right)^{n}\right]^{-1} = e^{-1} = \frac{1}{e} < 1 \qquad (\because 2 < e < 3)$$

Therefore, by Cauchy's root test, the given series, Σu_n is convergent.

2. *Test the convergence of the series* $\left(\dfrac{2^2}{1^2} - \dfrac{2}{1}\right)^{-1} + \left(\dfrac{3^3}{2^3} - \dfrac{3}{2}\right)^{-2} + \left(\dfrac{4^4}{3^4} - \dfrac{4}{3}\right)^{-3} + \ldots$

Sol. Here

$$u_n = \left[\left(\frac{n+1}{n}\right)^{n+1} - \frac{n+1}{n}\right]^{-n}$$

$$\therefore \quad u_n^{\frac{1}{n}} = \left[\left(\frac{n+1}{n}\right)^{n+1} - \frac{n+1}{n}\right]^{-1} = \left[\left(1+\frac{1}{n}\right)^{n+1} - \left(1+\frac{1}{n}\right)\right]^{-1}$$

$$= \left[\left(1+\frac{1}{n}\right)^{n}\left(1+\frac{1}{n}\right) - \left(1+\frac{1}{n}\right)\right]^{-1}$$

$$\therefore \quad \lim_{n\to\infty} u_n^{\frac{1}{n}} = \lim_{n\to\infty}\left[\left(1+\frac{1}{n}\right)^{n}\left(1+\frac{1}{n}\right) - \left(1+\frac{1}{n}\right)\right]^{-1}$$

$$= (e.1-1)^{-1} = (e-1)^{-1} = \frac{1}{e-1} < 1$$

$\therefore \lim_{n\to\infty} u_n^{\frac{1}{n}} < 1$, therefore, the given series, Σu_n is convergent (by Cauchy's root test).

3. *Test for the convergence of the series* $\Sigma\left(1+\frac{1}{n}\right)^{-n^2}$.

Sol. Here

$$u_n = \left(1+\frac{1}{n}\right)^{-n^2}$$

$$\therefore \quad u_n^{\frac{1}{n}} = \left(1+\frac{1}{n}\right)^{-n} = \left[\left(1+\frac{1}{n}\right)^{n}\right]^{-1}$$

$$\therefore \quad \lim_{n\to\infty} u_n^{\frac{1}{n}} = \lim_{n\to\infty}\left[\left(1+\frac{1}{n}\right)^{n}\right]^{-1} = \left[\lim_{n\to\infty}\left(1+\frac{1}{n}\right)^{n}\right]^{-1}$$

$$= e^{-1} = \frac{1}{e} < 1.$$

Hence, by Cauchy's nth rest test, the series Σu_n is convergent.

4. *Show that the series* $\Sigma\left(1-\frac{1}{n}\right)^{n^2}$ *is convergent.*

Sol. Here

$$u_n = \left(1-\frac{1}{n}\right)^{n^2}$$

$$\therefore \quad u_n^{\frac{1}{n}} = \left(1-\frac{1}{n}\right)^{n}$$

Then

$$\lim_{n\to\infty} u_n^{\frac{1}{n}} = \lim_{n\to\infty}\left(1-\frac{1}{n}\right)^{n} = \left\{\lim_{n\to\infty}\left[1+\left(-\frac{1}{n}\right)^{-n}\right]\right\}^{-1}$$

$$= e^{-1} = \frac{1}{e} < 1$$

Hence, by Cauchy's root test, the given series is convergent.

EXERCISE 4.6

Test the convergence of the following series:

1. $\Sigma \dfrac{1}{n^n}$

2. $\displaystyle\sum_{n=2}^{\infty} \frac{1}{(\log n)^n}$

3. $\Sigma \dfrac{(n-\log n)^n}{2^n . n^n}$

4. $\Sigma \left(1+\dfrac{1}{\sqrt{n}}\right)^{-n^{\frac{3}{2}}}$ [W.B.U.T, BCA. 2007]

5. $\Sigma \dfrac{(n+1)^n x^n}{n^{n+1}}$.

Answers

1. Convergent
2. Convergent
3. Convergent
4. Convergent
5. Convergent for $x < 1$, divergent for $x \geq 1$.

4.15 ALTERNATING SERIES

An alternating series is a series of the form $u_1 - u_2 + u_3 - u_4 + \ldots$
where $u_n > 0$ for all n.

Definition. If $u_n > 0$ for all n, the series $\displaystyle\sum_{n=1}^{\infty} (-1)^{n+1} u_n$ is called an *alternating series.*

Example. The series $1 - \frac{1}{2} + \frac{1}{3} - \frac{1}{4} + \ldots$, is an alternating series.

4.15.1 Leibnitz's test on alternating series

Statement. The alternating series $\Sigma(-1)^{n+1} u_n = u_1 - u_2 + u_3 - \ldots$ $(u_n > 0 \,\forall\, n)$ converges if

(*i*) each term is numerically less than its preceeding term, *i.e.*, $|u_n| > |u_{n+1}|$ for $n \geq 1$ and

(*ii*) $\lim\limits_{n \to \infty} u_n = 0$.

4.15.2 Illustrative Examples

1. *Examine the convergence of the series* $1 - \frac{1}{2} + \frac{1}{3} - \frac{1}{4} + \ldots$

Sol. Here $$|u_n| = \frac{1}{n}, |u_{n+1}| = \frac{1}{n+1}$$

Now $$|u_n| - |u_{n+1}| = \frac{1}{n} - \frac{1}{n+1} = \frac{n+1-n}{n(n+1)} = \frac{1}{n(n+1)} > 0, \forall n \geq 1$$

$\therefore$ $$|u_n| > |u_{n+1}|$$

and $$\lim_{n\to\infty} |u_n| = \lim_{n\to\infty} \frac{1}{n} = 0$$

Since, both the conditions of Leibnitz's test are satisfied, hence the series is convergent.

2. *Test the convergence of the alternating series* $\frac{1}{6} - \frac{2}{11} + \frac{3}{16} - \frac{4}{21} + \frac{5}{26} - \ldots$

Sol. Here $$u_n = (-1)^{n-1} \frac{n}{5n+1}, n \geq 1.$$

$\therefore$ $$u_{n+1} = (-1)^n \frac{n+1}{5(n+1)+1}$$

and $$|u_n| - |u_{n+1}| = \frac{n}{5n+1} - \frac{(n+1)}{5n+6} = \frac{5n^2 + 6n - 5n^2 - 6n - 1}{(5n+1)(5n+6)}$$

$$= \frac{-1}{(5n+1)(5n+6)} < 0, \text{ for } n \geq 1$$

hence $|u_n| < |u_{n+1}|$, for $n \geq 1$.

and $$\lim_{n\to\infty} |u_n| = \lim_{n\to\infty} \frac{n}{5n+1} = \lim_{n\to\infty} \frac{1}{5+\frac{1}{n}} = \frac{1}{5} \neq 0.$$

Since the given series does not satisfy the Leibnitz's test, so the given series is not convergent.
It is oscillatory.

3. *Test the convergence of the series* $\frac{x}{1+x} - \frac{x^2}{1+x^2} + \frac{x^3}{1+x^3} - \ldots,$ *(0 < x < 1).*

Sol. Here $$u_n = (-1)^{n-1} \frac{x^n}{1+x^n}, n \geq 1$$

$\therefore$ $$|u_n| = \frac{x^n}{1+x^n} \text{ and } |u_{n+1}| = \frac{x^{n+1}}{1+x^{n+1}}, \; 0 < x < 1$$

Now, $$|u_n| - |u_{n+1}| = \frac{x^n}{1+x^n} - \frac{x^{n+1}}{1+x^{n+1}}$$

$$= \frac{x^n(1-x)}{\left(1+x^n\right)\left(1+x^{n+1}\right)} > 0, \text{ for } 0 < x < 1.$$

i.e., $|u_n| > |u_{n+1}|$

and $\lim\limits_{n\to\infty} |u_n| = \lim\limits_{n\to\infty} \frac{x^n}{1+x^n} = 0, \quad (\because \lim\limits_{n\to\infty} x^n = 0, \text{ for } 0 < x < 1)$

Since, both the conditions of Leibnitz's test are satisfied, hence the given series is convergent.

4. *Examine the convergence of the series* $2 - \frac{3}{2} + \frac{4}{3} - \frac{5}{4} + \ldots$

Sol. The series is an alternating series.

Here $u_n = (-1)^{n-1} \frac{n+1}{n}, n \geq 1$

$\therefore$ $|u_n| = \frac{n+1}{n}$ and $|u_{n+1}| = \frac{n+2}{n+1}$

$\therefore$ $|u_n| - |u_{n+1}| = \frac{n+1}{n} - \frac{n+2}{n+1} = \frac{n^2+2n+1-n^2-2n}{n(n+1)}$

$$= \frac{1}{n(n+1)} > 0 \text{ for all } n.$$

Hence, $|u_n| > |u_{n+1}|$

Now $\lim\limits_{n\to\infty} |u_n| = \lim\limits_{n\to\infty} \frac{n+1}{n} = \lim\limits_{n\to\infty} \left(1+\frac{1}{n}\right) = 1 \neq 0$

i.e., the second condition of Leibnitz's test is not satisfied, hence the series is not convergent.

5. *Test the convergence of the series*

$$\frac{1}{2^3} - \frac{1}{3^3}(1+2) + \frac{1}{4^3}(1+2+3) - \frac{1}{5^3}(1+2+3+4) + \ldots$$

Sol. The series is an alternating series.

Here $u_n = (-1)^{n-1} \frac{1}{(n+1)^3}[1+2+3+4+\ldots+n], n \geq 1$

$$= (-1)^{n-1} \frac{1}{(n+1)^3}, \frac{n(n+1)}{2} = (-1)^{n-1} \frac{1}{2} \frac{n}{(n+1)^2}$$

$\therefore$ $|u_n| = \frac{1}{2} \frac{n}{(n+1)^2}$

and $$|u_{n+1}| = \frac{1}{2}\frac{n+1}{(n+2)^2}$$

$\therefore$ $$|u_n| - |u_{n+1}| = \frac{1}{2}\left[\frac{n}{(n+1)^2} - \frac{n+1}{(n+2)^2}\right]$$

$$= \frac{1}{2}\left[\frac{n^3+4n^2+4n-n^3-3n^2-3n-1}{(n+1)^2\,(n+2)^2}\right]$$

$$= \frac{1}{2}\left[\frac{n^2+n-1}{(n+1)^2\,(n+2)^2}\right] > 0 \text{ for } n \geq 1.$$

Hence, $$|u_n| > |u_{n+1}| \text{ for } n \geq 1.$$

Now $$\lim_{n\to\infty} |u_n| = \lim_{n\to\infty} \frac{1}{2}\frac{n}{(n+1)^2} = \lim_{n\to\infty} \frac{1}{2n\left(1+\frac{1}{n}\right)} = 0$$

Therefore, both the conditions of Leibnitz's test are satisfied, hence the given series is convergent.

4.16 ABSOLUTE CONVERGENT AND CONDITIONAL CONVERGENT

Definition. If a convergent series, whose terms are not all positive, remains convergent when all of its terms are made positive, then it is called an *absolutely convergent series*. In other words, the series Σu_n is said to be absolutely convergent if $\Sigma\,|u_n|$ is a convergent series.

Conditional Convergent Series

A series is said to be *conditionally convergent* if it is convergent but does not converge absolutely.

4.16.1 Illustrative Examples

1. *Test whether the series* $1 - \frac{1}{2^2} + \frac{1}{3^2} - \frac{1}{4^2} + \ldots$ *is absolutely convergent or conditionally convergent ?*

Sol. The given series is an alternating series.

(*i*) In this alternating series, each term is less than the preceeding term numerically,

(*ii*) Moreover, $u_n = \frac{1}{n^2} \to 0$ as $n \to \infty$.

Hence, the series satisfies both the conditions of Leibnitz's test and so the given series is convergent. Again, when all the terms of the series are made positive, the series becomes

$$\Sigma\,|u_n| = 1 + \frac{1}{2^2} + \frac{1}{3^2} + \frac{1}{4^2} + \ldots = \Sigma\frac{1}{n^2}$$

which is convergent. [since $\Sigma\frac{1}{n^2}$ is an auxiliary series and $p > 1$ by 4.18].

Thus the given series converges absolutely.

2. *Examine whether the series* $1 - \frac{1}{2} + \frac{1}{3} - \frac{1}{4} + \ldots$ *is absolutely convergent or conditionally convergent.*

Sol. The given series is an alternating series

Here $$|u_n| = \frac{1}{n},$$

$$\therefore \quad |u_{n+1}| = \frac{1}{n+1}$$

$$\because \quad \frac{1}{n+1} < \frac{1}{n} \text{ for } n \geq 1$$

$$\therefore \quad |u_{n+1}| < |u_n|$$

and $$\lim_{n\to\infty} |u_n| = \lim_{n\to\infty} \frac{1}{n} = 0$$

$\therefore$ $|u_n|$ satisfies both the conditions of Leibnitz's test.

Hence, the given series is convergent.

Again, when all the terms of the series are made positive, the series becomes $1 + \frac{1}{2} + \frac{1}{3} + \ldots$

which is divergent by the auxiliary series $\Sigma\frac{1}{n^p}$, since $p = 1$.

Hence, the given series is conditionally convergent.

3. *Examine whether the series* $\sum_{n=1}^{\infty} \frac{(-1)^{n-1}}{2n-1}$ *is absolutely convergent or conditionally convergent.*

Sol. The given series

$$\sum_{n=1}^{\infty} \frac{(-1)^{n-1}}{2n-1} = \Sigma u_n, \text{ say}$$

Putting $n = 1, 2, 3, \ldots$, the series is $1 - \frac{1}{3} + \frac{1}{5} - \frac{1}{7} + \ldots$

which is an alternating series.

Here $$u_n = (-1)^{n-1}\frac{1}{2n-1}$$

$\therefore$ $$|u_n| = \frac{1}{2n-1} \text{ and } |u_{n+1}| = \frac{1}{2n+1}$$

Here $$\frac{1}{2n+1} < \frac{1}{2n-1} \text{ for } n \geq 1$$

$\therefore$ $$|u_{n+1}| < |u_n|$$

and $$\lim_{n\to\infty} |u_n| = \lim_{n\to\infty} \frac{1}{2n-1} = 0$$

$\therefore$ $|u_n|$ satisfies both the conditions of Leibnitz's test, hence the series is convergent.

When all the terms of the series are made positive, the series becomes

$$\Sigma|u_n| = 1 + \frac{1}{3} + \frac{1}{5} + \frac{1}{7} + \dots$$

where $$u_n = \frac{1}{2n-1}$$

Taking $$v_n = \frac{1}{n}, \text{ we have } \frac{u_n}{v_n} = \frac{n}{2n-1}$$

$\therefore$ $$\lim_{n\to\infty} \frac{u_n}{v_n} = \lim_{n\to\infty} \frac{n}{n\left(2 - \frac{1}{n}\right)}$$

$$= \lim_{n\to\infty} \frac{1}{2 - \frac{1}{n}} = \frac{1}{2} \text{ which is finite.}$$

Therefore, by comparison test, Σu_n and Σv_n are of the same property. But $\Sigma v_n = \Sigma \frac{1}{n}$ is divergent (by p-test, since $p = 1$). Hence Σu_n is also divergent.

Hence, the given series is conditionally convergent.

4. *Show that the series* $1 + x + \frac{x^2}{2!} + \frac{x^3}{3!} + \dots$ *is absolutely convergent for all values of* x.

Sol. Here $$u_n = \frac{x^{n-1}}{(n-1)!} \text{ and } u_{n+1} = \frac{x^n}{n!}$$

$\therefore$ $$\left|\frac{u_n}{u_{n+1}}\right| = \frac{n}{|x|}.$$

Now $$\lim_{n\to\infty}\left|\frac{u_n}{u_{n+1}}\right| = \lim_{n\to\infty}\frac{n}{|x|} = \infty$$

whatever $x(\neq 0)$ may be.

Therefore, by D'Alembert's ratio test, the series $\Sigma\,|\,u_n\,|$ converges for all x(except $x\neq 0$).

when $x = 0$, the series converges (to 1).

Thus the series Σu_n converges absolutely for all values of x ($x\neq 0$).

We shall now state and prove a theorem on absolute convergent.

Theorem 4. Every absolute convergent series is convergent.

Proof. Let $\sum_{n=1}^{\infty} u_n$ be an absolutely convergent series.

Then $\sum_{n=1}^{\infty}|u_n|$ is convergent.

By Cauchy's general principle of convergence, given $\varepsilon > 0$, there exists a positive integer m such that

$|u_{m+1}|+|u_{m+2}|+...+|u_{m+p}| < \varepsilon$ ... (*i*), for all $n > m$,

whatever positive integer p may be. Again, by triangle inequality, we have

$$|u_{m+1}+u_{m+2}+\ldots+u_{m+p}| \leq |u_{m+1}| + |u_{m+2}| + \ldots + |u_{m+p}|,$$

or $|u_{m+1}+u_{m+2}+\ldots+u_{m+p}| < \varepsilon$, for all p [using (*i*)].

Therefore, by Cauchy's general principle of convergence, it follows that the series Σu_n converges.

Remark 1. Absolute convergence of a series implies its convergence, but convergence need not imply absolute convergence that is, the converse of the Theorem 4 need not be true.

As an example, the series

$$\Sigma\frac{(-1)^{n-1}}{n} = 1-\frac{1}{2}+\frac{1}{3}-\frac{1}{4}+\ldots$$

is convergent. [see Illustrative example 1]. But the series

$$\Sigma\left|\frac{(-1)^{n-1}}{n}\right| = \Sigma\frac{1}{n} \text{ is divergent.}$$

Remark 2. The divergence of $\Sigma\,|\,u_n\,|$ does not imply the divergence of Σu_n.

As an example,

$$\Sigma\left|\frac{(-1)^{n-1}}{n}\right| = \Sigma\frac{1}{n} \text{ is divergent, whereas } \Sigma\frac{(-1)^{n-1}}{n} \text{ is convergent.}$$

EXERCISE 4.7

1. Test whether the following series are absolutely convergent on conditionally convergent.

   (i) $1-\frac{1}{3}+\frac{1}{2}-\frac{1}{3^3}+\frac{1}{2^2}-\frac{1}{3^5}+\frac{1}{2^3}-\frac{1}{3^7}+...$

   (ii) $\sum_{n=2}^{\infty}\frac{(-1)^n}{n(\log n)^2}$.

2. Examine for the convergence of the following series :

   (i) $1-\frac{1}{\sqrt{2}}+\frac{1}{\sqrt{3}}-\frac{1}{\sqrt{4}}+...$

   (ii) $1-\frac{1}{2\sqrt{2}}+\frac{1}{3\sqrt{3}}-\frac{1}{4\sqrt{4}}+...$

   (iii) $\frac{1}{a}-\frac{1}{a+b}+\frac{1}{a+2b}-\frac{1}{a+3b}+...$ $(a>0, b>0)$

   (iv) $\frac{1}{\log 2}-\frac{1}{\log 3}+\frac{1}{\log 4}-\frac{1}{\log 5}+...$ (v) $\sum_{n=1}^{\infty}(-1)^n\frac{x^{n+1}}{n+1}$

3. Using Leibnitz's test, show that the series $\frac{1}{1+a^2}-\frac{1}{2+a^2}+\frac{1}{3+a^2}-...$ is convergent.

4. Prove that the series $\sum_{n=1}^{\infty}\frac{(-1)^{n-1}\cos^2 nx}{n\sqrt{n}}$ converges absolutely.

5. Show that the following series are convergent.

   (i) $1^{-p}-2^{-p}+3^{-p}-...$, when $p>0$

   (ii) $\frac{1}{x+1}-\frac{1}{x+2}+\frac{1}{x+3}-...$, except when x is a negative integer.

Answers

1. (i) Absolutely convergent (ii) Absolutely convergent
2. (i) convergent (ii) convergent (iii) convergent (iv) convergent (v) oscillatory.

4.17 RE-ARRANGEMENT OF TERMS OF AN ABSOLUTELY CONVERGENT SERIES

We shall state some properties of absolutely convergent series when its terms are re-arranged.

I. If the sum of an absolutely convergent series are re-arranged, the series remains convergent.

II. The sum of an absolutely convergent series is independent of re-arrangement of its terms.

III. By suitable re-arrangement of the given order of the terms and signs of the terms of a conditionally convergent series, we can make its sum equal to any given number including $+\infty$ and $-\infty$.

Remark. The sum of deranged series $= \frac{1}{2} \log 4k$, where $k = \frac{m}{n}$, m is the number of positive terms followed by n number of negative terms.

For derangement of a series, the ratio of the number of positive terms to the number of negative terms is given by $\frac{1}{4} e^{2S}$, where s is the sum of the series.

[Proofs of these results are omitted].

4.17.1 Illustrative Example

1. *Find the sum of the series* $1+\frac{1}{3}+\frac{1}{5}-\frac{1}{2}+\frac{1}{7}+\frac{1}{9}+\frac{1}{11}-\frac{1}{4}+\ldots$

Sol. The given series is the derangement of the series

$$1-\frac{1}{2}+\frac{1}{3}-\frac{1}{4}+\frac{1}{5}-\ldots \qquad \ldots(1)$$

which is an alternating series.

Here 3 positive terms are followed by one negative term,

so the sum of the deranged series is $\frac{1}{2} \log 4k$

$$= \frac{1}{2} \log 4 \times \frac{3}{1}, \text{(here } k = \frac{3}{1}\text{)}$$

$$= \frac{1}{2} \log 12$$

2. *Show that the sum of the series* $1+\frac{1}{3}+\frac{1}{5}-\frac{1}{2}-\frac{1}{4}+\frac{1}{7}+\frac{1}{9}+\frac{1}{11}-\frac{1}{6}-\frac{1}{8}+\ldots$ *is* $\frac{1}{2} \log 6$.

Sol. The given series is the derangement of the series $1-\frac{1}{2}+\frac{1}{3}-\frac{1}{4}+\frac{1}{5}\ldots \qquad \ldots(1)$

Here 3 positive terms are followed by 2 negative terms so that $k = \frac{m}{n} = \frac{3}{2}$.

$\therefore$ The sum of the given series $= \frac{1}{2} \log 4k = \frac{1}{2} \log 4 \times \frac{3}{2} = \frac{1}{2} \log 6$.

We now state two tests, without proof, which are known as (*i*) *Dirichlet's test* and (*ii*) *Abel's test.*

(*i*) **Dirichlet's test.** If Σu_n converges or oscillates between finite limits and $a_1, a_2, a_3, \ldots$ is a decreasing sequence of positive terms which tends to zero as a limit, then $\Sigma a_n u_n$ is convergent.

(*ii*) **Abel's test.** If Σu_n converges and $a_1, a_2, a_3, \ldots$ is a sequence of positive terms, tending to a finite limit, then $\Sigma a_n u_n$ is convergent.

4.17.2 Illustrative Examples

1. *Show that the series* $\sum_{n=1}^{\infty} \frac{(-1)^{n+1}}{\sqrt{n}}$ *is convergent.*

Sol. The given series is convergent by Dirichlet's test, since the sequence of partial sums $\{S_n\}$ of the series $\Sigma(-1)^{n+1}$ is bounded and the sequence $\left\{\frac{1}{\sqrt{n}}\right\}$ is a monotone decreasing sequences converging to 0.

2. *Show that the series* $\sum \frac{(-1)^{n-1}}{n} x^n$ *is uniform convergence in (0, 1).*

Sol. Let $v_n(x) = x^n$ and $u_n(x) = \frac{1}{n}(-1)^{n-1}$.

The sequence $\{v_n(x)\}$ is bounded and monotonic decreasing in (0, 1). Also the series $\sum \frac{(-1)^{n-1}}{n}$ is convergent.

Hence by Abel's test the given series is uniformly convergent in (0, 1).

3. *Show that the series* $\sum_{n=0}^{\infty} (-1)^{n-1} x^n$*, converges uniformly in* $0 \le x \le k \le 1$.

Sol. Let $u_n = (-1)^{n-1}, v_n(x) = x^n$.

Since, $S_n(x) = \sum_{n=1}^{\infty} u_r = 0$ or 1 according as n is even or odd. $S_n(x)$ is bounded for all n. Also, $\{v_n(x)\}$ is a positive monotonic decreasing sequence, converging to zero for all values of x in $0 \le x \le k < 1$. Hence, by Dirichlet's test, the given series is uniformly convergent in $0 \le x \le k < 1$.

4. *Prove that if* δ *is any fixed positive number less than unity, the series* $\sum \frac{x^n}{n+1}$ *is uniformly convergent in* $(-\delta, \delta)$.

Sol. Let $u_n(x) = x^n, v_n = \frac{1}{n+1}$.

Since $\quad | x | \le \delta < 1$, we have

$$| S_n(x) | = | x + x^2 + \dots + x^n | \le | x | + | x |^2 + \dots + | x |^n$$

$$\le \delta + \delta^2 + \dots + \delta^n = \frac{\delta\left(1-\delta^n\right)}{1-\delta} < \frac{\delta}{1-\delta}$$

Also $\{v_n\{$ is a monotonic decreasing sequence converging to zero. Hence, the given series is uniformly convergent by Dirichlet's test.

EXERCISE 4.8

1. Show that the series $\sin x + \frac{1}{2}\sin 2x + \frac{1}{3}\sin 3x + \dots$ converge uniformly in $0 < a \le x \le b < 2\pi$.
2. If $\sum a_n$ is a convergent series of positive constants, prove that the series $\sum a_n x^n$ is uniformly convergent in $0 \le x \le 1$.
3. Show that the series $\cos x + \frac{1}{2}\cos 2x + \frac{1}{3}\cos 3x + \dots$ converges uniformly in $0 < a \le x \le b \le 2\pi$.
4. Show that the series $\sin x + \frac{1}{3}\sin 3x + \frac{1}{5}\sin 5x + \dots$, converges uniformly in $0 < a \le x \le b < x$.

SOME USEFUL FORMULAE

1. Trigonometry :

(*i*) $2 \sin A \cos B = \sin (A + B) + \sin (A - B)$

(*ii*) $2 \cos A \sin B = \sin (A + B) - \sin (A - B)$

(*iii*) $2 \cos A \cos B = \cos (A + B) + \cos (A - B)$

(*iv*) $2 \sin A \sin B = \cos (A - B) - \cos (A + B)$

(*v*) $\sin C + \sin D = 2 \sin \frac{(C+D)}{2} . \cos \frac{(C-D)}{2}$

(*vi*) $\sin C - \sin D = 2 \cos \frac{(C+D)}{2} . \sin \frac{(C-D)}{2}$

(*vii*) $\cos C + \cos D = 2 \cos \frac{(C+D)}{2} . \cos \frac{(C-D)}{2}$

(*viii*) $\cos C - \cos D = 2 \sin \frac{(C+D)}{2} . \sin \frac{(D-C)}{2}$

(*ix*) $\sin 2A = 2 \sin A \cos A$

(*x*) $\cos 2A = \cos^2 A - \sin^2 A$

(*xi*) $2 \cos^2 A = 1 + \cos 2A$

(*xii*) $2 \sin^2 A = 1 - \cos 2A$

(*xiii*) $\sin 3A = 3 \sin A - 4 \sin^3 A$

(*xiv*) $\cos 3A = 4 \cos^3 A - 3 \cos A$

2. Hyperbolic Functions :

(*i*) $\cosh x = \frac{1}{2}\left(e^x + e^{-x}\right)$; $\sinh x = \frac{1}{2}\left(e^x - e^{-x}\right)$

(*ii*) $e^x = \cosh x + \sinh x$; $e^{-x} = \cosh x - \sinh x$

(*iii*) $\cosh^2 x - \sinh^2 x = 1$ (*iv*) $\text{sech}^2 x + \tanh^2 x = 1$

(*v*) $\coth^2 x - \text{cosech}^2 x = 1$ (*vi*) $\sinh 2x = 2 \sinh x \cosh x$

(*vii*) $\cosh 2x = \cosh^2 x + \sinh^2 x = 2 \cosh^2 x - 1 = 1 + 2 \sinh^2 x$

3. Complex Arguments :

(*i*) $(\cos \theta + i \sin \theta)^n = \cos n\theta + i \sin n\theta$ for all n.

(*ii*) $e^{iax} = \cos ax + i \sin ax$; $e^{-iax} = \cos ax - i \sin ax$

4. Algebra :

(*i*) $1 + 2 + 3 + \ldots + n = \frac{n(n+1)}{2}$ (*ii*) $1^2 + 2^2 + \ldots + n^2 = \frac{n(n+1)(2n+1)}{6}$

(*iii*) $1^3 + 2^3 + \ldots + n^3 = \left\{\frac{1}{2} n(n+1)\right\}^2$ (*iv*) $a + ar + ar^2 + \ldots + ar^n = \frac{a\left(r^n - 1\right)}{r-1}, (r \neq 1)$

(*v*) $a + ar + ar^2 + \ldots \infty = \frac{a}{1-r}$ $(r < 1)$

(*vi*) $\log_a m^n = n \log_a m$

(*vii*) $\log_a m = \log_e m \times \log_a e = \dfrac{\log m}{\log a}$

(*viii*) $\log_a b \times \log_b a = 1$

(*ix*) $e^{\log_e x} = x$.

(*x*) $^n P_r = \dfrac{n!}{(n-r)!}$ (*n* things are all different)

(*xi*) $^n C_r = \dfrac{n!}{r!(n-r)!}$

(*xii*) $(1+x)^n = 1 + nx + \dfrac{n(n-1)}{2!}x^2 + \dfrac{n(n-1)(n-2)}{3!} + \ldots\ldots + x^n$, when n is a positive integer.

(*xiii*) $(1+x)^n = 1 + nx + \dfrac{n(n-1)}{2!}x^2 + \dfrac{n(n-1)(n-2)}{3!} + \ldots\ldots \infty$, when n is negative or fraction.

(*xiv*) $(1 + x)^{-1} = 1 - x + x^2 - x^3 + \ldots\ldots \infty$

(*xv*) $(1 - x)^{-1} = 1 + x + x^2 + x^3 + \ldots\ldots \infty$

(*xvi*) $(1 + x)^{-2} = 1 - 2x + 3x^2 - 4x^3 + \ldots\ldots \infty$

(*xvii*) $(1 - x)^{-2} = 1 + 2x + 3x^2 + 4x^3 + \ldots\ldots \infty$

(*xviii*) $\log_e (1 + x) = x - \dfrac{x^2}{2} + \dfrac{x^3}{3} - \ldots\ldots + (-1)^{r-1}\dfrac{x^r}{r} + \ldots\ldots \infty, (-1 < x \le 1)$

(*xix*) $\log (1 - x) = -x - \dfrac{x^2}{2} - \dfrac{x^3}{3} - \ldots\ldots - \dfrac{x^r}{r} - \ldots\ldots \infty, (-1 \le x < 1)$

(*xx*) $e^x = 1 + \dfrac{x}{1!} + \dfrac{x^2}{2!} + \ldots\ldots + \dfrac{x^r}{r!} + \ldots\ldots \infty$; $e^{-x} = 1 - \dfrac{x}{1!} + \dfrac{x^2}{2!} + \ldots\ldots + (-1)^r \dfrac{x^r}{r!} + \ldots\ldots \infty$

5. Calculus :

(A) LIMIT

(*i*) $\lim\limits_{x \to 0} \dfrac{\sin x}{x} = 1$

(*ii*) $\lim\limits_{x \to 0} \dfrac{e^x - 1}{x} = 1$

(*iii*) $\lim\limits_{x \to 0} \dfrac{1}{x} \log(1 + x) = 1$

(*iv*) $\lim\limits_{x \to 0} \dfrac{(1+x)^n - 1}{x} = n$

(*v*) $\lim\limits_{x \to 0} (1+x)^{\frac{1}{x}} = e$

(*vi*) $\lim\limits_{x \to \infty} \left(1 + \dfrac{1}{x}\right)^n = e$

(*vii*) $\lim\limits_{n \to \infty} \dfrac{x^n}{n!} = 0$

(*viii*) $\lim\limits_{n \to \infty} x^n = 0 \; (-1 < x < 1)$.

(B) DIFFERENTIATION

(*i*) $D(x^n) = nx^{n-1}$

(*ii*) $D(c) = 0$

(*iii*) $D(e^{ax}) = ae^{ax}$

(*iv*) $D(\log_e x) = \dfrac{1}{x}$

(*v*) $D(\log_a x) = \frac{1}{x}\log_e a$

(*vi*) $D(\sinh x) = \cosh x$

(*vii*) $D(\cosh x) = \sinh x$

(*viii*) $D(\tanh x) = \text{sech}^2 x$

(*ix*) $D(\coth x) = -\text{cosech}^2 x$

(*x*) $D(\text{sech}\, x) = -\text{sech}\, x \,.\, \tanh x$

(*xi*) $D(\text{cosech}\, x) = -\text{cosech}\, x \,.\, \coth x$, where $D = \frac{d}{dx}$.

(C) INTEGRATION

(*i*) $\int x^n\, dx = \frac{x^{n+1}}{n+1}$ (n is a constant and $n \neq -1$)

(*ii*) $\int c\, dx = cx$

(*iii*) $\int \frac{dx}{x} = \log|x|$

(*iv*) $\int e^{ax}\, dx = \frac{1}{a}e^{ax}$

(*v*) $\int a^x\, dx = \frac{a^x}{\log_e a} = a^x \log_a e, \ (a > 0)$

(*vi*) $\int \sin mx\, dx = -\frac{1}{m}\cos mx$

(*vii*) $\int \cos mx\, dx = \frac{1}{m}\sin mx$

(*viii*) $\int \sec^2 mx\, dx = \frac{1}{m}\tan mx$

(*ix*) $\int \text{cosec}^2 mx\, dx = -\frac{1}{m}\cot mx$

(*x*) $\int \sec mx \tan mx\, dx = \frac{1}{m}\sec mx$

(*xi*) $\int \text{cosec}\, mx \cot mx\, dx = -\frac{1}{m}\text{cosec}\, mx$

(*xii*) $\int \sinh x\, dx = -\cosh x$

(*xiii*) $\int \cosh x\, dx = \sinh x\, dx$

(*xiv*) $\int \tanh x\, dx = \log|\cosh x|$

(*xv*) $\int \coth x\, dx = \log|\sinh x|$

(*xvi*) $\int \text{cosech}\, x\, dx = \log\left|\tanh\frac{1}{2}x\right|$

(*xvii*) $\int \text{sech}\, x\, dx = 2\tan^{-1}(e^x)$

(*xviii*) $\int \text{sech}^2 x\, dx = \tanh x$

(*xix*) $\int \text{cosech}^2 x\, dx = -\coth x$

(*xx*) $\int \frac{f'(x)}{f(x)}\, dx = \log|f(x)|$

(*xxi*) $\int \tan x\, dx = \log|\sec x|$

(*xxii*) $\int \cot x\, dx = \log|\sin x|$

(*xxiii*) $\int \frac{dx}{x^2 + a^2} = \frac{1}{a}\tan^{-1}\frac{x}{a}, (a \neq 0)$

(*xxiv*) $\int \frac{dx}{x^2 - a^2} = \frac{1}{2a}\log\left|\frac{x-a}{x+a}\right| (|x| > |a|)$

(*xxv*) $\int \frac{dx}{a^2 - x^2} = \frac{1}{2a}\log\left|\frac{a+x}{a-x}\right| (|x| < |a|)$

(*xxvi*) $\int \frac{dx}{\sqrt{x^2 + a^2}} = \log\left|x + \sqrt{x^2 + a^2}\right| = \sinh^{-1}\frac{x}{a}$

(xxvii) $\int \frac{dx}{\sqrt{x^2 - a^2}} = \log \left| x + \sqrt{x^2 - a^2} \right| = \cosh^{-1} \frac{x}{a}$

(xxviii) $\int \frac{dx}{\sqrt{a^2 - x^2}} = \sin^{-1} \frac{x}{a}, (| x | < | a |)$

(xxix) $\int \frac{dx}{x\sqrt{x^2 - a^2}} = \frac{1}{a} \sec^{-1} \frac{x}{a}$

(xxx) $\int uv \, dx = u \int v \, dx - \int \left(\frac{du}{dx} \int v \, dx \right) dx$

(xxxi) $\int e^{ax} \cos bx \, dx = \frac{e^{ax} (a \cos bx + b \sin bx)}{a^2 + b^2} = \frac{e^{ax}}{\sqrt{a^2 + b^2}} \cos \left(bx - \tan^{-1} \frac{b}{a} \right)$

(xxxii) $\int e^{ax} \sin bx \, dx = \frac{e^{ax} (a \sin bx - b \cos bx)}{a^2 + b^2} = \frac{e^{ax}}{\sqrt{a^2 + b^2}} \sin \left(bx - \tan^{-1} \frac{b}{a} \right)$

(xxxiii) $\int \sqrt{x^2 + a^2} \, dx = \frac{x\sqrt{x^2 + a^2}}{2} + \frac{a^2}{2} \log \left| \left(x + \sqrt{x^2 + a^2} \right) \right| = \frac{x\sqrt{x^2 + a^2}}{2} + \frac{a^2}{2} \sinh^{-1} \frac{x}{a}$

(xxxiv) $\int \sqrt{x^2 - a^2} \, dx = \frac{x\sqrt{x^2 - a^2}}{2} - \frac{a^2}{2} \log \left| \left(x + \sqrt{x^2 - a^2} \right) \right|$

$= \frac{x\sqrt{x^2 - a^2}}{2} - \frac{a^2}{2} \log \left| \left(x + \sqrt{x^2 - a^2} \right) \right|$

$= \frac{x\sqrt{x^2 - a^2}}{2} - \frac{a^2}{2} \cosh^{-1} \frac{x}{a}$

(xxxv) $\int \sqrt{a^2 - x^2} \, dx = \frac{x\sqrt{a^2 - x^2}}{2} + \frac{a^2}{2} \sin^{-1} \frac{x}{a}$

(xxxvi) $\int \text{cosec}\, x \, dx = \log \left| \tan \frac{1}{2} x \right| = \log | \text{cosec}\, x - \cot x |$

(xxxvii) $\int \sec x \, dx = \log \left| \tan \left(\frac{1}{4} \pi + \frac{1}{2} x \right) \right| = \log | (\sec x + \tan x) |$

EXAMINATION PAPERS
ENGINEERING & TECHNOLOGY EXAMINATIONS, JUNE-2004
BCA
MATHEMATICS
SEMESTER - 2

Time : 3 Hours] [Full Marks : 70

The figures in the margin indicate full marks.

Candidates are required to give their answers in their own words as far as practicable.

Answer Question No.1 and any five from the rest

1. Answer any five : 2 × 5

(*a*) Find the degree and order of the differential equation

$$\frac{d^2y}{dx^2} = 1 + 2\frac{dy}{dx} + \left[\frac{dy}{dx}\right]^3$$

(*b*) Form the differential equation by eliminating '*a*' and '*b*' from

$$y = a \sin mx + b \cos mx$$

where a and b are arbitrary constants.

(*c*) Show that the vectors (1, 1, 0), (1, 1, 1), (0, 1, −1) are linearly independent.

(*d*) Define the co-ordinate vector of an element v of a vector space V with respect to a basis $[v_1, v_2, ..., v_n]$ of V.

(*e*) When is a sequence $\{S_n\}$ said to converge to a real number l?

(*f*) Give an example of a monotonic increasing sequence, which is convergent.

(*g*) Examine if the sequence $\left\{\frac{n}{n+1}\right\}$ is bounded or not.

(*h*) If $\sum_{n=1}^{\infty} U_n$ is convergent, prove that $\lim_{n \to \infty} U_n = 0$.

2. Solve the differential equations :

(*a*) $xdy - ydx = \sqrt{x^2 + y^2}\,dx$ 6

(*b*) $y(1 + xy^2)\frac{dy}{dx} = 1$ 6

3. (*a*) Show that the solution of $\frac{d^2x}{dt^2} + K\frac{dx}{dt} + \mu x = 0$ is $x = e^{-Kt/2}$ (A cos nt + B sin nt)

if $K^2 < 4\mu$ and $n^2 = \mu - \frac{1}{4}K^2$ 6

(*b*) The differential equation for the current i in an electrical circuit containing an inductance L and a resistance R in series and acted on by an electromotive force

E sin ωt is $L\frac{di}{dt} + Ri = E \sin \omega t$. Find the current "$i$" at anytime "$t$" if initially there is no current in the circuit. 6

4. Solve any *two* : 2 × 6

(*a*) $\frac{d^2y}{dx^2} - 2\frac{dy}{dx} + y = xe^x$

(*b*) $(px - y)(py + x) = a^2p$ by using substitution $x^2 = u$ and $y^2 = v$, where $p = \frac{dy}{dx}$

(*c*) $x^2\frac{d^2y}{dx^2} - x\frac{dy}{dx} - 3y = x^2 \log x$

5. (*a*) Define a Sub-space of a Vector space. Show that intersection of two Sub-spaces of a Vector space is a Sub-space. 6

(*b*) Define a basis of a finite Dimensional Vector space V. If $\{\alpha, \beta, \gamma\}$ is a basis of a real vector space V, show that $(\alpha + \beta)$, $\beta + \gamma$, $\gamma + \alpha)$ is also a basis for V. 6

6. (*a*) Find a basis and the dimension of $S \cap T$ where S and T are sub-spaces of R^3 defined by

$$S = \{(x, y, z) \in R^3 : 2x + y + 3z = 0\} \text{ and}$$ 6

$$T = \{(x, y, z) \in R^3 : x + 2y + z = 0\}$$

(*b*) Determine the linear mapping $T : R^3 \to R^3$ which maps the basis vectors (0, 1, 1), (1, 0, 1), (1, 1, 0) of R^3 to the vectors (1, 2, 1), (1, 1, 2), (2, 1, 1,) respectively. Find dim ker T. 6

7. Examine the convergence of the following sequences : 3 × 4

(*a*) $\left\{\frac{n^n}{n!}\right\}$ (*b*) $\left\{\left(1 + \frac{2}{n}\right)^n\right\}$ (*c*) $\left\{\frac{(-1)^{n-1} n}{3^n}\right\}$

8. (*a*) Show that the sequence $\{S_n\}$ where

$$S_n = \frac{1}{1!} + \frac{1}{2!} + \frac{1}{3!} + \ldots + \frac{1}{n!} \text{ is convergent}$$ 6

(*b*) Test for convergence of the series $\frac{1}{2\sqrt{1}} + \frac{x^2}{3\sqrt{2}} + \frac{x^4}{4\sqrt{3}} + \frac{x^6}{5\sqrt{4}} + \ldots$ 6

9. Test the convergence of any *two* of the following series : 2 × 6

(*a*) $\sum_{n=1}^{\infty} \frac{V_n}{n^2 + 1}$

(*b*) $x + \frac{2^2x^2}{2!} + \frac{3^3x^3}{3!} + \frac{4^4x^4}{4!} + \ldots \infty \; (x > 0)$

(*c*) $\frac{1}{2} + \frac{2x}{3} + \left(\frac{3}{4}\right)^2 x^2 + \left(\frac{4}{5}\right)^3 x^3 + \ldots \infty, (x > 0)$

ENGINEERING & TECHNOLOGY EXAMINATIONS, JUNE-2005

BCA

MATHEMATICS

SEMESTER - 2

Time Allotted: 3 Hours | *Full Marks : 70*

The figures in the margin indicate full marks.

Candidates are required to give their answers in their own words as far as practicable.

Answer Question No.1 and any five from the rest

1. Answer *any five* of the following: 2×5

(*a*) Find the degree and order of the differential equation

$$\sqrt{\frac{dy}{dx}+y} = \sin x.$$

(*b*) Form a differential equation by eliminating A and B from

$$y = e^x (\text{A} \cos x + \text{B} \sin x)$$

(*c*) Write down a basis of R^3 with reasons.

(*d*) Examine if the set of vectors {(1, 2, 2), (2, 1, 2), (2, 2, 1)} is linearly independent.

(*e*) Define a basis of a finite dimensional vector space.

(*f*) Give an example of a sequence, which is bounded without being convergent.

(*g*) What is an oscillatory sequence? Give an example.

(*h*) If $\sum_{n=1}^{\infty} \text{U}_n$ is convergent, prove that $\lim_{n \to \infty} \text{U}_n = 0$.

2. Solve *any two* of the following: 6×2

(*a*) $\frac{d^2y}{dx^2} - 6y' + 13y = 0$

(*b*) $\frac{d^2y}{dx^2} - 24\frac{dy}{dx} + 144y = 0$

(*c*) $\frac{d^2y}{dx^2} - y = \cos 3x$

3. Solve *any two* of the following: 6×2

(*a*) $y = px + \frac{a}{p}$; $a > 0$ and $p = \frac{dy}{dx}$.

(*b*) $(x^2\text{D}^2 - x\text{D} + 2)y = \log x$.

(*c*) $\frac{d^2y}{dx^2} - 2\frac{dy}{dx} + y = e^3x$

4. (*a*) What do you mean by linear dependence of a set of finite number of vectors? Find the value of x such that the vectors (1, 2, 1), (x, 3, 1) and (2, x, 0) are linearly dependent. 6

(*b*) Show that the vectors (1, –2, 3), (2, 3, 1) and (–1, 3, 2) form a basis of R^3. 6

5. (*a*) Show that $W = \left\{ \begin{matrix} (x, y, z) \in R^3 : x + 2y - z = 0 \\ 2x - y + 3z = 0 \end{matrix} \right\}$ is a sub-space of R^3. Find a basis of W and its dimension. 6

(*b*) Let $T : R^3 \to R^2$ be a transformation such that T (1, 0, 0) = (1, 2), T (0, 1, 0) = (1, –1) and T(0, 0, 1) = (1, 1). Obtain a representative matrix for T. 6

6. Answer *any two* of the following: 6 × 2

(*a*) Find $\lim\limits_{n \to \infty} n^{\frac{1}{n}}$

(*b*) Show that $\lim\limits_{n \to \infty} \dfrac{x^n}{n!} = 0$ for all x.

(*c*) Find $\lim\limits_{n \to \infty} r^n$ for different values of r in $0 \le r < \infty$.

7. (*a*) Test the convergence of the series $\sum\limits_{n=1}^{\infty} \dfrac{n!\, 2^n}{n^n}$. 6

(*b*) Test the convergence of series $\dfrac{1}{1^p} + \dfrac{1}{2^p} + \dfrac{1}{3^p} + \dfrac{1}{4^p} + \ldots\ldots \infty$ for different values of p. 6

8. (*a*) Define Absolute and Conditional convergence of an infinite series. Show that the series $1 - \dfrac{1}{2} + \dfrac{1}{3} - \dfrac{1}{4} + \ldots$ converges conditionally. 6

(*b*) Test the convergence of the series 6

$$\left(\frac{1}{3}\right)^2 + \left(\frac{1.3}{3.5}\right)^2 + \left(\frac{1.3.5}{3.5.7}\right)^2 + \ldots\ldots \text{to} \infty.$$

9. (*a*) Let $S = \left\{ \begin{pmatrix} a, & b \\ c, & d \end{pmatrix} : a + b = 0 \right\}$, and $T = \left\{ \begin{pmatrix} a, & b \\ c, & d \end{pmatrix} : c + d = 0 \right\}$; Find Dimension (S), Dimension (T) and Dimension ($S \cap T$). 6

(*b*) Show that intersection of any number sub-spaces of a vector space V is a sub-space. Examine if union of two sub-spaces of V is a subspace. 6

ENGINEERING & TECHNOLOGY EXAMINATIONS, JUNE-2006

BCA

MATHEMATICS

SEMESTER - 2

Time : 3 Hours *Full Marks : 70*

The figures in the margin indicate full marks.

Note : *Answer* **Question No. 1** *which is compulsory and any* **five** *questions each from Q. Nos. 2 and 3.*

1. Choose the correct alternatives for any *ten* of the following : **(10 × 1 = 10)**

(*a*) Order of differential equation $\frac{d^2y}{dx^2}+1=2\frac{dy}{dx}+3\left(\frac{dy}{dx}\right)^2$ is

(*i*) 1 (*ii*) 2 (*iii*) 3 (*iv*) 0.

(*b*) If $\sum_{n=1}^{\infty} a_n$ is a convergent series of constant terms, then

(*i*) $\lim_{n\to\infty} a_n = 1$ (*ii*) $\lim_{n\to\infty} a_n = -1$ (*iii*) $\lim_{n\to\infty} a_n = 0$ (*iv*) $\lim_{n\to\infty} a_n$ does not exist.

(*c*) $\frac{dy}{dx} = x + y$ is a first order Linear equation

(*i*) True (*ii*) False.

(*d*) A solution of $\frac{d^2y}{dx^2} = 0$ is

(*i*) $y = 0$ (*ii*) $y = e^x$ (*iii*) $y = \sin x$ (*iv*) $y = \log_e x$.

(*e*) Eliminating A and B from $y = A\cos x + B\sin x$, the differential equation is

(*i*) $\frac{d^2y}{dx^2} = 0$ (*ii*) $\frac{d^2y}{dx^2} + y = 0$ (*iii*) $\frac{d^2y}{dx^2} - y = 0$ (*iv*) $\frac{d^2y}{dx^2} = 1$.

(*f*) $\lim_{n\to\infty} \frac{3n+1}{2n-3}$ is

(*i*) $\frac{1}{2}$ (*ii*) $\frac{3}{2}$ (*iii*) 1 (*iv*) $-\frac{1}{3}$.

(*g*) Three vectors (a_1, b_1, c_1), $(a_2, b_2, c_2,)$, (a_3, b_3, c_3) of Euclidean 3-space R^3 are linearly dependent, if

$$\begin{vmatrix} a_1 & a_2 & a_3 \\ b_1 & b_2 & b_3 \\ c_1 & c_2 & c_3 \end{vmatrix} = 0.$$

(*i*) True (*ii*) False.

(*h*) If $T : R^2 \to R^2$ is given by $T(x, y) = (x^2, y^2)$ for $(x, y) \in R^2$, then T is a Linear transformation.

(*i*) True (*ii*) False.

(*i*) If $T(x, y, z) = (x, y, 0)$ for all $(x, y, z) \in R^3$ is a Linear transformation, then Kernel T is

(*i*) $(0, 0, 0)$ (*ii*) X-axis (*iii*) Y-axis (*iv*) Z-axis

(*j*) Integrating factor of $\frac{dy}{dx} - y = 1$ is

(*i*) x (*ii*) $\frac{1}{x}$ (*iii*) $-x$ (*iv*) $-\frac{1}{x}$.

(*k*) The infinite series $\sum_{n=1}^{\infty} \frac{n}{n+1}$ is

(*i*) divergent (*ii*) convergent (*iii*) oscillatory (*iv*) none of these.

(*l*) The infinite series $\sum_{n=1}^{\infty} \frac{1}{n^p}$ is convergent if

(*i*) $p = 1$ (*ii*) $p \leq 1$ (*iii*) $p > 1$ (*iv*) $p \geq 1$.

2. Answer any *five* of the following : $5 \times 3 = 15$

(*a*) Find the values of x for which vectors $(1, 1, 1)$, $(1, x, x^2)$ and $(1, -x, x^2)$ shall form a basis of Euclidean 3-space R^3.

(*b*) Show that vectors $(1, 1, 0)$, $(1, 1, 1)$ and $(0, 1, -1)$ are linearly independent.

(*c*) Eliminate a and b from $y = ae^x + be^{-x}$ and obtain a differential equation.

(*d*) Obtain the equation of the curve that passes through origin and whose slope at any point equals to sum of abscissa and ordinate of that point.

(*e*) Solve $\frac{d^2y}{dx^2} = \cos x$.

(*f*) If $T : R^3 \to R^3$ is given by $T(x, y, z) = (x - y, y - z, z - x)$, show that T is a Linear transformation and obtain Kernel T.

(*g*) Let $W = \{(x, y, z) \in R^3 : x + y - 2z = 0\}$, show that W is a sub-space of R^3, and obtain a basis in W.

3. Answer any *five* of the following: $5 \times 9 = 45$

(*a*) (*i*) If $x_1 = \sqrt{2}, x_{n+1} = \sqrt{2x_n}, n > 1$, show that $\lim_{n \to \infty} x_n = 2$.

(*ii*) When $0 < r < 1$, show that $\lim_{n \to \infty} r^n = 0$. 5 + 4

(*b*) (*i*) State D'Alembert's ratio test for an infinite series of positive terms, and show that infinite series $\sum_{n=1}^{\infty} \frac{n^2 - 1}{n^2 + 1} x^n$ is convergent if $x < 1$ and divergent if $x > 1$.

(*ii*) Test for convergence the Hypergeometric series

$$\frac{\alpha}{\beta}+\frac{1+\alpha}{1+\beta}+\frac{(1+\alpha)(2+\alpha)}{(1+\beta)(2+\beta)}+\ldots \qquad 5+4$$

(*c*) (*i*) State Leibnitz test for convergence of an alternating series, and show that $1-\frac{1}{2}+\frac{1}{3}-\frac{1}{4}+\ldots$ is a convergent series.

(*ii*) Show that $1+\frac{1}{2}+\frac{1}{3}+\frac{1}{4}+\ldots$ is a divergent series. 5 + 4

(*d*) (*i*) If (α, β, γ) form a basis of Euclidean 3-space R^3, show that $(\alpha, \beta+\gamma, \gamma+\alpha+\beta)$ forms a basis of R^3.

(*ii*) If $T: R^3 \to R^3$ is given by $T(x, y, z) = (x-2y, y-2z, z-2x)$ for $(x, y, z) \in R^3$, obtain a representative matrix for linear transformation T. 4 + 5

(*e*) (i) Show that the family M_2 of all real square matrices of order 2 forms a vector-space over reals, and find a basis for M_2.

(*ii*) If $W_2 = \left\{\begin{pmatrix} a, & b \\ c, & d \end{pmatrix} : a, b, c \text{ and } d \text{ are reals with } a+b=0\right\}$, show that W_2 is a sub-space of M_2 and determine dimension (W_2). 4+5

(*f*) Solve any *two* of the following: 9

(*i*) $\dfrac{xdx+ydy}{xdy-ydx}=\sqrt{\dfrac{a^2-x^2-y^2}{x^2+y^2}}$

(*ii*) $\dfrac{dy}{dx}+\dfrac{y}{x}=\dfrac{y^2}{x^2}$

(*iii*) $y = px + \sqrt{a^2p^2+b^2}$ where $p=\dfrac{dy}{dx}$

(*g*) Solve any *two* of the following:

(*i*) $\dfrac{d^2y}{dx^2}-3\dfrac{dy}{dx}+2y=0$. Given that $x = 0$. $y = 0$ and $\dfrac{dy}{dx}=0$.

(*ii*) If $\dfrac{d^2V}{dx^2}+p^2V=0$, and $V = V_0$ at $x = 0$ and $V = 0$ at $x = l$, show that

$$V = V_0\frac{\sin p(l-x)}{\sin pl}.$$

(*iii*) $\dfrac{d^2y}{dx^2}-6\dfrac{dy}{dx}+9y=3x+1$.

(*h*) Answer any *two* of the following: 9

(*i*) Solve $\frac{d^2y}{dx^2} + 2\frac{dy}{dx} + 2y = 0$.

(*ii*) Reduce $x^2 (y - px) = p^2 y$ to Clairaut's form by making substitution $x^2 = u$ and $y^2 = v$ and then solve it. $\left(p \equiv \frac{dy}{dx}\right)$.

(*iii*) Solve $x^2 \frac{d^2y}{dx^2} - \frac{xdy}{dx} - 3y = x^2 \log x$.

ENGINEERING & MANAGEMENT EXAMINATIONS, JUNE-2007

BCA

MATHEMATICS

SEMESTER - 2

Time : 3 Hours] [Full Marks : 70

Group – A

(Multiple Choice Type Questions)

1. Choose the correct alternatives for any *ten* of the following : **10 × 1 = 10**

(*i*) Integrating Factor of the differential equation

$$\cos x \frac{dy}{dx} + y \sin x = 1 \text{ is}$$

(*a*) $\tan x$ (*b*) $\cos x$ (*c*) $\sec x$ (*d*) $\sin x$.

(*ii*) Family of curves $y = e^x$ (A $\cos x$ + B $\sin x$) is represented by the differential equation

(*a*) $\frac{d^2y}{dx^2} = \frac{2dy}{dx} - y$ (*b*) $\frac{d^2y}{dx^2} = \frac{2dy}{dx} - 2y$

(*c*) $\frac{d^2y}{dx^2} = \frac{dy}{dx} - 2y$ (*d*) $\frac{d^2y}{dx^2} = \frac{2dy}{dx} + y$.

(*iii*) Order and Degree of the differential equation,

$$\frac{d^2y}{dx^2} = \left\{ y + \left(\frac{dy}{dx} \right)^2 \right\}^{1/4} \text{ are}$$

(*a*) 4, 2 (*b*) 1, 4 (*c*) 2, 4 (*d*) 1, 2.

(*iv*) A particular integral of $\frac{d^2y}{dx^2} - \frac{2dy}{dx} + 4 = e^x \cos x$ is

(*a*) $e^x \cos x$ (*b*) $\frac{e^x \cos x}{2}$ (*c*) $\frac{e^x \sin x}{2}$ (*d*) none of these.

(*v*) T is a transformation from R_2 to R_3 defined by T $(x_1, x_2) = (x_1, x_1^2 + x_2, - x_1)$ then the image of (1, 2) is

(*a*) (1, 1, 1) (*b*) (0, 3, – 1) (*c*) (1, 3, – 1) (*d*) (0, 0, 0).

(*vi*) T is a linear transformation defined by T $(x_1, x_2) = (-x_1, -x_1)$ then the dimension of T (v_2) is

(*a*) 2 (*b*) 1 (*c*) 0 (*d*) none of these.

(*vii*) Which one of the following statements is true for a linear transformation T : V → W ?

(*a*) ker (T) may not be a vector space

(*b*) ker (T) is a subject of W.

(c) $\theta \in \ker(T)$ where θ is a null vector of V

(d) $\theta' \in \ker(T)$ where θ' is a null vector of W.

(viii) In V_3 the vectors (1, 0, 0), (0, 1, 0) and (1, 1, 1) are

(a) dependent

(b) independent

(c) none of these.

(ix) If $(3, 1) = x \cdot (1, 2) + y \cdot (0, 3)$ then the values of x and y are respectively

(a) $3, -5$ (b) $3, 1$ (c) $3, -\frac{5}{3}$ (d) $3, -\frac{5}{2}$.

(x) The infinite series $\sum_{n=1}^{\infty} \frac{n}{n+1}$ is

(a) divergent (b) convergent (c) oscillatory (d) none of these.

(xi) Find the lower bound of the sequence $\left\langle \frac{(-1)^{n-1}}{n!} \right\rangle$

(a) $-\frac{1}{2}$ (b) $\frac{1}{2}$ (c) 1 (d) 0.

(xii) Which one of the following statements is true ?

(a) Every convergent sequence is bounded

(b) Every divergent sequence is bounded

(c) Every bounded sequence is convergent

(d) Every bounded sequence is divergent.

(xiii) The series $1 - 1 + 1 - 1 + \ldots\ldots\ldots$ is

(a) convergent with sum equal to 0

(b) convergent with its sum equal to 1

(c) divergent

(d) oscillatory.

Group – B

(Short Answer Type Questions)

Answer any *three* questions. $3 \times 5 = 15$

2. Prove that the series

$$\frac{1}{3} + \frac{(1.4)}{(3.6)} + \frac{(1.4.7)}{(3.6.9)} + \ldots\ldots\ldots \text{ is divergent.}$$

3. Test the series $\sum_{n=1}^{\infty} \left(1 + \frac{1}{\sqrt{n}}\right)^{-n^{3/2}}$.

4. Show that the series $\sum \frac{1}{n^p}$ converges for $p > 1$.

5. Find the value of x for which the vectors (1, 2, 1), (x, 3, 1) and (2, x, 0) become linearly independent.

6. Solve any three of the following :

(*i*) $y = px + \frac{a}{p}$

(*ii*) $p^2 = ax^3$

(*iii*) $(D^2 - 4)y = e^{2x} + e^{-4x}$

(*iv*) $(D^2 + 9)y = \cos 3x$.

Group – C

(Long Answer Type Questions)

Answer any three questions. $3 \times 15 = 45$

7. (*a*) Find whether

$$W = \{(a_1, a_2, a_3) : a_1 + a_2 \geq 0 \,;\, a_1, a_2, a_3 \in R\} \text{ is a sub-space of } R^3.$$

(*b*) Give an example to show that union of 2 sub-spaces need not be a sub-space of V.

(*c*) Let $V = R^4 (R)$ and

$$W = \{(a, b, c, d) \in R^4 : a = b + c, c = b + d)\}.$$

Find a basis and dimension of W. 5 + 5 + 5

8. (*a*) Find the Kernel, Image set, Nullity and Rank of $T : R^3 \rightarrow R^2$ where

$T (1, 0, 0) = (2, 1)$

$T (0, 1, 0) = (0, 1)$

$T (0, 0, 1) = (1, 1)$. Hence verify Sylvester's law.

(*b*) Show that the mapping $T : R^2 \rightarrow R^3$ defined by

$T (x_1, x_2) = (x_1 + x_2, x_1 - x_2, x_2)$ is a linear transformation.

(*c*) Prove that the vectors (x_1, y_1) and (x_2, y_2) are linearly dependent, if and only if $x_1y_2 - x_2y_1 = 0$. 5 + 5 + 5

9. Test the convergence of any *three* of the following series : 3 × 5

(*a*) $\frac{6}{1.3.5} + \frac{8}{3.5.7} + \frac{10}{5.7.9} + \ldots\ldots\ldots$

(*b*) $1 + \frac{x}{2} + \frac{x^2}{5} + \frac{x^3}{10} + \ldots\ldots\ldots$

(*c*) $\left(\frac{2^2}{1^2} - \frac{2}{1}\right)^{-1} + \left(\frac{3^3}{2^3} - \frac{3}{2}\right)^{-2} + \left(\frac{4^4}{3^4} - \frac{4}{3}\right)^{-3} + \ldots\ldots\ldots$

(*d*) $1 + \frac{1}{2} + \frac{1}{2^2} + \frac{1}{2^3} + \ldots\ldots\ldots$

10. Solve any *three* of the following : 3 × 5

(a) $\frac{d^2y}{dx^2} - \frac{2dy}{dx} + y = xe^x$.

(b) $x^2(y - px) = yp^2$, taking $x^2 = u$, $y^2 = v$, $p = \frac{dy}{dx}$.

(c) $\frac{x^2 d^2y}{dx^2} - \frac{x\,dy}{dx} - 3y = x^2 \log x$.

(d) $\frac{d^2y}{dx^2} + y = \sec^2 x$.

11. (a) Show that the sequence $< a_n >$, where

$$a_n = 1 + \frac{1}{3} + \frac{1}{3^2} + \ldots\ldots + \frac{1}{3^{n-1}} \text{ converges. Find } \lim_{n\to\infty} a_n.$$

(b) Show that $< a_n >$,

$$a_n = \sqrt{n}\left(\sqrt{n+1} - \sqrt{n}\right), \text{converges.}$$

(c) Discuss the convergence of the series

$$\frac{x^2}{2\log 2} + \frac{x^3}{3\log 3} + \frac{x^4}{4\log 4} + \ldots\ldots\ldots$$ 5 + 5 + 5